# Habitat Management for Conservation

# Techniques in Ecology and Conservation Series

*Series Editor: William J. Sutherland*

**Bird Ecology and Conservation: A Handbook of Techniques**
William J. Sutherland, Ian Newton, and Rhys E. Green

**Conservation Education and Outreach Techniques**
Susan K. Jacobson, Mallory D. McDuff, and Martha C. Monroe

**Forest Ecology and Conservation**
Adrian C. Newton

**Habitat Management for Conservation: A Handbook of Techniques**
Malcolm Ausden

# Habitat Management for Conservation

*A Handbook of Techniques*

Malcolm Ausden

# OXFORD
**UNIVERSITY PRESS**

Great Clarendon Street, Oxford OX2 6DP

Oxford University Press is a department of the University of Oxford.
It furthers the University's objective of excellence in research, scholarship,
and education by publishing worldwide in

Oxford  New York

Auckland  Cape Town  Dar es Salaam  Hong Kong  Karachi
Kuala Lumpur  Madrid  Melbourne  Mexico City  Nairobi
New Delhi  Shanghai  Taipei  Toronto

With offices in

Argentina  Austria  Brazil  Chile  Czech Republic  France  Greece
Guatemala  Hungary  Italy  Japan  Poland  Portugal  Singapore
South Korea  Switzerland  Thailand  Turkey  Ukraine  Vietnam

Oxford is a registered trade mark of Oxford University Press
in the UK and in certain other countries

Published in the United States
by Oxford University Press Inc., New York

© Oxford University Press 2007

The moral rights of the author have been asserted
Database right Oxford University Press (maker)

First published 2007

All rights reserved. No part of this publication may be reproduced,
stored in a retrieval system, or transmitted, in any form or by any means,
without the prior permission in writing of Oxford University Press,
or as expressly permitted by law, or under terms agreed with the appropriate
reprographics rights organization. Enquiries concerning reproduction
outside the scope of the above should be sent to the Rights Department,
Oxford University Press, at the address above

You must not circulate this book in any other binding or cover
and you must impose the same condition on any acquirer

British Library Cataloguing in Publication Data

Data available

Library of Congress Cataloging in Publication Data

Data available

Typeset by Newgen Imaging Systems (P) Ltd., Chennai, India
Printed in Great Britain
on acid-free paper by
Biddles Ltd., King's Lynn

ISBN 978–0–19–856872–8   978–0–19–856873–5 (Pbk)

10 9 8 7 6 5 4 3 2 1

# Contents

*Acknowledgements*     vii

1. Introduction     1
2. Philosophical approaches to habitat management     13
3. Setting objectives and monitoring     27
4. General techniques and considerations     57
5. Dry grasslands     87
6. Dwarf-shrub habitats and shrublands     131
7. Forests, woodlands, and scrub     173
8. Freshwater wetlands and water bodies     229
9. Coastal habitats     301
10. Arable land     331
11. Gardens, backyards, and urban areas     355

*References*     371
*Index*     399

## Acknowledgements

I am very grateful to the following who commented on chapters of this book: Guy Anderson, Mark Bolton, Dave Buckingham, Dave Curson, Fiona Sanderson, John Sharpe, Bill Sutherland, Ken Smith, and Geoff Welch. I also thank Ian Sherman at Oxford University Press for his enthusiasm and Bill Sutherland and Graham Hirons for their inspiration and encouragement.

Special thanks to Siobhan.

Dedicated to Liam and Callum.

Please send any comments, relevant additional information, or corrections (e-mail: malcolm.ausden@rspb.org.uk; postal address: RSPB, The Lodge, Sandy, Beds SG19 2DL, UK).

# 1
# Introduction

This book is about managing land to enrich it for wildlife and provide places of beauty and inspiration for people. Habitat management is often perceived as merely seeking to conserve, or 'preserve', habitats and species assemblages created through past, so-called traditional land management. Whereas there will inevitably be an element of maintaining existing assemblages of species and landscapes, managing habitats for conservation is really a process of *managing* inevitable *change*. This should involve maintaining the best of what we have inherited, while also making the most of future opportunities.

## 1.1 What is habitat management and why is it necessary?

Managing habitats for wildlife mainly involves influencing the successional stage and physical structure of vegetation to benefit particular species, or assemblages of species, considered to be of high conservation or other intrinsic value (Figure 1.1). Succession is the process by which assemblages of plants and animals change over time in the absence of disturbance. Habitat management can also include:

- manipulations to specifically increase the abundance and accessibility of prey;
- provision of nest sites;
- control of unwanted plants, which are often alien or exotic species;
- minimizing the effects of damaging human activity.

Habitat management can in some cases also include planting vegetation, although this is more usually carried out during habitat (re-)creation or restoration.

Many species are dependent on early successional habitats: those that have been subject to recent disturbance. In fully natural systems the rate and direction of succession is influenced by vegetation removal and other physical disturbance. The main causes of these are:

- grazing, browsing, trampling and other physical disturbance by larger herbivores (Figure 1.2);

**Fig. 1.1** Habitat management and rare species. Habitat management can be undertaken to increase populations of specific, rare species, such as this red-cockaded woodpecker, *Picoides borealis*. These woodpeckers are restricted to open, pine-dominated forests in the south-eastern USA. Prescribed burning and thinning are used to remove broad-leaved trees to provide an open understorey and maintain dominance by pines. Photograph by Andy Swash.

- physical-disturbance events, particularly fires, storms, drought and variations in water levels;
- periodic large-scale herbivory by insects;
- outbreaks of disease in plants.

However, in parts of the world such as temperate Europe, virtually all natural habitats have been highly modified by past human activity. The remaining fragments of modified vegetation are usually referred to as being semi-natural. The natural processes of vegetation removal and other physical disturbance usually either no longer operate in these semi-natural habitats or, if they do, only at an inappropriate scale or frequency to maintain the desired assemblages of species. In these situations, habitat management can be used to mimic the *effects* of these natural processes to provide suitable conditions for species that depend on these forms of disturbance.

Most of the remaining fragments of semi-natural habitat have only survived until now because they have been subject to a long history of human resource use. Examples include grasslands managed by grazing and mowing for hay, forests managed for wood production, peatlands cut for fuel and swamps and fens cut

**Fig. 1.2** Effects of wild, large herbivores. Large herbivores can have profound impacts on habitat conditions, but are now absent from many small, isolated patches of habitat. Examples of their effects include:

(a) grazing and browsing by large herds of ungulates maintaining open habitats by arresting succession (Masai Mara National Reserve, Rift Valley Province, Kenya);

(b) damming of rivers by American beavers, *Castor canadensis*, creating ponds and drowning trees, the latter providing dead wood habitat for invertebrates and woodpeckers (Jug Bay Wetlands Sanctuary, Maryland, USA); and

(c) soil disturbance by rooting wild boar, *Sus scrofa*, providing bare ground in which seeds can germinate (Rabinówka, Podlaskie, Poland).

to provide reed and sedge for thatching. Habitats have also been managed to provide harvestable quantities of wildfowl, game birds, deer and other quarry species. In many cases this management has happened to produce habitats that have not only retained a valuable and characteristic subset of natural biodiversity, but which are also of high cultural and aesthetic value. The habitats that have been

**Fig. 1.3** Cultural habitats. The savannah-like wooded dehesas (Spanish) and montados (Portugese) of the western Mediterranean are cultural habitats that support a characteristic and valuable flora and fauna. They consist of scattered live oaks, *Quercus rotundifolia*, or cork oaks, *Quercus suber*, set amidst grazed semi-arid grassland, some of which is cultivated periodically. The oaks are pruned (lopped) to provide firewood and charcoal, and to maximize the crop of acorns that is used to fatten pigs in winter and to flavour spirits. Cork oak is periodically stripped of its bark to provide cork.

(b) The wooded dehesas and montados are especially important for their birdlife. Large concentrations of common cranes, *Grus grus*, feed on fallen acorns in winter and the habitat also supports a wide range of wintering and breeding birds. The dehesas and montados are used by  hunting Spanish imperial eagle, *Aquila adalberti*, and other raptors, and the large herds of livestock provide carrion for important populations of vultures (top: near Embalse del Tozo; bottom: near Retamosa; both in Extremadura, Spain).

derived from, maintained by or heavily impacted by human management are known as cultural habitats. An example is the wooded dehesas (in Spanish) and montados (in Portugese) of the Iberian Peninsula (Figure 1.3). The high value afforded to maintaining cultural habitats is largely a European phenomenon.

Lack of suitable natural processes is usually most acute in small, isolated fragments of habitat. In particular, small, isolated patches of semi-natural habitat rarely contain more than a tiny proportion, if any, of their native large herbivores, let alone any larger carnivores to prey on them. Large-scale disturbances caused by fire or flood are largely prevented in small (and often even in very large) areas of habitat as a matter of policy. If they do occur, and the habitat is small and isolated, these forms of catastrophic disturbance can impact upon the whole of the remaining fragment of habitat. Such events can make the entire habitat patch temporarily unsuitable for most of its existing compliment of species. If this habitat is isolated from sources of potential re-colonization, then many of its former species are unlikely to return, even if the habitat eventually becomes suitable for them again. This is particularly the case for many invertebrate species with limited powers of dispersal (Figure 1.4). In these situations, habitat management can be used to mimic the effects of these natural processes over only a *proportion* of the habitat and, in so doing, retain a viable population in the unmanaged area to re-colonize the managed area as it becomes suitable again.

Ancient habitats, which have existed in a similar form for a long time, can be particularly valuable. Examples include ancient grasslands, forests and woodlands. Ancient habitats can support specific, important features for wildlife, such as veteran trees, that are rare or absent from more recently created habitat. They can also contain a range of species associated with relatively stable conditions which are often poor at dispersing. The ancient habitats that we have inherited are a product of past events and are consequently impossible to fully recreate/restore.

In some cases the main priority for conservation management is to prevent or minimize deleterious effects of *existing* human activities, rather than to necessarily introduce or maintain active habitat management. Examples include reducing levels of nutrients in water entering a wetland, minimizing the effects of human physical disturbance on fragile plant communities and reducing herbicide and pesticide use on farmland.

## 1.2 Principles of habitat management

There are five main ways of removing vegetation to influence the **successional stage, vegetation composition** and **structure** of terrestrial habitats. These are by:

- grazing and browsing;

**Fig. 1.4** A contradiction: insects that are restricted to temporary habitats but are poor at dispersing. A range of insect species on the cool, wet edges of their climatic range are dependent on ephemeral, early successional habitat, but are remarkably poor at dispersing. An example of this is the heath fritillary butterfly, *Mellicta athalia*.

This is widespread in most of continental Europe, but on the north-west edge of its climatic range in south-east England it has persisted (without re-introduction) in just one complex of woodland. Here it is largely restricted to the first 4 years of re-growth and adjacent rides that contain its sole larval foodplant there, common cow-wheat, *Melampyrum pratense*.

It would be expected that species characteristic of early successional habitat would have good powers of dispersal to allow them to colonize new areas of early successional habitat as existing areas become unsuitable. The heath fritillary, though, rarely colonizes suitable habitat further than 600 m away from existing colonies (Warren 1987).

Two explanations have been proposed to explain the apparent contradiction of relatively immobile species confined to ephemeral habitat (Thomas 1994). The first is that under current conditions only cleared areas provide the warm microclimates that these species require, but that they colonized these regions when summer temperatures were higher and they could persist in a wider range of natural conditions. The second explanation is that former management has resulted in selection for individuals with low powers of dispersal. This might occur if management usually created areas of suitable ephemeral habitat in close proximity to one another, and habitat fragmentation meant that any individuals with better powers of dispersal had a negligible chance of finding suitable habitat.

- cutting and removing vegetation;
- burning;
- soil disturbance, such as by ploughing and rotovating;
- removal of the topsoil, usually known as sod cutting or turf stripping (Figure 1.5).

The timing and frequency of these methods will influence the structure and composition of the vegetation at any given time and its suitability for any animals living within it.

# Principles of habitat management

**Fig. 1.5** Sod cutting/turf stripping. (a) Removal of the topsoil can be used to set back succession by removing accumulated soil nutrients and exposing the buried seedbank.

(b) This area had been arable land for 40–50 years. Its topsoil was removed 6 years before this photograph was taken and the area now consists of pioneer heathland.

Seedbank studies are useful in identifying the most abundant seeds in the soil. However, seeds of many plants are difficult, or take time, to germinate. It only requires a tiny number of seeds to survive and germinate following sod cutting/turf stripping for a species establish. Many rare and scarce wet heathland plants have apparently re-appeared from the seedbank after having been buried for 40–50 years (Langdonken, Flanders, Belgium).

Key considerations when grazing and browsing are the type of livestock, grazing pressure and timing of grazing. With cutting, burning and soil disturbance the principle considerations are their timing and frequency, together with the size of area cut, burnt or disturbed at any one time.

All of these methods of vegetation removal and disturbance have the potential to make conditions temporarily unsuitable for the species that they are intended to benefit. Therefore, it is important to time the management so that its immediate, damaging effects are minimized and only carry out the management on a proportion of the habitat at any one time. This can be done through **rotational management**. While some species might require large areas of habitat of the same or similar successional stage, others only require small patches of habitat, or a closely knit **mosaic** of different vegetation types and structure. In practice, creating an intimate mosaic of conditions through small-scale patchy cutting and burning is usually more logistically complicated and expensive than undertaking it on fewer, larger areas. Thus, the size and number of patches managed is usually a compromise between the desirable and the practical.

Habitat management in wetlands also involves **manipulation of water levels** and in some cases influencing **water quality**. Manipulation of water quality usually aims to reduce nutrient levels to prevent dominance by algae and other more competitive plant species. Manipulation of **salinity** can also be important in saline wetlands.

A further concept worth introducing at this stage is that of transitions and boundaries between habitats, often referred to as **edge** (Figure 1.6). Many species exploit so-called soft edges between habitats, for example where woodland gradually merges into scrub and the scrub merges into grassland. These soft edges can support especially diverse assemblages of species. Many habitats in intensively managed landscapes have 'hard' edges, where one habitat abruptly ends and another one starts without any gradation in conditions between them.

Finally, the suitability of habitat for individual species can also be influenced by introducing **variation in topography**. This creates differences in hydrology and variations in aspect and therefore sunlight and temperature. Small-scale variations in sunlight and temperature can be particularly important in influencing conditions for many insect and reptile species.

## 1.3 The aims and scope of this book

There are a number of excellent books describing the principles and techniques of habitat (re-)creation and restoration (e.g. Gilbert and Anderson 1998; Middleton 1999; Whisenant 1999; Perrow and Davy 2002a, 2002b; van Andel

**Fig. 1.6** Edge. Management often aims to maximize the length of edge, or the boundary between different habitat types. The edges of swamp and open water provide important habitat for many waterbirds and wetland songbirds (Lake Kvismaren, Örebro, Sweden).

and Aronson 2006). As far as I am aware, though, there is not a single overview discussing the range of techniques available for ongoing management of habitats for conservation. This book aims to provide such an overview of techniques for *actively* managing habitats. In so doing, it aims to both inform management of existing habitats, but also provide information on how future management might influence the design of newly created habitat. For example, if you are going to create a common reed *Phragmites australis*-dominated wetland, then you do not necessarily have to design one that will have to be managed by rotational reed cutting, as many have historically been. Instead, you could create one that will be managed by grazing and using periodic fluctuations in water levels. However, if you decide upon this second option, then it will necessary to design in a suitable area of high, dry ground for grazing animals to lie up in, and ensure that the site includes enough suitable plants to provide sufficient foraging for these grazing animals during winter. It will also be sensible to design the site so that it has different hydrological units, thus allowing water levels to be periodically lowered across only one part of the site at a time. Conversely, if you were intending to maintain the area of reed bed by cutting, then you would need to ensure suitable access to the areas mown and that the topography is flat enough to ensure safe use of mowing equipment.

Methods other than habitat management are also used to benefit particular populations of species, particularly those at critically low population levels, or to maximize the harvestable surplus of game. These methods include re-introduction and translocation of species, supplementary feeding and control of predators. They are often called species management (Sutherland 2000). This book generally avoids species management and will not cover techniques of establishing vegetation during habitat (re-)creation, most of which are covered well elsewhere. It will also generally not discuss planting vegetation in existing semi-natural habitats. Exceptions to this will be when considering addition of seeds to diversify species-poor, agriculturally improved grasslands, forestry planting regimes, crop types on arable land and planting in gardens/backyards and urban areas.

The techniques described have a geographical bias towards those used in temperate Western Europe. This is due to both the author's own experience, and also the wide range of cultural habitats, conservation aims and methods employed in this region. In addition, in the cool, wet Atlantic climate of Western Europe there has been a particular emphasis on often fairly intensive intervention management aimed at providing warm, open, early successional habitats to support southern, warmth-loving plants and animals. Despite this, an attempt has been made to include additional techniques used outside of this region, and to emphasize the general principles of all management techniques, so they can be applied in other situations. For example, there are several management methods commonly used in the USA that are used rarely in Europe. Examples include moist-soil management to provide food for wintering wildfowl (Section 8.4.1), and the use of regulated tidal exchange in coastal impoundments (Section 9.3.1). Conversely, there is a range of techniques used to benefit farmland birds in parts of Western Europe but not in the USA.

Cultural or intensively managed habitats of high conservation value are rare in, or absent from, tropical regions. Here the primary focus of conservation is usually on maintaining areas of near-pristine, so-called wild nature. Specific management of tropical vegetation is outside the scope of this book. Similarly, habitats in which conservation actions are largely restricted to minimizing damaging human activities, rather than specific, interventionist management for conservation, are also excluded. These habitats include deserts and other arid land, mountain tops, rocky shores and areas below the low-water mark.

## 1.4 Outline of the book

The rest of this book is divided into chapters discussing issues and techniques common to all habitats (Chapters 2–4), and those detailing methods of habitat management used in specific habitats (Chapters 5–11).

Chapter 2 discusses different approaches to habitat management. It concentrates on the extent to which management focuses on individual species, groups of species, habitats and ecological processes, and the extent to which habitat management for conservation is integrated with other interests. Chapter 3 describes decision-making at a site level. It discusses management/site action planning: the process for deciding what you want to achieve, how to achieve it, and how to monitor whether you are achieving it.

Chapter 4 includes a range of issues and considerations common to all, or most, habitats. It discusses general principles of managing habitats for different groups of plants and animals. It includes general principles of grazing, control of unwanted plants and the effects of climate change. The chapter also highlights landscape factors to take account of when undertaking habitat management.

The final seven chapters each concentrate of the main considerations and methods for managing individual habitats. For practical reasons, it has been necessary to discuss management of different habitats in separate chapters. However, it is important to emphasize that many areas consist of mixtures and transitions of habitat that are managed together, particularly through grazing. Indeed, it is often very desirable to create and maintain these valuable gradations and mosaics.

# 2
# Philosophical approaches to habitat management

There are a range of philosophical approaches to habitat management. Fundamental differences involve whether management aims to create and maintain cultural habitats or those considered closer to habitats that existed under more natural conditions, and the level of human intervention and control. At one extreme, habitat management may involve very specific actions aimed primarily at benefiting just one or a small suite of species. At the other, it may involve introducing key natural processes with little further intervention and expectation of the outcome.

The approach to habitat management also depends on the extent to which conservation is integrated with other interests, such as recreation, wider resource use and provision of other ecological services. A further consideration at a higher level is the extent to which resources are focused on achieving conservation objectives in protected areas and nature reserves, compared with on farmed land in the wider landscape (Figure 2.1).

## 2.1 Preserving cultural habitats or managing change?

In many areas of the world, particularly much of Europe, the only existing areas of near-natural habitat are cultural habitats that have been subject to long periods of human resource use. The usual starting point when considering management of these cultural habitats is to continue, or reinstate, similar management to that which created and maintained them. This is based on the assumption that if the desired assemblage of species already co-existed under a particular management regime, then the best way to perpetuate it is to continue, or reinstate, a similar regime. Introduction of different management might provide even better conditions for some of the species already present. In fact, they might have persisted more *despite* existing management, rather than because of it. Introduction of different management might also provide suitable conditions for species not currently

## 14 | Philosophical approaches

**Fig. 2.1** Conserving farmland wildlife. In parts of Europe low-input arable land is considered of high conservation value for its farmland birds and 'arable weeds' (ruderal plants). This uninspiring arable landscape in Hungary supports a number of Europe's rarest birds, whose European population in largely confined to farmland. These include saker falcon, *Falco cherrug*, and imperial eagle, *Aquila heliaca*, the first visible here just before this photograph was taken and the other nesting nearby (Poroszló, Heves County, Hungary).

present and allow potentially desirable dynamism in the system. However, there is also the possibility that it might result in the loss of existing species from that particular patch of habitat. This is therefore a potentially dangerous approach if the habitat is isolated from sources of potential re-colonists, since the lost species may be slow or unable to re-colonize (Figure 2.2). Similarly, isolation will probably also mean that the species that could potentially take advantage of the new conditions will also be slow or unable to colonize (Thomas and Jones 1993). This will not be the case for species with good powers of dispersal. However, species with good powers of dispersal, by the very fact that they are easily able to exploit new areas of habitat, are unlikely to be of high conservation priority.

Despite these strong arguments for continuing or reinstating traditional management in cultural habitats, there are number of reasons why this may not always be the best method of conserving species in them. In some cases, what is currently considered to have been traditional management may in fact have varied quite significantly over time. For example, at Wicken Fen, one of the UK's oldest

# Preserving habitats or managing change? | 15

**Fig. 2.2** Maintenance of traditional management. Mickfield Meadow is an isolated 1.7-ha damp meadow in Suffolk, England, surrounded by hundreds of hectares of intensively farmed arable land. The meadow is managed by cutting for hay and aftermath grazing to maintain its beautiful display of rare fritillaries, *Fritillaria meleagris*, and an array of other damp-meadow species. Fritillaries are found at only three other sites in Suffolk.

A change in traditional management might benefit a different range of species, but how will any other species found in damp meadows colonize the site? If the change in management resulted in the loss of the fritillaries or other damp-meadow species, then they would be unlikely ever to re-colonize naturally.

nature reserves, the timing of so-called traditional harvesting of great fen-sedge *Cladium mariscus* for thatching (see Figure 8.15) has varied historically. During different periods sedge has been harvested between April and July, between July and December, or in a piecemeal fashion at varying times throughout most of the year (Friday 1997).

The most important reason, though, why reliance on traditional-type management may not be the best option, is because of more recent human-induced changes in the wider environment. The most important of these are:

- changes in climate (Section 4.7);
- increases in nutrient levels influencing plant growth (Section 4.1.1);
- acidification of soils and water bodies (Section 6.5.2);
- changes in the assemblages of species within a given area due to past extinctions and recent colonization, especially by alien or exotic species (Sections 4.2, 4.5 and 4.6);
- changes due to population processes and numbers of generalist predators in the wider landscape (Section 4.2).

In some cases, additional management may now be needed to mitigate these effects.

## 16 | Philosophical approaches

A further reason why reliance on traditional management may not always be the best option is due to its often high costs. Most traditional management of cultural habitats of high conservation value is now uneconomic, unless supported by grants (it is important to note that most conventional agriculture is also heavily subsidized).

For the reasons discussed, it may be more useful to consider management of cultural habitats more as a process of *managing* inevitable *change*. This involves maximizing the positive effects of these changes and minimizing their negative effects, rather than merely seeking to conserve or preserve assemblages of species that happened to be present when the decision to conserve the area was made.

If it is intended to introduce new forms of management, it is still prudent to trial this new management over just a portion of the site. If the new management proves unsuitable for existing species, then they should at least persist in nearby habitat not subject to this change. It is also important to monitor changes taking place under both the new and existing management regimes to help understand the effects of the new management (Section 3.2).

### 2.2 Recreating former cultural habitats or creating new ones?

There is less of a case for relying on traditional management techniques when designing the creation and management of new areas of habitat on land of low or negligible conservation value. In these situations there is no need to use traditional management to perpetuate assemblages of species of conservation value already present. Instead, there is greater freedom to create conditions not provided in cultural habitats, and thereby cater for different assemblages of species. There is also greater potential to manage such areas less intensively, with a greater reliance on more natural processes of vegetation removal and physical disturbance. This has the advantages of providing a greater sense of wilderness and will in many cases be cheaper than maintaining them using traditional management. A frequently cited example of this approach is management of the Oostvaardersplassen in the Netherlands (see Figure 4.6). In some cases, though, there may be other cultural or historic reasons for creating cultural habitats, including increasing the size and potential viability of existing fragments.

### 2.3 The level of intervention: focusing on individual species or encouraging natural processes?

A fundamental decision when undertaking habitat management is the level of intervention. In some situations there is the potential to manage habitats specifically to benefit just one or a small suite of species. In practice, this invariably involves maintaining the habitat in a suitable condition for its characteristic assemblages of species, but refining the details of this management to specifically benefit one or a group of them. This approach has been successful at reversing declines of species of high conservation value, but its success is obviously dependent on a correct diagnosis of the reasons for decline (Green 1995). The more single-species-led approach to habitat management has been most often used to benefit birds and some other popular groups, such as butterflies.

Focusing on the requirements of individual species, rather than the habitat or ecosystem as a whole, is viewed by some people as being too interventionist and controlling, and far removed from their philosophical view of what wild nature is really about. Focusing on single species also means it is necessary to decide, through valued judgement, which species to benefit, and in some cases which species to disadvantage, through management. The latter can be contentious, especially when involving control of predators.

Even if the main aim of management is to benefit particular species or groups, a decision still needs to be made regarding the level of intervention. This is well illustrated by the range of management options available for increasing food supply for wintering seed-eating wildfowl. The quantity of seed available to them can be increased by artificially manipulating water levels to maximize the growth of abundant seed-producing ruderal vegetation, using moist-soil management (Section 8.4.1). Although this is clearly a management intervention, it still largely mimics the natural process of the seasonal drawing down of water levels. A further intervention would be to sow favoured wildfowl foods within these wetland areas to increase the quantity of suitable seed. More interventionist still would be to grow and leave unharvested favoured wildfowl foods on arable land using conventional farming methods, so-called sacrificial crops (Section 10.5). The most extreme end of the naturalness–unnaturalness continuum would be to simply feed wildfowl with grain (Figure 2.3). Individuals and organizations differ in the level of intervention they consider acceptable.

There are a number of important practical advantages in focusing habitat management to benefit individual species. It results in the setting of clear, unambiguous targets for management, and clear actions to achieve these targets, providing that the key species' habitat requirements are well understood. It is therefore a strong tool for focusing minds and resources. This approach can be used to define targets for species (and habitats) at national and other levels that can be cascaded down to those at individual sites. This type of planning approach has the advantages of providing a standard 'language' with which to define priorities and allocate resources. In the event of these targets not being achieved, organizations and

## 18 | Philosophical approaches

**Fig. 2.3** Naturalness. An important consideration when deciding what habitat management to undertake is the appropriate level of intervention. Artificial feeding probably represents the most unnatural extreme of the natural–unnatural continuum. However, as in this case, it can be spectacularly successful in attracting wildlife close to people and, in so doing, reaching out beyond traditional conservation audiences. This photograph shows the feeding of wild whooper swans, *Cygnus cygnus*, and other wintering wildfowl in front of a viewing station (Ouse Washes, Cambridgeshire, England).

governments can be lobbied to provide the extra resources required to achieve the agreed targets.

The other extreme to focusing on the requirements of favoured individual species is to focus on restoring or maintaining key *natural processes*. These may include introducing naturalistic, large-scale grazing and flooding regimes, and then allowing them to operate with minimal further intervention. Encouragement of so-called keystone species also falls into this category. These are species which have an important influence on ecosystem function and biological diversity disproportionate to their numerical abundance. Examples include beavers through their damming activities, prairie dogs, *Cyonomys spp.*, and European rabbits *Oryctolagus cuniculus* because of their effects on the vegetation and importance as prey for a wide range of species, and large carnivores that influence the numbers and grazing patterns of large herbivores (Figure 2.4). The main advantages of such an approach are the high level of perceived naturalness, lower ongoing maintenance costs and the potential to support species rare or absent from more intensively managed habitats.

**Fig. 2.4** The importance of large carnivores. No ecosystems in temperate areas can be considered to operate in very close to a truly natural way, because they contain different assemblages of large mammals compared to that in a natural state. The full suite of native large carnivores is rarely present even in the largest protected areas.

The importance of large carnivores in influencing habitat conditions for other species is well illustrated by changes in tree regeneration following the re-introduction of gray wolves, *Canis lupus*, to Yellowstone National Park in Wyoming/Montana/Idaho, USA. Wolves prey largely on elk (red deer), *Cervus elaphus* (shown here) in the park. Since wolves have been re-introduced elk have altered their behaviour, now spending a lower proportion of time grazing areas where they are at high risk from wolf predation (Creel *et al.* 2005, Hernández and Laundré 2005). This has allowed formerly heavily grazed saplings in areas of high wolf-risk to start growing into taller trees, while stands in low-wolf-risk areas remain heavily browsed (Ripple and Beschta 2003). Photograph by Bill Sutherland.

Discussions over the level of control and intervention to apply when managing habitats for conservation often focus on the level of control over grazing regimes (Section 4.4.1). This can vary from, at one extreme, grazing for discrete periods of the year at prescribed stocking levels, to naturalistic grazing by free-ranging mixtures of large herbivores. A logical continuation of this decreasing level of intervention is the process of rewilding. This aims to restore large, strictly protected core wilderness areas, connected by corridors and supporting free-ranging large herbivores and large carnivores and other keystone species (Noss and Soulé 1998). Some proponents of rewilding even suggest the introduction into North America of African and Asian megafuana, such as lions *Panthera leo*, cheetahs *Acinonyx jubatus*, Asian elephants *Elaphus maximus* and African elephants *Loxodonta africana*, to perform similar ecological roles to their now extinct North American equivalents (Donlan *et al.* 2005).

Introducing natural processes and letting them operate with minimal intervention is in many cases unlikely to create systems very similar to those that occurred naturally, because land use has changed. An example would be the re-naturalization of river floodplains (Figure 2.5). Simply removing artificial river walls to recreate a naturally functioning floodplain is unlikely to recreate similar conditions to those that occurred under natural conditions. The land in the floodplain will usually have been levelled to allow agricultural activities, thus removing its natural variation in topography and hydrology, and consequently reducing the variation in habitat conditions that will subsequently develop. Artificial drainage within the rest of the catchment will increase the rate of runoff following storm events, and create a higher and more short-lived flood peak. If the re-naturalized area forms only part of the floodplain, then it will receive a larger quantity of floodwater than if the water was able to spread over the entirety of the original floodplain. The fertility of the floodplain will also be higher than under natural conditions, due to anthropogenic nutrient inputs within the catchment. There will also be less than a full compliment of large herbivores to graze the vegetation. The way to make the resulting habitats more similar to those that existed under natural conditions would be to use further intervention to recreate variation in topography through land-forming, and introduce management by livestock to mimic the effects of now absent large herbivores. Often, leaving areas to nature will produce conditions quite different to those that occurred in an original, natural state.

The concept of naturalness is frequently used when discussing habitat re-creation/restoration and management. In practice, there is no single natural state of a site. There are, though, a number of different forms of naturalness, as described by Peterkin (1981). These are:

- **original-naturalness**: the state that existed prior to human influence;
- **present-naturalness**: the state that would exist now if there had been no human influence;
- **past-naturalness**: the quality associated with sites whose components have been inherited directly from the original state that existed prior to human influence;

**Fig. 2.5** Re-naturalization of river floodplains. The Skjern River Restoration Project in Western Jutland, Denmark, is one of the largest habitat-restoration projects in Northern Europe. It involved re-excavation of the meandering channel of the River Skjern to close to its original course before it had been straightened as part of a major land drainage scheme. The quantity of soil moved was enough to fill an unbroken line of dumper trucks stretching from Skjern to Rome! An excavated section of the river is shown in (a), and part of the flooded floodplain in (b). The restored 2200-ha river floodplain is now a wonderful wetland. However, because of land shrinkage, changes in the quantity and quality of water the floodplain receives and the lack of native large herbivores, the floodplain will be very different to its original-natural state.

- **potential-naturalness**: the state that would in theory exist if human influence was removed now and the site instantaneously developed into the successional state that it would ultimately achieve following this removal of human influence;

- **future-naturalness**: the state that would eventually develop if human influence was removed now and remained permanently removed.

It is important to recognize that reinstatement of natural processes and removal of further human influence will never result in sites achieving their original-naturalness. Some species will be no longer present or able to re-colonize. Others, particularly alien or exotic species, may have colonized since its original-natural state. Changes in climate and soils will also prevent sites from returning to their original-natural state.

## 2.4 Integrating habitat management with other interests and values

Like it or not, nature conservation *per se* is a minority interest. Even for people who regularly visit non-urban areas, the most highly valued features are usually the landscape, presence of livestock or sense of wilderness, rather than their specific biodiversity value. For nature conservation to maximize its potential, it therefore needs to be integrated with other interests including health and well-being (Figure 2.6) and its economic case made in terms of the entire range of benefits provided by more sustainable land management (Figure 2.7).

At a site-based decision-making level, consideration needs to be given to the extent that habitat management for nature conservation is integrated with, or even compromised by:

- recreation;
- education and research;
- landscape and aesthetic considerations;
- cultural history;
- resource use;
- provision of wider environmental benefits, like ecosystem services such as flood protection.

Habitat management also needs to consider the wider environmental damage that management might cause, especially through increasing carbon emissions. Methods for taking account of these interests when decision-making at the site level, including setting different objectives for different areas of a given site through zoning, are discussed in Chapter 3.

## 2.5 Conclusions

As we have seen, there is a range of considerations when deciding how to approach habitat management. Although it is important for organizations and individuals

## Conclusions | 23

**Fig. 2.6** Integrating nature with other recreational interests. Conservationists are often excellent at providing facilities for people already interested in conservation, but can be less good at reaching out to new audiences.

In the National Park de Hoge Veluwe in Gelderland, The Netherlands, wildlife, landscape, art, architecture, and outdoor recreation are fully integrated. The park houses the Kröller-Müller Museum, containing many works by Van Gogh, Picasso, and others and a sculpture park set amidst large areas of heathland, acid grassland, and woodland. There are also 42 km of cycle paths and 1500 white bicycles available for free use by visitors.

**Fig. 2.7** Making the economic case for nature conservation. Large areas of formerly high conservation value peatlands in Belarus have been drained for agriculture and forestry, but subsequently left unmanaged. Once drained, unmanaged peatlands are vulnerable to fire. In 2002, US$1.5 million was spent extinguishing peatland fires in Belarus. Re-wetting the peatlands (a) reduces the incidence of fires and the large sums of money spent extinguishing them. It also has considerable wildlife benefits, including restoring suitable habitat for the vulnerable aquatic warbler, *Acrocephalus paludicola* (b), while the re-wetting and reduction in fires will provide economic benefits through the developing carbon market by restoring the functions of these peatlands as carbon sinks (a, Bortenicha, Minsk Region, Belarus; photograph by Alexander Kozulin; b, photograph by RSPB IMAGES).

to formulate a philosophy regarding the extent to which they focus on species, habitats and natural processes, these principles have to be applied in a pragmatic way. In practice, differences in the level of intervention usually largely reflect differences in the type and size of sites inherited.

In the case of cultural habitats of high conservation, cultural or landscape value, the emphasis is on maintaining these invariably open, early successional habitats and their associated species through continuation, reinstatement or modification of former land use. Introduction of other management that relies on the reinstatement of near-natural ecological processes will inevitably lead to an impoverishment of their existing suites of species and loss of cultural and existing landscape value. Where more pristine areas of habitat remain, there will be a greater emphasis on maintaining these large, near-natural areas through minimal human intervention. It would be absurd to introduce intensive, small-scale management to large areas of near-natural wilderness. The interesting area of debate, though, concerns how to manage medium-sized areas of less highly valued cultural habitat, and in particular how to design and manage newly created habitat within otherwise intensively managed landscapes: whether to recreate cultural habitats or landscapes or something different.

Introducing and maintaining key natural processes might have many benefits, particularly in terms of its perceived level of naturalness and the potentially low ongoing management costs. However, it is also clear that no sites can ever be considered fully natural because of their past modification, the influence of wider human activities and because they are unlikely to be large enough to operate in a truly natural way. We have also seen that there is no all-embracing, natural state of a habitat. The best we can hope to achieve when seeking to create 'natural' habitats is to choose some desirable past-, present- or future-natural state to aim at.

Any desire to allow natural processes to take precedence has therefore to be tempered with pragmatism. The approach will usually be to let natural processes and functions operate as far as practical, while recognizing that on occasions more interventionist management may be necessary to conserve species considered to be of high value. It is also important to distinguish between encouraging natural processes to create conditions closer to what we consider to be original-naturalness, and viewing the introduction of key natural processes more as an end in itself.

Management of an existing site may often involve extending its area to increase the long-term viability of populations of desired species, and decrease unit management costs. Where this is the case a combination of traditional and more naturalistic management may be appropriate. The existing fragment of habitat may have to continue to be managed through traditional management in the short term to maximize the chance of maintaining its existing species compliment. Meanwhile, the newly created habitat surrounding it could be managed less intensively at lower unit cost. The intention would be that the highly valued species maintained by intensive management will in time colonize the surrounding, less intensively managed land. Once this has happened, the whole area could then be managed less intensively with little risk of extinction of the then more widespread, highly valued species.

# 3
# Setting objectives and monitoring

Habitat management has the potential to both benefit and damage the conservation value of an area. It is therefore important to think out clearly:

- what you intend to achieve;
- how you intend to achieve it;
- how you will know whether you are achieving it.

The best way to decide these is through management planning, which is sometimes referred to as site action planning.

## 3.1 Management planning

A management (or site action) plan is a document that helps to ensure that you follow a logical decision-making process. There are many different formats for management plans. Despite this, all good management plans have the same basic logic. This is described in Section 3.1.1.

Once produced, the management plan will inform people what needs doing at the site. It can also be used as a bid for resources by demonstrating that the proposed actions are the product of a logical decision-making process. The plan will also help facilitate communication, by providing a compendium of information about the site, and by setting out the objectives for the site and how it is intended that they will be achieved.

It is important to recognize that the management plan is only a *part* of an *ongoing process* of decision-making, monitoring and re-evaluation (Figure 3.1). It is especially important that the effects of management are monitored, and that the results of this monitoring are used to inform future site management. Many management plans fail because the plan itself is seen as the end-product.

It is worth noting that 'management' is simply the terminology used to describe what is done (or not done) at a site. Management does not necessarily imply *active* management of a habitat, such as grazing or control of water levels.

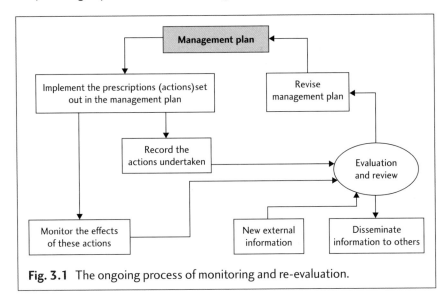

**Fig. 3.1** The ongoing process of monitoring and re-evaluation.

Instead, it may refer to lobbying for change in land use, consulting with local people, or even managing an area by non-intervention.

Key decisions before starting to produce a management plan are which stakeholders (the people involved in managing the site and other landowners and resource users) to involve and how to involve them. Nobody likes being told what to do, without having been involved in deciding why. It is, though, often best if one person *co-ordinates* the process of producing the management plan. It is also important to agree a timescale for plan production.

Another issue worth considering at an early stage is ensuring that monitoring is afforded a suitably high priority and is adequately funded. Monitoring requirements should be decided through the management process, and not overly led by personal interests.

### 3.1.1 The format of the management plan

The first decision is how complex and 'glossy' the plan should be: is the plan primarily for internal use, or will it be needed to communicate to wider audiences? It can be tempting to include everything that you can possibly think of in a plan, particularly when designing a generic plan format for use by an organization. However, including too much information can slow down production and make the plan less accessible. The ideal format is short, simple, and focused on the key decisions. The main body of plans can be kept short and concise by including more detailed information in appendices. Figure 3.2 summarizes the sections and logic of a typical management plan.

Management planning | 29

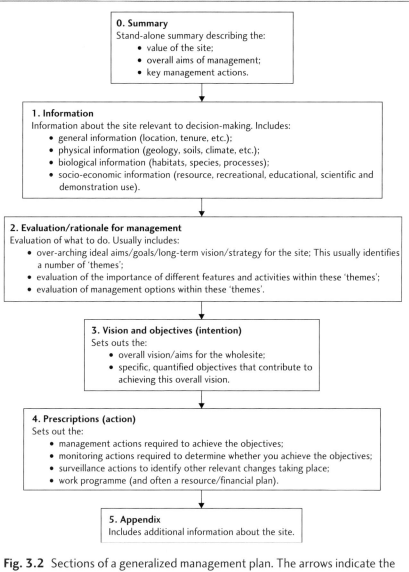

**Fig. 3.2** Sections of a generalized management plan. The arrows indicate the progression of the decision-making process.

## 0. (Executive) Summary

This is used to communicate the key elements of the plan. It usually includes the following:

- a description of the site; that is, why it is special;

- the main objectives for the site written in a user-friendly language; that is, what we want the site to be like;
- the main prescriptions for the site written in a user-friendly language; that is, what management will take place.

If the summary is also used to inform the public, then it might also be worth including a map, pictures, information on access arrangements and details of where to obtain further information. The style and detail in the Summary will depend on the site's key audiences.

## 1. Information section

The first main section of any management plan comprises a collation of information about the site. At one extreme it can be a fairly exhaustive repository of information, and may serve a useful purpose in being so. At the other it might include the minimal quantity of information required for the decision-making in the rest of the plan. This section should also indicate where further information about the site is held. Collation of information for this section is useful in identifying knowledge gaps.

Another decision is the extent of the use of maps. It is essential to include at least some maps in the plan. Annotated maps are excellent for displaying information about the site, although electronic versions of maps can be time-consuming to produce.

Table 3.1 shows the types of information usually included in the Information section of management plans. Not all will be relevant to every site. Some management plans have a separate introduction/context section at their beginning, in addition to the (Executive) Summary. This usually includes information on the legal basis for the plan, its process of development and procedures for its modification, updating and implementation. Sometimes, the system for reviewing the plan is included as a separate section.

## 2. Evaluation/rationale for management section

This section begins the decision-making process. It evaluates the information in Section 1 and provides the rationale for the objectives and management described in Section 3.

Evaluation/rationale for management is the section of management plans that varies most between formats. Most, though, broadly follow the logic described below.

**Table 3.1** *Typical contents of the Information sections of a management plans*

| Main section | Sub-section | Information required | Relevant to all types of site? |
|---|---|---|---|
| General | Organizational policy and legal framework | Statement of organizational policy and legal framework governing management of the site | Yes, but not always included. Can be included in separate Introduction/context section instead. |
| | Location of site | Map showing location, site boundaries and details of relevant local authorities and area | Yes |
| | Area | Area of the site in hectares or square kilometres | Yes |
| | Map and aerial-photograph coverage | List of relevant maps and aerial photographs and where they are held | Yes, but not always included. |
| | Designations | List of designations and how they potentially influence land management | Yes |
| | Land tenure and rights | Maps and details of which areas of land are freehold, leasehold, common land, under management agreement, other legal rights, planning permissions, etc. Details of lessors, etc., with renewal dates of agreements, etc. | Yes |
| | Conditions of land purchase, grants and gifts, etc. | Any conditions of land purchase that might influence land management | Yes, where relevant, but rarely in management-plan formats. |
| | Management resources and infrastructure | Details of buildings and other structures and of current staffing. | Yes, although information on management resources is not given in all formats. |
| Physical | Geology, topography, geomorphology, and soils | Information on these that aids understanding of the site and of potential management options | Yes |
| | Hydrology | Description of the hydrology (including map) and of water quality | Yes |

**Table 3.1** (Continued)

| Main section | Sub-section | Information required | Relevant to all types of site? |
|---|---|---|---|
| Physical | Climate | Mean rainfall and temperature and long-term trends | Yes |
| Biological | Recording areas | Maps of recording areas for different groups, sampling points and transect routes, etc. | Yes, but not in all management-plan formats. |
| | Habitats | List of habitats, their areas, and often brief descriptions | Yes |
| | Vegetation communities | List of vegetation communities, their areas, and often brief descriptions | Yes |
| | Flora | Lists and status at the site's key species and information on changes in their abundance over time. Full lists of species are sometimes included in an appendix. | Yes |
| | Fauna | As above | Yes |
| | Ecological processes | Information on key processes that might influence management | Not in all formats. |
| Socio-economic | Archaeological, historic, and socio-cultural | Details of historic and socio-cultural uses of the site | Yes, but specific information required varies greatly between sites. |
| | Resource use, local communities, human threats and pressures | Details of resource use (farming, timber extraction, water use, etc.) and human pressures (e.g. deforestation) | Yes, but specific information required varies greatly between sites. |
| | Recreational use | Details of access arrangements and recreational use, information on landscape | No |
| | Educational/demonstration/scientific use | Details of educational, demonstration and scientific use | No |
| | Previous/ongoing conservation activities | Summary of previous and ongoing conservation activities and who implemented them | Yes, but not always included. |
| | Stakeholders and their roles | List of current stakeholders and their roles | Yes, but not always included. Sometimes included in an Introduction/context section. |

The first step is to identify the **overall aims** for the site *during* and *beyond* the life of the management plan being written. Different plan formats may refer to these as the ideal aims, goals, long-term vision, or strategy. Defining the overall aims involves defining the main themes for the site and the relative priorities of potentially competing themes; for example, whether the main overall aims are to benefit biodiversity or to provide an educational resource. Commonly used themes for sites are listed below. This process might also include dividing the site into zones, including so-called buffer zones, and identifying different themes and aims for these.

### Biological, environmental, and landscape themes

- Protection or enhancement of biodiversity.
- Maintenance of environmental/ecosystem services.
- Protection of natural/cultural features, including landscape.

### Human themes

- Maintenance/promotion of recreation/tourism.
- Maintenance/promotion of education and awareness.
- Maintenance/promotion of sustainable use.
- Maintenance of cultural/traditional attributes.
- Support of community development.
- Promotion of scientific research.
- Institutional administration/Health and Safety, etc.

The next stage is to examine these individual themes in more detail to inform the process of objective setting in Section 3.

### Biological, environmental, and landscape themes

The procedure for evaluating biodiversity, environmental/ecosystem services, and natural/cultural themes involves first identifying what is important about the site. There is usually a presumption that management will aim to maintain or enhance the site's existing most important features, where feature is a generic term used to describe individual species, assemblages or communities of species, habitats, processes, and environmental/ecosystem services. Methods of assessing the relative importance of individual features vary between plan formats. The importance of biodiversity features is usually assessed using the criteria shown in Table 3.2. There are no universally used criteria for evaluating the importance of environmental services and natural/cultural features.

**Table 3.2** *Criteria commonly used to evaluate the importance of biodiversity features*

| Biodiversity feature(s) | Criteria |
| --- | --- |
| Ecosystems, habitats, and plant communities and assemblages | Rarity |
| | Size |
| | Condition |
| | In some formats also: naturalness, fragility, typicalness, diversity |
| Individual plant and animal species | Rarity |
| | Population size |
| | Importance of their role in the ecosystem: keystone species |

In some cases it is useful to also identify which features have the potential to become important in the future—potentially important features—for example, where habitat (re)-creation/restoration is taking place. It is also possible to identify important features that are thought likely to be present, but for which there is currently no or only limited information: possible important features. This is most often the case for poorly recorded groups of invertebrates. The management plan will usually include an action to survey at least some of the key, under-recorded groups, focusing on those whose presence might influence management.

Once the important features of the site have been identified, the next stage is to identify the main factors that will determine whether each important feature achieves the desired state set out in the management plan's overall aims. For example, the overall aims may involve increasing the populations of particular rare species. The main 'factors' influencing whether this can be achieved might be the extent of suitable habitat or freedom from human persecution. Negative factors are often called threats or pressures. Factors that cannot be overcome are often referred to as constraints. Recognizing these factors helps us identify what actions, if any, are necessary to attain the desired condition of the important feature.

### Human themes

There are a variety of approaches for evaluating information and developing a rationale for management for various human themes. Typical steps in these processes are outlined below.

### Maintenance/promotion of recreation/tourism

Identify the:
- main features of interest for visitors;
- numbers and profile of current and potential visitors;
- types of recreation suitable for the area;
- current and potential impacts of tourism;
- existing and potential sources of conflict;
- ways that tourism can benefit the area and key communities.

### Maintenance/promotion of education and awareness

Identify the:
- suitability of the site for various education and awareness activities;
- target groups and key messages.

### Maintenance/promotion of sustainable use

Identify:
- the main positive and negative impacts of the most important forms of land-use in the area;
- any opportunities for continuing/promoting sustainable land use.

### Maintenance/promotion of cultural/traditional attributes

Identify:
- key cultural and traditional attributes;
- potential benefits and conflicts between these and achievement of biological objectives;
- any opportunities for co-existence of cultural/traditional attributes and biological objectives.

### Support of community development

Identify:
- the main positive and negative interactions between local communities and the area;
- any opportunities for local people to gain benefits from the site.

### Promotion of scientific research

Identify the suitability of the area for research.

## 36 | Setting objectives and monitoring

*Institutional administration/Health and Safety, etc.*

Identify what is needed to satisfy the requirements of administration and Health and Safety.

## 3. Objectives and prescriptions

This is the key part of the plan that sets out:

- *what* you want to achieve (your **objectives**);
- *how* you intend to achieve them (usually called **management prescriptions**);
- what *monitoring* you need to undertake to determine whether you are achieving your objectives (usually called **monitoring prescriptions**).

The objectives and prescriptions are the product of the process that has taken place in Section 2, which has already identified what is important about the site, what state we want it to be in, and what main factors we consider to be affecting whether it attains this desired state. It is also important to decide the extent to which the intended actions are limited by current resources, or used as a bid to gain *additional* resources.

The objectives need to be sufficiently clear and detailed to ensure that everybody involved with the site knows what you are collectively aiming to achieve. Objectives should therefore be as SMART (i.e. specific, measurable, assessable, realistic, and time-specific) as possible.

The objectives should clearly set out the desired condition of the important features. This is best done by describing the best measures of the condition of these features, and then setting numerical targets, or target ranges, for these measures. These measures of condition are often called attributes. The idea of an attribute is that it is:

- a general condition of a feature;
- practical to measure;
- ideally informative about something other than itself;
- ideally an indicator of the future rather than the past.

Attributes and their targets are called a variety of names in different plan formats. Sometimes they are just called targets, sometimes just limits of acceptable change, and sometimes indicators or performance indicators. They are all more or less the same. Examples of useful attributes for different types of feature are shown in Table 3.3. An example of an objective setting out attributes and targets is shown below. The attributes are in italics and their targets shown in bold.

**Management planning** | 37

**Table 3.3** *Examples of useful attributes for different biological features*

| Feature | Attribute(s) |
|---|---|
| Wintering population of a bird species | Mean (or maximum) winter count derived from standardized monthly counts |
| Population of a butterfly species | Summed annual weekly or two-weekly counts along a fixed transect |
| Population of a plant species | Mean numbers of individuals of the plant in quadrats in a defined area |
| | Extent of mapped distribution of the plant |
| Species-rich grassland | Frequency of positive indicator plant species in quadrats |
| | Frequency of negative indicator plant species in quadrats |
| | Scrub cover |
| | Sward height at a given time of year (cm) |
| | Litter cover |
| Woodland | Tree species composition |
| | Canopy cover |
| | Understorey composition |
| | Indices of the quantity of dead wood |
| | Tree regeneration potential (scores of regenerating seedlings and saplings) |

Objective 1: To maintain the lowland wet grassland primarily to maintain its wintering and passage waterfowl, breeding wader populations, and important ditch flora and fauna.

### *Species attributes and targets*

- Increase the 5-year mean number of breeding pairs of:
  - northern lapwings from 132 to >170
  - common redshank from 93 to >130.
- Maintain *5-year September-to-March summed monthly counts* of wintering waterfowl at >5500.
- Increase the numbers of ditches supporting:
  - frogbit from 23 to >35
  - common bladderwort from 10 to >15.

Etc.

The level of detail and precision that it is useful to include in the objectives will vary, especially according to the amount of information available about key species and habitats and the level of management intervention and control. Whatever the case, objectives still need to be sufficiently quantified to enable progress towards achieving them to be monitored effectively. Without this ability, it is impossible to determine whether you are achieving what you intended. In a wet grassland that is managed intensively to provide suitable conditions for specific groups of species, it might be useful to set quite specific, often single-species, targets.

It is meaningless to set meaningful attributes for most invertebrates, assemblages of invertebrates, and other species that are difficult or impossible to monitor realistically. An alternative is to use a measure of habitat condition as a surrogate. This is not ideal, since the species may still fare badly, despite the habitat being in apparently suitable condition. It will also be less useful setting targets for individual species, or groups of species in more complex and less intensively managed habitats. It will be difficult or impossible (and not necessarily very useful) to set precise, quantified, targets for the functioning of many natural processes.

Often, there is more than one level of targets. These are sometimes referred to as ends or means objectives. The idea of these definitions is to differentiate between what we *ultimately* want to achieve (our ends objective), from our means of achieving it (our means objective). A common situation is where our ends objective is to conserve a particular species, but our means of doing so is by providing particular habitat conditions (our means objective). In some cases there may be more than one level of means objective.

In the wet grassland example, the ends objective might be to increase numbers of breeding waders and meadow birds or wintering waterfowl, but our means objectives might be to provide a particular sward height or flooding regime to benefit them. An alternative to referring to ends and means objectives is to instead refer to them as *species* attributes and targets and *habitat* attributes and targets. Habitat attributes for the wet grassland example are shown below:

### Habitat attributes and targets

- Median height of the sward at the beginning of April: **< 5 cm**.
- Rush cover: **< 20%**.
- Percentage of shallow winter flooding over the site: **25–100%**.
- Percentage shallow spring flooding over the site: **15–20%**.
- Height of ditch-water levels relative to field level at the end of May: **< 30 cm**.
- Percentage of ditches in the following successional stages:
  - early successional (i.e. > 80% open water): **10–40%**.
  - late successional (i.e. choked): **10–40%**.

- Mean annual nutrient levels in ditch water:
  - total oxidized nitrogen: **< 2.0 mg/l**.
  - soluble reactive phosphorus: **< 0.2 mg/l**.

Etc.

Even though there is a danger of becoming too esoteric when setting objectives, it is important to differentiate between means and ends objectives and between targets for species and habitats. It is a common mistake to only focus on your means objective/habitat target, but lose sight of your ends objective/species target. This can result in focusing solely on, for example, grazing a sward to a particular height (achieving the means objective/habitat target) without recognizing that this is not achieving (or not knowing whether it is achieving) the ends objective/species target of maintaining the population of a particular species that the grazing is aimed at benefiting.

Once you have agreed your objectives and defined your attributes and targets, the next stage is to set out what management you need to undertake to achieve these targets, and what monitoring you need to do to determine whether this management is proving successful. Monitoring is the use of repeat surveys using **standardized methodology** to determine progress towards a **target**, compliance with a **pre-determined standard**, or the degree of deviation from an **expected norm**. In this case, we have already defined our targets or target ranges.

Monitoring is often set up and undertaken *without* the intention of comparing the results with pre-determined targets, target ranges, or limits of acceptable change. Instead, it is often used solely to determine change. This process is termed **surveillance**. This is the use of repeat surveys using standardized methodology to determine whether, and to what extent, something changes over time. However, in practice many surveillance actions are commonly (although wrongly) referred to as monitoring. We return to surveillance later on.

A commonly used method of preserving the decision-making logic and maintaining clarity is to list the individual management and monitoring prescriptions (or at least a summary of them; the detail is usually held later on in the plan) under their relevant objectives. This makes it difficult to include management prescriptions that do not contribute to achieving specific objectives, and difficult to include monitoring prescriptions that do not measure progress towards achieving these objectives. The objective and prescriptions for the wet grassland example are shown below.

**Objective 1: To maintain the lowland wet grassland primarily to maintain its wintering and passage waterfowl, breeding wader populations, and important ditch flora and fauna.**

### Species attributes and targets

- Increase the 5-year mean number of breeding pairs of:
  - northern lapwings from 132 to >**170**
  - common redshank from 93 to >**130**.
- Maintain *5-year September-to-March summed monthly counts* of wintering waterfowl at >**5500**.
- Increase the numbers of ditches supporting:
  - frogbit from 23 to >**35**
  - common bladderwort from 10 to >**15**.

### Habitat attributes and targets

- Median height of the sward at the beginning of April: **< 5 cm**.
- Rush cover: **< 20%**.
- Percentage of shallow winter flooding over the site: **25–100%**.
- Percentage shallow spring flooding over the site: **15–20%**.
- Height of ditch-water levels relative to field level at the end of May: **< 30 cm**.
- Percentage of ditches in the following successional stages:
  - early successional (i.e. >80% open water): **10–40%**.
  - late successional (i.e. choked): **10–40%**.
- Mean annual nutrient levels in ditch water:
  - total oxidized nitrogen: **< 2.0 mg/l**.
  - soluble reactive phosphorus: **< 0.2 mg/l**.

Etc.

### Management prescriptions

- Summer grazing by cattle.
- Top to maintain a short sward.
- Create and maintain winter floods.
- Create and maintain scattered pools in spring.
- Maintain high ditch-water levels.
- Carry out rotational ditch management.

Etc.

### Monitoring prescriptions

- Monitor breeding waders.
- Monitor wintering waterfowl.
- Monitor ditch plants.

- Monitor ditch-water levels and extent of surface flooding.
- Monitor sward height.
- Monitor water quality.

Etc.

Setting objectives for non-biological objectives follows the same principles, although it can be more difficult to set realistic targets and practically monitor progress towards them. In the example below, the ultimate measure of whether or not the objective has been achieved is whether the target audiences have increased their awareness of the reserve and awareness and understanding of environmental issues. In practice, this is rarely monitored effectively. It is more common to just set targets for the means objectives, in this example called the people targets.

**Objective 5: To increase awareness of the reserve and awareness and understanding of environmental issues in general.**

*People targets*

- Engage with >1000 schoolchildren and students per annum during the next 5 years.
- Produce ten or more media releases per year.
- Run four or more open days per year.
- Give more than three presentations to workshops or conferences on upland management during the next 5 years.

So far, we have only discussed monitoring to determine progress towards achieving specific site-management objectives. Monitoring can also be undertaken to contribute information to wider monitoring schemes. It is often also informative to track changes in other biological and environmental features that we do not currently aim to influence through management intervention. Measuring changes in these can help:

- interpret the rest of your monitoring results, for example by recording precipitation to interpret changes in hydrology;
- detect changes that might result in you taking action at a later date, for example declines in water quality, spread of alien/exotic species, vegetation succession, or in populations of species that you are not currently seeking to influence.

We may set targets or limits of acceptable change for some of these variables, even though we are not currently seeking to influence them. An example would be setting maximum acceptable levels for water-quality parameters, which, if exceeded, would trigger some intervention. In many cases, though, we may simply want to track changes in a variable, without setting any targets or levels of acceptable change. This constitutes **surveillance**, as described above. It is useful to include these monitoring and surveillance prescriptions under a separate objective, to separate them from prescriptions that are directly monitoring the effectiveness of specific management actions. An example is shown below.

> Objective 7: To inform management of the site and contribute to national recording schemes.

### Survey, surveillance, and monitoring projects

- Survey under-recorded groups.
- Record daily precipitation.
- Undertake surveillance of alien/exotic plants.
- Carry out national Wetland Bird Survey counts.

## 4. Projects

Management and monitoring prescriptions are generally broken down into individual projects, which are included elsewhere in the plan.

Examples of individual projects are given below:

### Summer grazing by cattle

> Manage habitat, grassland, by controlled grazing.
> SUMMER GRAZING BY LIVESTOCK
> Grazing livestock to be the principal means of achieving predominately short and varied sward including tussock-forming species. Cattle are preferred, though a limited number of sheep and ponies may be used. Early season grazing to start from mid-May and then concentrated where fewest nesting birds to avoid nest trampling. Earlier grazing may be possible in some areas. Usual grazing season ends at end of October, though can be extended to December if conditions suitable, and poaching is avoided. Stocking rates between 150 and 300 livestock unit days/ha per year.

> Liaise, owners/occupiers
> LIAISE WITH GRAZIERS
> Site manager to maintain good relations with all graziers taking rights on the reserve.

Collect data, management, by owners/tenants/public bodies/neighbours.
RECORD GRASSLAND MANAGEMENT
Collect data on grazing regimes, recorded as livestock numbers present in each field within reserve area on a daily basis from May to November. Results to be summarized in annual report.

### Topping to maintain a short sward

Manage habitat, grassland, by mowing
TOPPING TO MAINTAIN SHORT SWARD
Topping is necessary to maintain predominately short sward over most of reserve where late and light grazing leaves tall, coarse vegetation. Fields to be topped from centre out to reduce risk of killing chicks and mammals.

### Create and maintain winter floods

Manage habitat, swamp/fen/inundation, by water-level control.
CREATE AND MAINTAIN WINTER FLOODS
To flood shallowly a minimum of 130 ha to provide feeding and roosting areas for wintering wildfowl. Natural floods draw down to leave extensive bodies of water trapped in low-lying fields. In absence of natural flooding the 30 ha of artificial floods can be achieved by use of water-level control structures and pumping.

Manage habitat, swamp/fen/inundation, by water level control
MAINTAIN INFRASTRUCTURE FOR HYDROLOGICAL CONTROL
Maintain bunds and sluices necessary to control water levels.

Etc.

## 5. Work programme

Management plans always include a prioritized work programme for the period covered by the plan, which can be broken down into an annual work plan. This usually details:

- what you intend to do;
- in what year it will be done;
- what priority it is.

### 6. Resources/budgeting/financial plan

Some plans also include the cost of agreed actions in the form of a financial plan.

## 3.2 Monitoring and surveillance

Unfortunately, monitoring is too often regarded as an unaffordable luxury, with little or no benefits compared to immediate conservation action. Without monitoring, though, we have no way of:

- determining whether our management is achieving its aims, or, if not, requires modifying;
- detecting long-term changes in conditions, some of which may require intervention at some stage to maintain the site's interest.

Without monitoring, we will be unable to communicate the effects of our management to others, so they can learn from our successes and avoid repeating our failures. As we shall see, the costs of learning whether or not actions are proving successful can be surprisingly small compared to the total costs of carrying out habitat management.

An example of the importance of monitoring is provided by the *inadequate* monitoring of many European agri-environment schemes (Kleijn and Sutherland 2003). This has made it difficult, or impossible, to determine the effectiveness of many of the measures included in them, and has thus hindered the development of more effective schemes, especially those introduced to European Union accession countries. When applying for grants for conservation management, also seek funding to monitor the effectiveness of these actions. Unfortunately, many grant-giving bodies do not fund monitoring, even though it would presumably be useful for them to know whether the schemes they are funding are actually achieving what they purport to.

Communicating the results of monitoring is usually considered an even lower priority than the monitoring itself. It is particularly valuable to record and disseminate negative results. Unfortunately, this is rarely done and negative results are especially difficult to publish. The results of monitoring which re-confirms previous findings, but under different conditions and at different sites, are also informative. The Internet provides excellent opportunities for communicating the results of habitat management (see www.conservationevidence.com).

In the following sections I emphasize some key principles and considerations to help in the design of monitoring and surveillance projects. The process for deciding *which* variables to monitor has already been described in the previous section. Details of specific methods that can be used are described in a number of practical guides (e.g. Hill *et al.* 2005; Hurford and Schneider 2006; Sutherland 2006a).

### 3.2.1 Recording management actions

It is impossible to interpret the results of monitoring without also recording what management was undertaken. If the monitoring shows that an objective is not being achieved, then there are two possible reasons for this: either the agreed actions were not carried out, or they were carried out but did not have the desired effect. If the latter, you need to re-consider the link between the agreed actions and their predicted effects.

### 3.2.2 Making the most of existing information

The first step when designing a monitoring project should be to search for any *existing* monitoring information that can be put to good use. It is often tempting to design a new, well-thought-out monitoring project, while wasting data that have already been collected as part of a previous, albeit less-well-designed, project.

### 3.2.3 Reverse planning and deciding what level of evidence is required

It is a common mistake to go out and collect lots of monitoring information (usually the fun bit), often for years, and then eventually wonder how it is going to be analysed and whether it was useful to collect in the first place. Instead, it is best to design the monitoring project backwards; that is, reverse planning (Sutherland 2006b). First, think about what you expect the changes to be and how the information collected will be analysed and used. Unless you know what the results will be used for, you will not know the level of evidence required, and therefore how to design your data collection.

A common reason for the failure of monitoring or surveillance projects is choosing the wrong level of evidence. For example, undertaking rigorous, complicated sampling to determine something that is blindingly obvious. Alternatively, taking too few samples to enable the detection of small, but important, changes.

An example of where a low level of evidence is required may be where you are monitoring the extent of an unwanted, invasive species, such as rushes, *Juncus* spp., on our previously described wet grassland. In this case, the aims of the monitoring might be to:

- provide a check of whether rushes are spreading over too great an area and whether they need controlling;

- enable changes in abundance of breeding waders/meadow birds on the grassland to be interpreted, thus allowing the effects of differences in rush cover to be considered, or dismissed, as potential causes.

There is no point in collecting lots of data on the cover of rushes in a large number of quadrats in each field to determine subtle differences in cover between fields and over time. Instead, all you want to know is whether the extent of rushes is increasing so much that it might have a detrimental effect on breeding waders/meadow birds and require control. In this example, taking some quick, fixed-point photographs each year should be sufficient to answer this question. Using a more complicated method would be an inefficient use of time. There might even be an argument for not carrying out any formal monitoring at all. Memories can be deceptive, though. In this case, taking some photographs is so quick and easy, that its benefits far outweigh its costs.

An example of where a higher level of evidence would be required is if monitoring whether management is providing a suitable sward height and abundance of a food plant for the larvae of a rare moth. In this case, the aims of the monitoring might be similar to those already described:

- to provide a check of whether the desired conditions have been met, and if not, whether the grazing management needs to be adjusted;
- enable changes in abundance of the moth to be interpreted, so that the effects of differences in sward conditions can be considered, or dismissed, as potential reasons.

In this example more detailed monitoring than using fixed-point photography is needed to obtain the required information. It would, though, also be an inefficient use of time to take a large number of sward measurements per field to enable a statistical comparison of small differences in sward height. Instead, a quick visual assessment of whether the sward at different points in each field falls into one of several height bands (e.g. 0–5, 5–10, or 10–20 cm) should suffice, together with a count of how many of the places across the field the larval food plant occurs in. Again, there may be an argument for not undertaking formal monitoring at all. However, even though the current state of the site might appear obvious to those managing the site, it might not in a few years time, particularly if there are changes in personnel. The benefits of a few hours each year spent monitoring the results of management should again far outweigh its costs.

An example of where higher levels of evidence might be required is where the results are genuinely less predictable, where small-scale changes may have important implications and are less easy to detect visually, and where the results are likely to be most useful for others (i.e. novel management). In these cases, quite detailed, and thereby more expensive monitoring will be required to detect the relevant changes. However, if the results are used to inform others, then the benefits to *wider conservation* are likely to far outweigh the costs of monitoring.

So, before you design a sampling strategy consider what type of measures are practical to collect, what densities of the particular species you expect to occur, how they will be distributed, and, importantly, what magnitude of differences over time or between areas you want to be able to detect. When setting up monitoring or surveillance to detect long-term changes, think of the changes that you expect to occur over the relevant timescale. It is also worth thinking back in time: what monitoring would it have been really useful for people to have set up 20 years ago?

### 3.2.4 Using indices and sampling

A common misconception is that the best method for monitoring a species is the one that produces the highest numbers. Practitioners frequently bemoan methods that produce lower counts of a species than they 'know' to be there. As already discussed, though, the primary aim of monitoring is to provide a *measure of change*, not necessarily to produce the highest population count or estimate. Ideally, the method used will do both. In reality, some methods might be good at detecting change without necessarily providing a very close estimate of the true population size or environmental variable. Often, *indices* of population size provide a more efficient way of detecting change, and can be carried out more quickly and be easier to standardize.

If you are recording the whole population of a species in an area, then you can simply count or obtain an index of all of them and compare changes in these total counts or indices. In many other cases, though, it is not practical to measure the whole population. For example, if you want to determine the abundance of invertebrates in the mud of a wetland, then it will be impossible to count *all* of the invertebrates in the entire expanse of mud. The only way to measure change will be to *sample* the population, and use this information to *estimate* the total population.

If you decide to sample a population, then it is crucial to carry out statistical tests to determine the likelihood that any differences in numbers recorded are due to real differences in the numbers present, or just due to chance. It is tempting to believe that if you record more of something in your samples one year, then this reflects a real increase in the population you are sampling. For example, if you record an average cover of 24% of a particular plant species in 15 quadrats in a field in one year, and an average cover of 35% of the plant in 15 quadrats the following year, it is tempting to conclude that the plant has increased within the field. Unfortunately, though, if you had sampled the same field twice on the same day, this would almost certainly have produced different average cover values for the plant: it is unlikely that the average from your 15 samples would be exactly the same. Without carrying out statistical tests, you are unfortunately deluding yourself that real changes have taken place when they have probably not.

It is a common misconception that if you are sampling a larger area of habitat, then you need to take more samples; for example, that if you have two similar fields of 5 and 10 ha, you should take twice as many samples from the 10-ha field compared to the 5-ha one. What you should instead be considering when deciding how many samples to take are the:

- *variation* in the samples you are taking;
- magnitude of the difference you need to detect.

Methods for determining the number of samples you need to take to obtain a desired level of precision in your estimate are described in books on ecological monitoring, for example by Greenwood and Robinson (2006). In this example, if the vegetation throughout the 5-ha field is more varied than that throughout the 10-ha one, then you need to take more samples from the 5-ha field to estimate its composition with a similar level of confidence to the 10-ha field.

For samples to be truly representative of the area they are taken from, then they need to be distributed *randomly*. This ensures that each part of the area has an equal likelihood of being sampled. Sampling randomly often means that you end up sampling lots of uninteresting areas. It is crucial, though, that your samples remain truly random, however tempting it may be to adjust them slightly. In the example already given, these biases might involve, even at a subconscious level, dropping the 'random' quadrat so that it encloses a rare plant, rather than just misses it. Once you start subjectively choosing the locations of samples even slightly to include areas you consider to be 'more representative', then you will completely bias and invalidate your results. The results will instead be more a product of how, and the extent to which, you have subjectively chosen your sampling points, rather than be a true sample of what is actually there. Importantly, the next person to carry out the monitoring may subjectively alter the positioning of the sampling points in a slightly different way, making your and their results incomparable.

A method for reducing the number of 'uninteresting' areas sampled is to stratify your random sampling. This involves defining different sampling areas at the outset and then taking different numbers of random samples from these different areas.

Although it is important to choose a suitable level of evidence, it is nevertheless often sensible to take a large number of samples, at least to start off with. It can be frustrating to repeat the monitoring after several years, only to find that so few samples were taken during the first round of monitoring to stand much chance of detecting statistically significant changes. Often, when sampling, the time taken to carry out the fieldwork is relatively small compared to the time and cost of organizing it, travelling to and from the site, perhaps staying overnight somewhere and writing up and analysing the results. Doubling the number of samples might take another few hours, but make the difference between the project providing useful information in years to come or having been a waste of time.

Repeating sampling at the *same* points greatly increases the precision of your estimate, because you are not introducing variation caused by carrying out sampling in different areas. The introduction of hand-held global positioning systems (GPSs) has revolutionized the re-location of sampling points. Even so, since GPSs are only accurate to within a few metres, it is still usually necessary to mark on the ground the precise sampling location. Locating sampling points along permanently located belt transects is useful for monitoring changes across a gradient of environmental conditions.

## 3.2.5 Use of controls

In most situations it is sufficient to just monitor changes in the area you are managing, to determine whether or not the management is proving successful. In some cases, though, it is useful to determine whether the observed changes are due to the management itself, or would have happened irrespective of this management. This can be done by comparing changes in the managed area with those in a similar, unmanaged, control area over the same period of time. An excellent way to determine the effects of grazing is by monitoring changes inside and outside grazing exclosures.

It is still possible, though, that any observed differences between the managed and unmanaged areas are simply due to chance. The most rigorous level of monitoring is therefore to monitor changes taking place in several managed and unmanaged areas. This involves setting up a randomized, replicated experiment. It is rarely practical to set up such experiments on a large-enough scale to investigate the effects of management on habitat use by vertebrates. It is, though, often feasible to use randomized, replicated experiments to determine the effects of management on other factors, such as vegetation height and composition, that are thought to be important in influencing habitat use by them.

### 3.2.6 Frequency of monitoring

Another important consideration is the frequency of monitoring. Often people invest a lot of time in monitoring at frequent intervals, for example annually, only to become disillusioned that the monitoring is not showing any meaningful changes. They often then cease doing it. The starting point should be to review existing information from the site or elsewhere to estimate the likely timescale of detectable changes. If it is still not clear how frequently monitoring should be undertaken, then a good principle is to start monitoring frequently at first to obtain an idea of the likely rate of change. The frequency of monitoring can subsequently be reduced if changes are only taking place slowly.

In some cases it may be best to undertake different levels of monitoring at different frequencies. An example would be monitoring water quality at frequent intervals using a cheap, hand-held conductivity meter. This is easy to use but only provides a crude measure of dissolved plant nutrients. More rigorous, but costly, chemical analysis of levels of key nutrients can be carried out at the start of the monitoring programme, and then repeated should the results from the conductivity monitoring indicate that significant changes are taking place.

A familiar problem with many donor-funded projects is that they take place over too short a period of time to expect to detect measurable biological effects. The ideal is to use some of the donor money to set up detailed and rigorous biological monitoring at the start of the project. Subsequent monitoring can then be used to determine whether appropriate management systems and actions are in place. It can also be used to train staff in the use of simple, practical monitoring methods, whose results can be used to directly inform management. The ideal is to also secure funding to repeat the detailed biological monitoring at an appropriate time in the future when measurable changes would be expected. Any short-term project can only be considered successful if it has measurable conservation benefits well after its period of funding has ceased.

### 3.2.7 Maintaining consistency of methods

The most important attribute of a monitoring method is that it provides a *consistent* measure of the relevant attributes of the species or habitat feature. It is important to realize that even quite small changes in methods can change the relationship between what you record and what is really there, and thereby invalidate the results of the monitoring. No monitoring method is perfect. All are subject to some biases.

Any decision to change methods has to compare the benefits of using the new method with the disadvantage of potentially being unable to compare your future results with those collected previously. If it is decided to change methods, then the best option is to carry out monitoring using both methods for a period, to allow calibration of the results produced by the two of them.

Changes in recording areas can make it impossible to determine long-term changes in populations of species. If the recording area increases, then it is important to record numbers on the new area *separately* from those on the original recording area.

Other issues that commonly cause inconsistencies of methods are:

- ambiguities in recording areas and details of methods;
- inclusion of *ad hoc* records and those from outside the defined recording area;
- differences in the abilities of recorders.

Ambiguities in recording areas and methods can be largely overcome by clearly writing them down and including a map of the recording areas. Standardized recording forms are particularly useful. Always describe the methods and recording areas in sufficient detail that somebody completely unfamiliar with the site and methods can repeat them without introducing unavoidable biases. It is worth trialing methods and standardized recording forms on colleagues first. It is surprising how instructions that appear completely clear to the person writing them can be interpreted in numerous ways by those unfamiliar with the technique.

Addition of *ad hoc* records to otherwise standardized surveys is often an issue. Adding information from *ad hoc* records increases figures by increasing recording effort. This variation in recording effort will again make it impossible to differentiate whether a change in numbers recorded is due to changes in numbers actually present, changes in recording effort, or both.

There is often a strong, and sometimes irresistible, urge among some recorders to not want to 'waste' interesting records, even if these are from just outside the recording area. Common examples include moving the registration of an interesting breeding bird from just outside the recording area to just inside it, and recording interesting butterflies that are a few metres outside of the recording area along a standard butterfly transect. The justification is often 'after all, they probably uses the land within recording area sometimes' or 'they could have just as well been inside the recording area when I actually saw them'. Tempting though this may be, it introduces bias into the results unless undertaken consistently from year to year. It will also provide misleading information about what is actually within the pre-defined recording area. For example, if you are recording breeding birds within a 1-km square, but also add in records that outside of this 1-km square but within 100 m of its boundary, then you increase the recording area (and estimated density of breeding birds) by 44%! The urge to include records from outside the recording area is often strongest when there are few, and particularly no, records for that particular species *within* the recording area. This problem can be largely overcome by designing the recording form so it includes a section for recording species of interest from outside the recording area. This prevents the record from being 'lost' by the observer (hence no need to move it to within the recording area), but means that records from outside the standard recording areas can still be differentiated from those within it.

Differences in the abilities of recorders can sometimes be difficult to overcome, even with training. It is therefore better to use the same recorder as far as possible. It is helpful for the new recorder to go out with the old one prior to the changeover, to learn how the monitoring has previously been carried out.

### 3.2.8 'Quick and dirty' site audits

A useful form of monitoring to improve site management is to carry out periodic, site-based audits. These can comprise a 'quick and dirty', semi-quantified assessment of habitat condition by specialists or advisers not closely involved with the site. As with many other monitoring methods, the cost of these audits can appear high when viewed in isolation. They are, though, likely to be small compared to the total costs of habitat management (Table 3.4).

**Table 3.4** *Cost-effectiveness of site audits*

| Activity | Cost |
|---|---|
| Staff time for visit plus follow-up workshop (4.5 person days) | £500 |
| Travel | £120 |
| Total cost of audit plus follow-up workshop | £620 |
| Total cost of audit plus follow-up workshop as a percentage of annual site-management costs | =£620/£47,000=1.3% |

Quick and simple 'expert' assessments of habitat condition by specialists can provide a very cost-effective way of providing fresh insights and solutions not necessarily apparent to those more intimately involved at the site. In the habitat audit of a wetland site shown here, so long as the results of the annual audit increase the efficacy of site management by more than 1.3%, then it will be money well spent. If the audit is only conducted once every 5 years, then it need only increase the effectiveness of site management by more than 1.3%/5=0.26% to have been worthwhile.

## 3.3 Achieving conservation objectives in the wider countryside: agri-environment schemes and conservation programmes

An important mechanism for achieving biodiversity conservation, landscape, and broader environmental protection objectives in the wider landscape is through agri-environment schemes and conservation programmes. These involve private landowners entering into management agreements in return for payments from government.

A major distinction is between schemes that require the land to remain in agricultural production, such as most European agri-environment schemes, and those that remove it from agriculture, such as the USA's largest environmental programme for private lands, the Conservation Reserve Program (CRP), and its offshoot the Conservation Reserve Enhancement Program (CREP; Figure 3.3). Another difference is the extent to which schemes are spatially targeted. Several designs of agri-environment scheme have been tried, as described in the following sections.

### 3.3.1 Prescription-led

There are two main types of prescription-led management agreement. The first involves offering a selection of pre-set management agreements, often called tiers, in return for fixed payments. Each tier contains a set of measures (prescriptions) that the landowner agrees to adhere to. In practice, the prescriptions usually largely involve restrictions on farming operations, such as maximum permissible levels of fertilizer use and earliest permissible dates for particular management operations. The second type of agreement involves landowners *choosing* from a *menu* of prescriptions, again in return for fixed payments. The process is usually competitive and applications are selected that offer the greatest potential benefits. This menu-based system is generally considered better. This is because it offers landowners greater flexibility, while also allowing management for wildlife to be more closely tailored to the needs of individual sites.

### 3.3.2 Outcome-led

This involves paying landowners for the wildlife that they produce, rather than compensating them for restrictions on farming practices. This type of scheme requires monitoring of the biological outcomes of the farming, and is therefore only practical where there are clear, short-term, predicted management outcomes. The results of small-scale experiments in The Netherlands have demonstrated the

**Fig. 3.3** Conservation programs. The USA's Conservation Reserve Enhancement Program (CREP) is a partnership scheme between landowners and tribal, state and federal governments. It aims to provide environmental benefits by taking private land out of agricultural production to help decrease erosion, safeguard ground and surface water, and restore wildlife habitat.

Measures to achieve the schemes aims include restoring wetland habitat (a; Leipsic, Delaware USA) and planting filter strips of native warm-season grasses adjacent to watercourses (b; Pickering Creek Audubon Center, Maryland USA).

benefits of this approach. Landowners were paid per clutch of breeding waders/meadow birds on their land. Payment per clutch resulted in higher breeding success and cost 60–90% less per clutch (including the costs of monitoring) than similar, prescription-led schemes. It also resulted in better monitoring of scheme effectiveness and was considered to foster greater cooperation between landowners and conservationists (Musters *et al.* 2001).

### 3.3.3 Auctioning conservation contracts

In this type of scheme landowners offer sealed bids for contracts to manage their land. These bids contain a package of actions intended to benefit the environment. Individual bids are assessed in terms of their value for money, by evaluating the likely benefit of the proposed actions in relation to the price offered to undertake them. This auctioning process has offered large potential cost savings to the body offering the contracts, although a number of design problems were identified which would need to be overcome (Stoneham *et al.* 2003).

# 4

# General techniques and considerations

This chapter covers some general techniques and considerations common to managing most habitats. These include principles of managing for different groups, landscape factors, and disturbances, different approaches to grazing, and methods of controlling unwanted plant species. Eradication of rats and cats on islands is also covered, since this affects a number of different habitats, and can be of critical importance in conserving some island endemics and important seabird colonies. The chapter also includes a discussion of the likely effects of climate change on species and habitats and potential ways to mitigate and compensate for its damaging effects.

## 4.1 Management for different taxa

This section describes some overriding principles applicable to managing all habitats for different groups. Conflicts between the requirements of different groups can occur, but are usually rare, principally because most habitat management involves maintaining groups of species that have already co-existed under previous management.

### 4.1.1 Plants

Habitat management for higher plants usually focuses on maintaining or increasing plant species richness and maintaining populations of rare or otherwise highly valued plants. Habitat management rarely focuses on specifically benefiting lower plants. Many highly valued assemblages of bryophytes (mosses and liverworts) are associated with later successional habitat, particularly old-growth forest that should be left with minimal or no intervention.

The major factors important in maintaining or creating high species richness of higher plants in terrestrial habitats are soil fertility and the amount of vegetation removal and other disturbance. In many vegetation types plant species richness shows a hump-shaped relationship with these variables; that is, it is highest at intermediate fertility and intermediate levels of disturbance.

Nutrients in terrestrial and aquatic habitats have in many cases been significantly raised through human activities, especially in the lowlands of intensively managed countries. High levels of nutrients, especially of phosphorus, are a major constraint in maintaining or restoring high plant species richness. Hence, vegetation types typical of *low* nutrient conditions are usually considered to be of high conservation value, especially in otherwise nutrient-enriched lowlands.

The hump-shaped relationship between plant species richness and the amount of vegetation removal and disturbance results in another general principle of vegetation management. This is that management of grasslands, fens and other herbaceous vegetation aimed at increasing plant species richness usually involves periodic vegetation removal or disturbance. This prevents larger, more competitive plant species from out-competing smaller ones, thereby enabling a wider range of plant species to co-exist in the same area.

A further general principle of vegetation management involves providing suitable bare ground for the germination of seedlings (germination gaps), and subsequent freedom from competition from other plants to allow these seedlings to grow. Even though many plants are long-lived and many spread vegetatively, all need to regenerate from propagules at some stage. Many plant species in freshwater and brackish wetlands require damp, bare mud on which to germinate and establish. This can be provided by periodically lowering water levels.

A final important point regarding vegetation management is that plants are better able to survive periods of adverse conditions than most animals. They can withstand adverse conditions during their reproductive period by persisting as vegetative rootstock, and longer periods by surviving as seeds or other propagules. However, plant species without wind-borne propagules can be poor at dispersing compared to many animals.

### 4.1.2 Fungi

Fungi are important components of most habitats other than wetlands. There are two types of fungus: saprotrophic and mycorrhizal. Saptrotrophic fungi obtain their nutrients from dead organic matter. Mycorrhizal fungi form mutually beneficial associations with the roots of higher plants.

The requirements of fungi are rarely taken into account during habitat management. Exceptions are the creation and retention of decaying wood and the recognition of the value of a small number of fungal assemblages, notably waxcap fungi of the genus *Hygrocybe* in grasslands in Northern Europe. Small-scale sod cutting and litter stripping have been used in woodland to increase species richness of ectomycorrhizal fungi, but are not used widely. The effects of management on most groups of fungi, though, are poorly understood.

Mycorrhizal fungi influence the species composition of vascular plants. They do this by enabling some otherwise slower-growing plants to exploit resources otherwise unavailable to them, and by influencing soil nutrient recycling. Nitrogen is recycled more slowly in soils dominated by mycorrhizal fungi, compared to by bacteria, and these conditions favour slower-growing vascular plant species. Mycorrhizal fungi tend to dominate nutrient recycling in low-fertility soils, particularly more acidic ones. Addition of inorganic fertilizer reduces fungal biomass in at least some grassland types (Smith *et al.* 2003). Soil disturbance and cultivation also lead to bacterially dominated nutrient recycling. Fungal hyphae are important in binding soil particles and thereby influencing soil structure. Methods of influencing vascular plant species composition via soil microbial processes are poorly understood (Pywell 2006).

## 4.1.3 Lichens

Lichens typically prefer open conditions. Particularly rich assemblages are associated with ancient habitats. These include ancient trees, stable dunes and shingle, short, open grassland, montane habitats including flushes, rock exposures and late-lying snow patches and rocks in lowland areas, especially in old churchyards. Of these, calcareous substrates tend to be richer than acidic ones. An exception is the characteristic assemblages of lichens associated with metal-rich environments, especially metal-rich mine workings. The main considerations are to identify important lichen assemblages, and, if necessary maintain suitably open conditions for them. However, because most lichens are relatively slow to recover from disturbance, any management needs to be carried out especially sensitively. Habitat management for lichens is discussed by Fletcher (2001).

## 4.1.4 Vertebrates

For birds, the major factor influencing habitat use is usually habitat structure (MacArthur and MacArthur 1961), particularly through its influence on the abundance and accessibility of food and potential nest sites. Even though habitat structure will obviously be influenced to some extent by plant species composition, high plant species richness *per se* can often be irrelevant to bird conservation. It is often possible to provide habitat for birds of high conservation priority on land with high nutrient levels and botanically species-poor vegetation. A prime example of this is shallow, nutrient-rich wetlands. Most species of birds are relatively mobile and good at colonizing new habitat. They can move elsewhere when conditions become unsuitable.

There are few general principles concerning managing habitats for mammals. Densities of many larger mammals have been greatly reduced by human persecution, and many occur at far lower levels than could be supported by the current habitat. An important exception to this is the 'artificially' high densities of deer in parts of Western Europe and North America.

Habitat management is often undertaken to specifically benefit reptiles and amphibians, particularly towards the cooler edges of their climatic range. Management for reptiles in these areas usually involves provision of open ground, particularly on sunny slopes, for basking and egg laying, and suitable warm, well-drained banks for hibernation. Many amphibian species only breed successfully in temporary water bodies free of predatory fish and high densities of large, predatory invertebrates. Habitat management for amphibians often involves providing these conditions.

### 4.1.5 Invertebrates

Principles of managing habitats for invertebrates, especially insects, differ in a number of fundamental ways from that aimed at benefiting plants and vertebrates. These differences are mainly related to:

- the small size of invertebrates and often extreme habitat specialization;
- the importance of the surrounding temperature in influencing habitat suitability (similar to reptiles and to a lesser extent amphibians);
- the different habitat requirements of the larval and adult stages of many insect species;
- the often limited mobility of many invertebrates;
- the annual life cycles of many invertebrates, including most insects.

The small size and specialization of many invertebrates, especially insects, means that they can exploit, and are consequently often restricted to, very small features within larger areas of habitat. These are called microhabitats. There are many microhabitats and other features of particular value for invertebrates, but of little or no relevance to other groups. Examples are shown in Figure 4.1 and Table 4.1. Important invertebrate faunas are more likely to occur where these features have been continually present over a long period of time. An example would be where there has been a long, continual supply of tidal refuse. It is important to identify any specific features of potentially high value for invertebrates, and ensure that these are retained and, if possible, enhanced. Kirby (1992b) provides an excellent guide to managing temperate habitats for invertebrates.

The body temperature of invertebrates is heavily dependent on that of their immediate surroundings (the surrounding microclimate). The microclimate can vary greatly depending on habitat structure. For example, under hazy summer sunshine at midday in southern England there is a 7°C difference in temperature

**Fig. 4.1** Important features for invertebrates. Many habitats and smaller features (microhabitats) can be of particular value for invertebrates, but of little or no conservation value for other groups.

This unremarkable-looking roadside verge (a) in Breckland, England, supports a richer fauna of ground beetles characteristic of heathland and sandy grassland than do most heathland nature reserves in this region (Eversham and Telfer 1994). The bank on the left is maintained by piling up of sand that blows onto the road from adjacent fields. The continual supply of sand maintains a continuity of early-successional bare ground and annual plants, which these species require. Such highly disturbed conditions are rare on most nearby heathland nature reserves.

(b) Old wooden fence posts with beetle holes in them can provide valuable nesting habitat for solitary bees and wasps in areas with little or no other suitable nesting habitat. When upgrading fencing it is worth checking to see whether the old fence posts are of potential value, and, if so, retaining them next to the new ones.

**Table 4.1** *Important features for invertebrates*

| Feature(s) | Importance for invertebrates |
|---|---|
| Bare and disturbed ground and annual plants, especially on southerly facing slopes | Important for a range of warmth-loving species, especially those at the cold and wet margins of their range and species that require open ground. Also important for species that feed on the seeds of annual plants and those that favour plants stressed by drought, pollutants, or mineral deficiency. |
| Decaying wood | Supports a large and specialized fauna, especially of flies and beetles. Different types of decaying wood support their own specialized faunas. |
| Nectar sources | Many adult insects feed on nectar, especially bees, butterflies, moths, flies, and some beetles. Suitable nectar sources can be in short supply in some habitats at particular times of year. |
| Springs, freshwater and saline seepages, damp peat, bare mud and 'ancient' temporary waterbodies | Can support a wide range of often specialized species, including small wetland features that are botanically uninteresting. These features are often threatened by plans to create larger, more permanent water bodies for conservation in their place. |
| Tidal refuse | Contains a distinctive fauna dominated by flies and beetles. |
| Soft rock cliffs | Supports a range of species associated with the mosaic of different successional stages, including cracked and soft ground and ruderal vegetation, together with wet areas caused by seepage. |
| River shingle | Contains a distinctive fauna. |
| Dung | Supports a specialized fauna comprising mainly flies and beetles. The fauna varies with the type of dung and surrounding habitat. |

at the soil surface beneath vegetation 1 and 7 cm high (Thomas 1983). This is equivalent to the difference in mean maximum July temperature between London and Naples. Therefore, as with birds, habitat structure can be of fundamental importance to invertebrates, but in the case of invertebrates it operates on a very small scale. Differences in the microclimate provided by different habitat structure and aspect can be critical in determining their presence or absence towards the cooler edges of their range.

The overall habitat requirements of many insects can be even more demanding where adults and larvae of the same species exploit different habitats. For example, many beetle larvae feed on decaying wood, while the adult beetles require suitable nectar sources. For these species, the often exacting requirements of both stages of their life cycle have to be present for the species to persist. Some insect species, such as many true bugs and the larvae of moths and butterflies, specialize in feeding on only a limited range of plant species. These often not only require the presence of the particular plant species, but also for it to be in the correct growth form. Some only feed on stressed plants. Habitat structure or architecture is important in influencing suitability for spiders, especially web-spinning species (see review by Bell *et al.* 2001).

The limited mobility of many invertebrates can make providing suitable habitat conditions even more challenging. Flightless species are particularly poor at dispersing, but even those that can fly often move only short distances. Limited powers of re-colonization mean that management needs to be even more sensitive in ensuring that it does not inadvertently eradiate an entire population of a species at a particular site, because it will be unlikely to re-colonize. Limited mobility also means that where a species requires different conditions during different stages of its life cycle, then these also need to be in very close proximity. A few groups range more widely. On arable land bumblebees regularly forage several hundred metres from their nests, with some species ranging up to 750 m from them. Bumblebee nests only occur at low densities and they therefore require the presence of suitable conditions at more of a landscape scale (Osborne *et al.* 1999, Darvill *et al.* 2004, Knight *et al.* 2005).

Finally, the annual life cycles of many species, especially most insects, place even greater demands on providing suitable conditions. Most populations of long-lived vertebrates can survive periods of low or zero productivity, because the adults will still survive. For species with annual life cycles, suitable conditions need to be present for the species to successfully complete its life cycle *every* year for it to persist. On the positive side, the high reproductive rate of many invertebrates means they can increase rapidly when conditions are suitable.

These combinations of factors mean that habitat management for invertebrates has to be undertaken more sensitively, and on a smaller scale, than for most other groups. The key principles are listed below.

- To minimize the risk of local extinctions never carry out catastrophic management such as cutting, burning, or heavy grazing over a large proportion of the site in any one year.
- Maintain a diversity of habitats, microhabitats and transitions between these. You will need to maintain suitable conditions for species that you do not know are present and whose ecology is unknown.

## 4.2 Landscape factors

There is a range of factors other than habitat conditions at a site that can influence its fauna. It is important to be aware of these when planning site management. These are discussed in the following sections. In addition, high levels of pesticides and hunting may reduce populations below those that the habitat is otherwise capable of supporting.

### 4.2.1 Area of habitat

Many species require relatively large areas of habitat and species can occur at higher densities in larger blocks of habitat. This is sometimes referred to as area sensitivity (e.g. Robbins *et al.* 1989). Species with high area sensitivity are those very intolerant of habitat fragmentation. For larger animals such as birds, there may be a minimum area needed to support a breeding pair of a particular species. For example, in grasslands in Maine in North America breeding upland sandpipers, *Bartramia longicauda*, only reach 50% occurrence in patches of grassland of greater than about 200 ha, whereas savannah sparrows, *Passerculus sandwichensis*, reach 50% occurrence in patches of grassland of only 10 ha (Vickery *et al.* 1994). The minimum area required by a species will vary according to the quality of habitat.

Size is an important consideration when designing habitat creation. In general, larger sites will be better for maintaining high-priority species than smaller ones. However, the benefits of creating larger blocks of habitat also have to be balanced against those of creating small-scale mosaics of habitats, transitions and edge habitat, especially for many invertebrates. From a visitor perspective, large sites, particularly more natural-looking ones, can provide a greater sense of wilderness. Conversely, providing a larger number of smaller sites will increase the feeling of localness and decrease the distance people need to travel to visit them. An overriding reason for creating larger blocks of habitat is that they are far cheaper to manage per unit area (Figure 4.2).

### 4.2.2 Connectivity

Connectivity influences the ability of species to disperse between habitats. This ability varies between species. There has been much debate on the value of habitat corridors (also known as conservation or wildlife corridors) in facilitating movement of species between habitat patches. Habitat corridors are linear strips of semi-natural habitat that connect otherwise isolated habitat fragments. The potential benefits of habitat corridors vary between species. If it is good at dispersing across unsuitable habitat, then providing a habitat corridor is unlikely to provide significant additional benefits. If it is very poor at dispersing, then it is unlikely to make significant use of a linear corridor of habitat. Hence, corridors will only be of value to species of intermediate dispersal ability. A decision has to be made whether resources are better spent creating habitat corridors between patches of habitat or increasing the size of these patches.

**Landscape factors** | 65

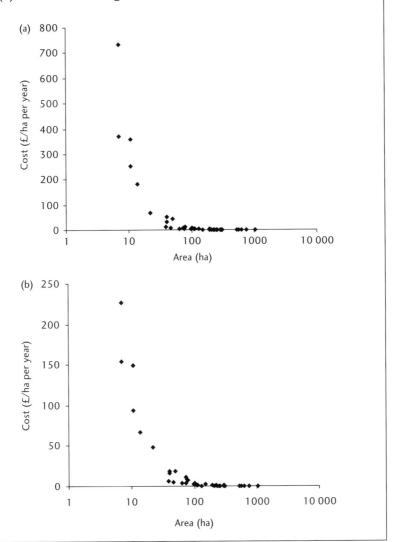

**Fig. 4.2** The costs of managing different-sized reserves. The unit costs of managing reserves decrease dramatically with reserve size. The figures shown are for the Royal Society for the Protection of Birds' staffed wetland reserves in the UK (2005/2006 prices). Note the logarithmic scales. For this set of reserves, sites that are smaller than about 100 ha cost up to 13 times as much to manage per hectare than sites larger than 100 ha. The increase in unit management costs with reserve size is a product of both higher unit (a) staff and (b) other habitat management costs.

## 66 | General techniques and considerations

Roads and other development can form barriers to the movement of large mammals and be a significant cause of mortality for some vertebrates. There are, though, few studies investigating the effect of this mortality on overall population size. Structures for controlling water levels can impede regular movement and migration of fish. The effects of roads in preventing movement of animals can be reduced by providing green bridges (Figure 4.3) and underpasses. Green bridges are also known as ecoducts, wildlife bridges, biobridges, or wildlife overpasses. Fencing is also required to prevent animals from crossing the road elsewhere and to funnel them towards these safe crossing points. Bridges are used mainly for larger mammals. Underpasses are aimed at smaller mammals and, especially, to

**Fig. 4.3** Barriers to movement of large mammals. Seasonal movements and dispersal of mammals can be prevented by roads and development.

In Nairobi National Park in Kenya, residential development to the south of the park has prevented natural movement of large herbivores into and out of the park, thereby reducing prey for lions within it. This lack of wild prey has resulted in the lions preying on domestic cattle in the surrounding area, bringing them into inevitable conflict with people. Most of the lions in the park have now been killed by people to protect their livestock.

Green bridges can be built across busy roads to allow animals to safely cross them. The grassland over which the people are walking is actually a bridge over the busy A50 motorway between Arnhem and Apeldoorn in The Netherlands. This green bridge was built to allow deer, wild boar and free-ranging Highland cattle to cross the motorway that separates adjacent areas of heathland. The bridge is covered in soil and planted with grass, with hedgerows to screen its edges. Small ponds were dug either side of the bridge to encourage animals to visit and use it. Monitoring of footprints in strips of sand showed that the bridge was quickly used by the intended species.

allow amphibians to travel to and from their breeding pools. Further information on reducing adverse effects of roads is provided by Spellerberg (2002). Passes can be installed to help fish move past water-level control structures (e.g. Knights and White 1998).

Increasing connectivity between areas can, though, be damaging if it facilitates movement of undesirable species. This can occur where islands free of ground predators are connected to the mainland. For example, connection of the island of Tautra in North Trøndelag in Norway to the mainland via a stone causeway has allowed American mink, *Mustela vison*, red foxes, *Vulpes vulpes*, European badgers, *Meles meles*, and martens to cross to the island. Connection of the island has been followed by a large decrease in numbers of ducks, particularly of common eiders, *Somateria mollissima*. This is thought to have been caused by both predation and disruption of feeding conditions caused by a reduction in tidal currents around the island. Attempts have been made to deter predators from using this causeway and other crossings by broadcasting of a variety of sound frequencies from loudspeakers, although the deterrent effect of these appears to decline over time (but see also Section 11.1.5). The effectiveness of these sound devices is possibly greater in tunnels than on open bridges and causeways.

Where unwanted introduced predatory mammals cannot be eradicated throughout an entire landmass, an option is to enclose an area with fencing and maintain this as a predator-free 'island' within it (e.g. Moseby and Read 2006).

Lack of colonization can be overcome through species introductions. This is commonly undertaken to introduce propagules of plants at creation sites, particularly keystone plant species that form the fabric of the habitat, such as common reeds, in reedbeds, trees in woodlands, and heather in heathlands. More stringent criteria are usually applied to (re-)introducing animals. There is, though, a strong argument for carrying out more species introductions to aid the spread of species to suitable habitat in response to climate change.

## 4.2.3 Edge effects

Processes taking place on the margins of the habitat patch, so-called edge effects, can diminish its overall conservation value. A frequent example is where a wetland is surrounded by more intensively drained land, and the surrounding drainage lowers water levels on the margins of the wetland. Open habitat surrounding the edges of forest can reduce the humidity on its edges, thereby affecting conditions for some groups, such as forest bryophytes. Habitats surrounded by agricultural land can be affected by pesticide drift and fertilizer runoff. Negative effects of surrounding habitats can be reduced by creating a buffer zone between the core habitat and the surrounding habitat.

### 4.2.4 Surrounding habitats

Populations of species within a particular patch of habitat can be influenced by population processes taking place in surrounding habitats. Numbers of migratory and eruptive species will be influenced by conditions elsewhere in their range. Even though these considerations can rarely be overcome, they can nevertheless be useful in interpreting changes taking place within the habitat patch. Potential effects of surrounding habitats will, though, be important when deciding the location and design of (re-)creation sites.

For some more generalist species there can be considerable interchange of individuals between populations in the remaining block of semi-natural habitat and the surrounding habitat. Maintenance of the population within the specific habitat patch may be dependent on immigration of individuals from outside of it.

The abundance of some more generalist predators, or less commonly parasitic species, in surrounding habitats may also influence populations of habitat specialists within it. A probably widespread example of this involves increased nest predation of birds on the edges of woodland surrounded by farmland, caused by habitat-generalist nest predators such as corvids and mammalian predators whose high population densities are a product of habitat conditions in the surrounding farmland (Andrén 1995). Another example is brood parasitism by brown-headed cowbird, *Molothrus ater*, in North America. In practice, only resource-hungry long-term population studies are likely to detect whether these processes are taking place.

The proximity of other suitable habitat can also be beneficial for groups, especially birds, which can easily commute between areas. Locating a shallow wetland near to an estuary will provide additional feeding foraging for waders/shorebirds at high tide. Provision of safe, daytime shallow-water roosts for wildfowl will enable them to feed on surrounding habitats at night, which are too disturbed for them in daytime.

## 4.3 Disturbance

Human disturbance can reduce habitat use by species, although just because a species more strongly avoids human presence does not necessarily mean it is more disadvantaged by it (e.g. Gill *et al.* 2001). Damaging effects of disturbance can be limited by controlling specific, high-disturbance activities (e.g. Bregnballe and Madsen 2004) and by limiting access, for example by closing tracks (e.g. Summers *et al.* 2004). Disturbance can also be used to provide conservation benefits, an example being its use to discourage large gulls from competing with terns for nest space (Figure 4.4).

**Fig. 4.4** Blow-up scarey man.

This inflatable scarey man is here being tested prior to being deployed on an island where it is used to discourage herring gulls, *Larus argentatus*, and lesser black-backed gulls, *Larus fuscus*, from settling to nest in areas suitable for later-nesting terns. Scarey man is attached to an electric motor fan which, once activated by remote control from a hide, inflates and deflates it five times every 18 minutes. There is also a light and a siren attached to him, both of which can be operated independently of each other. Scarey man is re-positioned regularly while deployed to maximize its effect, and removed once the terns arrive.

## 4.4 Grazing

The effects of grazing will form a significant proportion of Chapters 5–11. There are, though, some overriding principles of grazing that are common to a range of habitats. These are discussed below.

### 4.4.1 Levels of control over grazing

Grazing has been important in creating and maintaining many highly valued cultural habitats. Conventional, commercial regimes in the lowlands usually involve grazing relatively uniform blocks of land with one type of livestock at a set stocking density. The livestock are also usually of the same sex and of similar age. The aim of this grazing is to maximize farming income.

The effects of grazing uniform areas of habitat with one type of animal at a set stocking rate is relatively predictable. An alternative approach is to manage livestock less intensively, so their effects more closely mimic those of now-absent wild, large herbivores (WallisDeVries 1995, Kampf 2000). Two types of approach can be recognized: *extensive* and *naturalistic* grazing. The characteristics of these are summarized in Table 4.2. Conventional, extensive, and naturalistic grazing

**Table 4.2** Characteristics of conventional, extensive, and naturalistic grazing regimes (modified from Hodder et al. 2005).

| | Grazing system | | |
|---|---|---|---|
| | Conventional | Extensive | Naturalistic |
| Breed type | Usually 'conventional' | Usually primitive and other hardy ones | Typically breeds considered most similar to their wild ancestors. |
| Herbivore density | Pre-determined by specific aims of grazing. Usually based on, or modified from, former agricultural practice. | Pre-determined by aims of grazing | Self-regulated and therefore prone to periodic fluctuations (although animal-welfare issues will prevent large-scale death by starvation or disease). |
| Herbivore demography | Determined by grazier. Invariably single-sex, single-age, single-species herds. | Determined by grazier. Can involve more than one sex, age, and stock type. | Mixed-sex and -age herds and often involving more than one stock type. Herbivores allowed to breed and form family groups. |
| Seasonality of grazing | Usually seasonal | Seasonal or year-round | Year-round |

regimes represent a continuum of:

- decreasing levels of control/predictability of their effects;
- increasing levels of naturalness.

Extensive and naturalistic grazing are usually used in areas containing a mixture of habitats. Naturalistic grazing is only possible on very large areas of land.

The effects of extensive grazing can be far more difficult to predict than in conventional systems because of the combinations of mixed types, ages, and sex of livestock. Different livestock types, ages, and sexes can exhibit quite different feeding behaviour, and their preferences for different plants and habitats often varies throughout the year. Where herbivores are allowed to graze a mixture of habitats, then this will make prediction even more difficult. This is because the vegetation selected by livestock at any one time will vary according to their preferences for, and the availability of, alternative food. The grazing animals will themselves alter the nature of the vegetation, and thereby alter the choices available and subsequent grazing patterns. A number of highly valued cultural habitats have been created and maintained by extensive grazing regimes. An example is the New Forest in England (Figure 4.5).

The effects of grazing in naturalistic grazing systems are even more difficult to predict. This is because numbers of herbivores are also allowed to be

**Fig. 4.5** Extensive grazing in the New Forest. The 20000-ha New Forest in Hampshire, England, has been maintained since medieval times, originally as a royal hunting forest. It has been managed by year-round grazing of free-ranging large herbivores, gathering of wood and bracken, and digging of peat and marl. The large herbivores include New Forest ponies, cattle, and five species of wild and introduced deer. Pigs are also let out in autumn to feed on fallen acorns and beech mast.

Numbers of all these large herbivores, including the deer, are managed, but have been subject to large fluctuations over time. The combination of extensive grazing and other past management have created a unique mixture of heavily grazed lawns, heaths, mires, scattered scrub, groves of trees, and wood-pasture. The heavy grazing supports many species, notably those of heavily grazed and trampled areas, that are rare or absent elsewhere in the UK, but also makes the area unsuitable for many others.

self-regulating and thereby subject to potentially large variation. In this way, naturalistic grazing regimes can be considered even more similar to natural grazing regimes. However, the lack of large carnivores which can regulate both herbivore numbers and grazing behaviour will probably mean that the effects of naturalistic grazing regimes will still be quite different to that of an original-natural system. The most well-known example of naturalistic grazing in Europe is in the Oostvaardersplassen in The Netherlands (Vulink and Van Eerden 1998, Kampf 2000; Figure 4.6).

**Fig. 4.6** Naturalistic grazing in the Oostvaardersplassen. This area was claimed from the sea in 1968 and, after unsuccessful attempts to drain it followed by a downturn in the economy, was allowed to develop into a nature area. It consists of 3600 ha of wetlands, mainly reed-dominated marshes, shallow open water, mud and ruderal vegetation, and 2000 ha of drier habitats now managed by naturalistic grazing. Three types of large herbivore have been introduced to these drier areas: Konik ponies, Heck cattle, and red deer. These breeds of ponies and cattle are considered to be, respectively, the closest present-day equivalents to now-extinct tarpan, *Equus ferus ferus*, and aurochs, *Bos taurus primigensis*. Roe deer, *Capreolus capreolus*, colonized naturally.

The large herbivores in the Oostvaardersplassen are allowed to live more or less natural lives and to increase in population until natural food supply begins to limit their population. At this point, cattle and ponies in poor condition are culled. The drier habitats managed by naturalistic grazing mainly comprise open grassland with or without mainly scattered trees and bushes and areas of tall-herb vegetation and dry reedbed.

The economics of extensive and naturalistic grazing will differ to those of conventional systems. If conventional systems are profitable, then naturalistic grazing will be less economic, because the grazing will not provide any income. If a suitable conventional grazing can only be achieved by paying graziers, then both extensive and near-natural systems will usually be cheaper to run.

Both extensive and naturalistic grazing systems offer great potential for use in newly created habitat. They are unlikely, though, to be realistic options for managing existing cultural habitats, because of their potential to severely damage their existing interest. Any such damage would in many instances contravene conservation legislation. Most existing sites would in any case be too small for naturalistic grazing.

Introducing extensive or naturalistic grazing to uniform areas of species-poor habitat, such as on former arable land, is unlikely to provide great conservation benefits. The relative uniformity of topography and soil conditions created through arable farming will result in relatively uniform vegetation, whatever the grazing regime. Natural landscapes would have dry, south-facing slopes, colder and wetter north-facing slopes, wet hollows, areas of thin and deep soil, etc., each supporting different assemblages of plants and animals. This variation is lost through ploughing. Maximizing the positive effects of the grazing will depend on careful design. This may include land-forming to recreate suitable topographic variation, consideration of how the design of the site will influence grazing patterns and, in the case of wetlands, designing the hydrology to provide spatial and temporal variation in habitat conditions.

There is debate over the extent to which the habitats created by extensive and naturalistic grazing in Western Europe resemble those that occurred under original-natural conditions. Vera (2000) argues that these open grasslands, savannah-like habitats, and wood-pasture are the closest analogues to original-natural habitats. There is, though, also evidence that most drier ground in Western Europe was largely forest, with a small proportion of temporary and permanent glades probably created by extreme weather events, fire, and grazing by large herbivores (Hodder *et al.* 2005).

As with other types of grazing, naturalistic grazing still needs to take full account of animal-welfare issues. No grazing system can be considered natural, and no areas of grazed habitats in developed countries will have their full natural compliment of large predators. Hence, numbers of large herbivores will eventually increase under naturalistic grazing until animals succumb to disease and starvation. Naturalistic mortality of large herbivores is considered unacceptable in terms of animal-welfare legislation in many countries. As with overall conservation philosophy, a pragmatic view is to allow naturalistic grazing to operate in as natural a way as possible within pre-defined limits, until a point where intervention is required. This may involve allowing densities of large herbivores to reach levels that precipitate fundamental and interesting changes in vegetation, and then removing or culling animals before numbers exceed their carrying capacity and mass mortality takes place. Alternatively, fluctuations in herbivore density could be created to produce dynamism in habitat conditions by periodically culling or removing animals at irregular intervals before they reach close to this carrying capacity. In both instances, culling or removing young, sick, and old animals would probably most closely mimic the changes in demography that would take place under original-natural conditions.

### 4.4.2 Breed type

There is, unfortunately, limited objective information on the relative merits of different livestock breeds used for conservation grazing. There is widespread interest in the use of traditional, non-commercial breeds, and a general belief that these are better at grazing coarser vegetation. However, it appears that where differences in feeding behaviour do exist, these are usually more related to differences in body size, rather than to breed type *per se* (Rook *et al.* 2004). Beef cattle are, though, generally considered better at grazing coarse vegetation than dairy cows. Younger and smaller livestock are generally poorer at grazing coarser vegetation. Smaller breeds of cattle may be more able to graze on soft ground in wetlands without becoming stuck. Age and past experience are undoubtedly important in influencing grazing behaviour. There is evidence that exposure to different foods during early life has at least some influence on grazing preferences later on. It is not uncommon, though, for 'hardy', breeds to be brought up in rather pampered conditions. So-called primitive sheep are especially good at browsing and preventing the establishment of scrub. They are intermediate in their feeding habits between conventional sheep and goats.

It is also widely considered that some hardy breeds are more tolerant of insect bites and wounds and better able to withstand adverse conditions than modern breeds. Thus, they are considered better suited for wintering outside in cold, wet, and windy conditions. Some breeds seem better able to gain weight when food is plentiful, and to use these reserves during periods of shortage. Examples of hardy and primitive breeds of cattle, ponies, sheep, and goats commonly used in conservation are described in Figure 4.7. There are no particularly hardy breeds of horse. Donkeys, which are thought to be derived from wild asses, *Equus africanus*, from north-east Africa, are more suited to hot, arid conditions. Breeds can also vary in temperament and ease of handling, although this can be modified by how they are managed.

**Fig. 4.7** Hardy and primitive breeds commonly used in conservation grazing.

**Cattle:** hardy breeds used in harsh environments and suitable for out-wintering include Galloways, Highlands (shown here), and Hecks. The first two are medium-sized, old-fashioned breeds widely used in conservation grazing. Heck cattle were bred in the early twentieth century by the brothers Heck in Germany, in an attempt to recreate the extinct aurochs, the ancestor of European cattle.

**Ponies:** hardy breeds of pony include the Exmoor, New Forest (Fig. 4.5), Welsh Mountain, Shetland, Fjord, Icelandic, Carmargue, Przewalski, and the Konik Polski or Konik (Fig. 4.6). Koniks were created by Polish farmers crossing the Eurasian wild horse, the tarpan, with domestic horses. Tarpans became extinct in 1875 and Koniks are considered their closest living relative.

**Sheep:** primitive sheep breeds include Hebrideans (Fig. 6.9), Soays, Manx Loaghtans, and Shetlands. Soays and Manx Loaghtans are rare. Hebrideans are the most widely used primitive sheep in conservation grazing.

**Goats:** hardy breeds of goat used in conservation grazing are feral goats and dwarf/pygmy goats. The small size of dwarf/pygmy goats makes them slightly easier to handle and contain.

Another commonly used term is rare breed. This is a specific term used to describe a breed's conservation status. Because knowledge of breeding to create desirable features only fully developed from the late eighteenth century onwards, most rare breeds are of relatively recent origin. They do not necessarily possess some of the advantageous traits of some more hardy and primitive breeds. Their rarity also means they can be difficult to obtain.

### 4.4.3 Use of anti-parasitic drugs

A wide range of invertebrates are associated with large-herbivore dung, particularly that of cattle (Figure 4.8). There has been concern over the effects of one particular group of modern anti-parasitic drugs, avermectins, on this fauna. Avermectins are used to control internal and external parasites in cattle, equines, sheep, and pigs. They produce residues in the animal's dung that prevent the

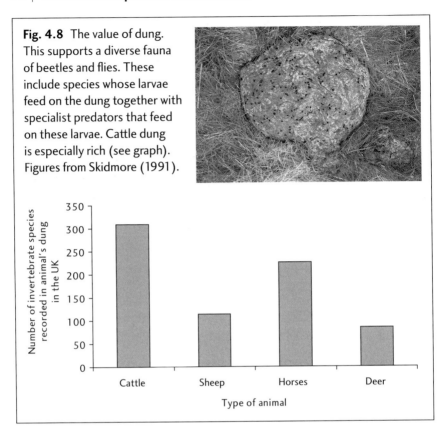

**Fig. 4.8** The value of dung. This supports a diverse fauna of beetles and flies. These include species whose larvae feed on the dung together with specialist predators that feed on these larvae. Cattle dung is especially rich (see graph). Figures from Skidmore (1991).

development of beetle and fly larvae (e.g. Wall and Strong 1987, Madsen *et al.* 1990). The following options should be considered where there are concerns over the potential effects of anti-parasitic drugs on the dung fauna and species that feed on dung invertebrates.

- Only treat livestock on the basis of need, rather than as a matter of routine. Have a sample of dung checked for parasite eggs and only treat if there is a problem. An exception to this is where livestock are grazing areas with liver flukes, *Fasciola hepatica*. Infestations of liver flukes are associated with particular areas of marshy land, and are less dependent on previous or current stocking regimes, as are most other parasites. Livestock also lose condition very quickly when infested with liver fluke.
- Reduce the length of time that avermectin residues are excreted in the dung, especially by not administering it using a bolus.
- Treat livestock with avermectins before they graze the site of interest and do not release them onto it until about 35 days after the last treatment. This will

mean that their dung no longer contains significant residues. Alternatively, only treat animals after they have left the area.
- Use an alternative anti-parasitic drug that is less harmful to the dung fauna. This is, though, also likely to be less effective at treating parasites.

## 4.5 Controlling unwanted plants

A common form of habitat management is reducing the abundance of unwanted plant species, particularly alien/exotic plants that can, or are threatening to, out-compete native vegetation or cause other problems (Figure 4.9). The main techniques used are:

- modifying existing management;
- cutting the above-ground parts of undesired species;
- digging up and removing the plants;
- treating with herbicide.

These methods can also be combined, as described below.

It is important to evaluate whether the predicted benefits of control will outweigh its costs. Issues to consider are whether the unwanted plant species is spreading, what vegetation is likely to replace it following removal, and the likelihood of the unwanted species re-colonizing. For example, it may not be worth removing a group of alien/exotic mature trees in a woodland that are not regenerating and which provide mature tree habitat for a variety of species, albeit of lesser value than provided by similar-aged native species. Conversely, there will be enormous benefits in eradicating a small patch of a recently arrived highly invasive plant that threatens to out-compete native vegetation of high conservation value.

The first option to consider is whether this can be achieved simply by modifying the *existing* management. For example, whether reducing the grazing pressure might reduce the availability of germination sites for the unwanted species or whether increasing grazing pressure or changing its timing or the type of livestock might reduce its abundance.

The next option is to repeatedly cut the above-ground growth of the unwanted species to reduce its vigour relative to that of surrounding vegetation. Trees and shrubs can be felled and even quite large-diameter woody vegetation destroyed using a forestry mulcher. Cutting above-ground vegetation is most effective at reducing the vigour of plants if undertaken during periods of the year when least of their reserves are stored below ground. Grazing the re-growth following cutting will further reduce the plant's competitive ability. Some emergent plants, for example common reed and sea club-rush/alkali bulrush, *Scirpus maritimus*,

**Fig. 4.9** Alien/exotic plants. Some alien/exotic plant species can become dominant in semi-natural habitat, such as this water hyacinth, *Eichhornia crassipes*, in Nepal.

Water hyacinth has probably one of the longest lists of detrimental effects of any alien plant. It affects the ecology of water bodies by reducing light and oxygen levels in the water beneath it, restricts water flow and so interferes with irrigation and can increase flooding, reduces water supply by increasing evapotranspiration rates by up to 13 times that from a free water surface, and can restrict access to fishing grounds and cause damage to hydro-electric installations. Originally a native of South America, water hyacinth now occurs in many tropical and subtropical areas of Central and North America, Africa, India, Australia, and New Zealand. Millions of dollars are spent controlling it each year in North America alone.

and are most effectively controlled by cutting and immediately flooding their stems, or cutting them underwater. This prevents oxygen being transported from the plant's leaves above the water to its roots in the anoxic mud. Although repeated cutting is often useful in preventing plants from growing large and setting seed, and in preventing or reducing their vegetative spread, it rarely kills them. Herbicides are more effective at this.

If none of the previous strategies proves successful, then in some cases plants can be dug up or else treated with herbicide. Digging up, pulling, and otherwise removing unwanted plants will only be feasible on a small scale, and may be followed by regeneration of the unwanted species from seed in the freshly disturbed soil. Herbicides can be very effective, but are also dangerous to people and other non-target groups. They should not be used close to water bodies.

There are several methods of herbicide application. Boom-spraying involves spraying herbicide over all the vegetation in a given area. This has the disadvantage of killing non-target species as well, unless a very specific herbicide is used. Bare ground created by die-off of non-target plant species can provide competition-free gaps for any seeds of the unwanted plant to germinate and grow. Herbicide can also be sprayed by helicopter. Selective methods of application include:

- spot-spraying: spraying the herbicide on individual plants, often in combination with a dye to show which plants have been sprayed;
- weed-wiping: wiping a systemic/translocated herbicide (i.e. a herbicide that is applied to the foliage and translocated throughout the rest of the plant) against tall, unwanted vegetation without touching shorter, desirable vegetation below it;
- drilling holes into cut tree stumps and pouring herbicide into them;
- using ecoplugs: small, herbicide-containing plugs that are hammered into small holes bored into a tree trunk. The hammering action breaks open the plug to and releases a capsule of systemic/translocated herbicide.

Vegetation is usually heavily grazed prior to weed-wiping to reduce the height of non-target, palatable vegetation and thereby increase the height differential between this and the taller, unpalatable vegetation at which the herbicide is aimed. Uptake of systemic/translocated herbicides can be increased by cutting the plants first and applying the herbicide to the fresh, re-growth, instead of to older, less actively growing foliage. Manufacturer's instructions and best practice should always be followed.

## 4.6 Eradicating introduced cats and rats on islands

Introduced predators have been responsible for a large number of declines and extinctions of native birds, mammals, and reptiles on islands. Predators that have had the greatest impacts are domestic cats, *Felis catus*, rats, *Rattus* spp., and small Indian mongooses, *Herpestes auropunctatus* (Lever 1994). Introduced rats are thought to have been responsible for the global extinction of at least 12 bird species, and the local extinction and decline of many more (Atkinson 1985). In one extreme case the entire world population of a bird, the Stephen Island wren, *Xenicus lyalli*, in New Zealand, was exterminated by just one cat, Tibbles, which was kept by the lighthouse keeper on the island (Fuller 2000). Domestic cats and rats have also seriously depleted many seabird colonies.

It is useful to consider the following questions when deciding whether to attempt eradicating domestic cats or rats.

## 1 Is there a problem?

The first step is to determine whether species of high conservation value are at risk, whether this is due to the presence of introduced predators and, if so, which species are causing the problem. The impact of a predator can vary between islands. For example, introduced black rats *Rattus rattus* on islands in the Mediterranean have less effect on breeding seabird densities on larger islands and those made of limestone. The effect of substrate is probably because the limestone rocks provide more suitable cavities and caves for seabirds to nest in, and make these nest sites less accessible to rats (Martin *et al.* 2000). Even if it is considered desirable to eradicate the predator, then knowledge of the severity of their effects on different islands will still be important in at least allowing prioritization of the often limited resources.

## 2 Is eradication feasible?

One of the primary considerations regarding whether eradication is likely to be feasible is the size of the island. It is easier to eradicate introduced mammal species from smaller islands. Of the 48 islands listed by Nogales *et al.* (2004) from which domestic cats have been successfully eradicated, 36 were less than 5 $m^2$. The largest was the 290 $m^2$ Marion Island in the Republic of South Africa. Rat eradication has been attempted on at least 135 islands, of which more than 90 are in New Zealand. Ninety-three per cent of attempted rat eradications on islands have been successful. The largest island from which rats have successfully been eradiated is the 113 $m^2$ Campbell Island in New Zealand. Other considerations are the terrain and logistics.

## 3 What are the risks to non-target species?

Non-target species can be killed through direct ingestion of the poison (primary poisoning) and through ingestion of poisoned prey (secondary poisoning). The potential effects of the eradication methods on other key species on the island should be evaluated thoroughly.

## 4 What is the likelihood of re-colonization by the predator?

Domestic cats are unlikely to re-colonize by accident. Rats, though, can easily re-colonize from boats unless specific precautions are taken. The risk of re-colonization by rats will depend on the:

- distance from the nearest source of rats;
- type and frequency of boat landings;
- type of cargo carried.

Re-colonization by rats can be minimized by placing cargo in quarantine in areas with rat-elimination procedures and by fixing rat baffles to ship mooring ropes to prevent rats from running along them. Quays can also be designed so that if rats do come ashore, they are funnelled by walls into areas containing rat poison. On inhabited islands it is important to undertake environmental education programs (see Jacobson *et al.* 2006) to explain the issues and thereby help reduce the risk of re-colonization.

The only method for controlling rats is by poisoning. A wider range of methods can be employed to eradicate domestic cats, usually in combination. The most commonly used methods have been:

- trapping (82% of islands), commonly using gin traps and less frequently cage traps;
- hunting (51% of islands) including with dogs in daytime and using headlamps at night;
- poisoning (27% of islands; Nogales *et al.* 2004).

Poisoning is commonly used at the beginning of cat-eradication programmes. This is often followed by trapping and hunting to remove any remaining cats. There are three ways of presenting poison (Table 4.3).

It is essential to monitor the success of eradication. There are two methods used for rats—chew sticks and snap traps. Chew sticks are sticks of about 20 cm long by 1–2 cm square soaked in peanut oil, vegetable oil, or lard and pushed vertically about one-third their length into the ground. They should be set over a period of at least 2 years following poisoning and checked periodically for signs of gnawing by rats. If rats are detected, then another round of poisoning should be undertaken. Eradication can be considered successful if no rats are detected for 2 years.

It can be difficult to determine the success of cat eradication. Presence of remaining cats can be detected by placing cat food in areas of smoothed sand to check for footprints and by finding scats and prey remains.

Following successful eradication of rats it is obviously important to minimize the risk of them re-colonizing the island by taking the precautions described above. The risk of rats establishing themselves can be further reduced by running permanent bait stations on the island. An emergency plan should also be in place should rats be detected.

**Table 4.3** *Methods of presenting poison to introduced predators*

| Method | Advantages | Disadvantages |
|---|---|---|
| Bait stations* | Bait less available to birds. | Time-consuming |
| | Bait does not degrade quickly in poor weather. | Logistically complicated |
| | Smaller quantities of bait required compared to the other two methods. | |
| | Possible to monitor the quantity of bait taken. | |
| Hand broadcasting | Faster than setting bait stations (1 day for islands <30 ha), and so suitable for islands with difficult access. | Baits available to non-target species, including livestock. |
| Aerial broadcasting | Covers large areas far more rapidly than hand broadcasting. More suitable for larger islands and those with poor access by sea. | Baits available to non-target species, including livestock. Requires specially equipped aircraft and specially trained pilot. Consequently the most expensive method. |

*Bait stations for rats are typically laid out on a 25 m × 25 m or 50 m × 50 m grid.

It is also essential to formulate a plan for how the programme will be communicated to the public. Public consultation and awareness also have to be handled sensitively. It is better to name the programme a Seabird-Recovery Project, than a Cat-Eradication Programme.

## 4.7 Taking account of climate change

### 4.7.1 Impacts

Climate change will have enormous impacts on both people and wildlife. The main effects on wildlife are likely to be:

- changes in climatic conditions at existing sites;
- changes in habitat requirements of species within a given area;
- changes in the timing of biological events, such as migration and flowering, known as phenology;
- effects of adverse weather during migration;
- loss of coastal habitats due to sea-level rise.

Climate change will change the suitability of habitats for species already present in them by influencing temperature, precipitation patterns, and storms, which will in turn affect hydrology and the frequency of wildfires. Climatic warming will benefit warmth-loving species that are best able to colonize new areas as they become climatically suitable for them; that is, highly dispersive and competitive ones. Species that are less mobile and which require cooler conditions will be most disadvantaged. A woodland snail cannot disperse across hundreds of metres of unsuitable farmland, let alone the tens of hundreds of kilometres that might be needed to reach its new, so-called climate envelope. The inability of many species to move to suitable areas will inevitably result in a widespread impoverishment in biodiversity at a local scale. A particularly insidious threat is from the increased potential for mobile, competitive, alien/exotic species to spread and out-compete native species in semi-natural habitats.

Changes in climate will also affect the range of conditions that a species is able to occupy in a given area. These changes will be most rapid for insects in areas where they are currently restricted to specific habitat conditions because of the warm microclimate they provide. For example, insects currently restricted to warm, open vegetation on sunny slopes will be able to exploit longer vegetation on slopes facing away from the sun (Thomas *et al.* 2001, Davies *et al.* 2006).

Changes in phenology have the potential to 'de-couple' interactions between species. Examples might include the timing of emergence of insects and flowering of plants that they pollinate, or the timing of nesting of birds and of peak abundance of food they require for successful breeding (Gordo and Sanz 2005, Laaksonen *et al.* 2006). Changes in weather conditions, especially increased storm activity during migration, might result in increased mortality of birds.

Rising sea levels will have profound impacts on coastal habitats (Figure 4.10). In an original-natural situation, although sea-level rise would have destroyed intertidal, brackish, and coastal freshwater habitats in some areas, it would also have created similar habitats further inland. Nowadays, formation of new intertidal habitat further inland is in most cases prevented by the presence of hard coastal defences, a process known as coastal squeeze (Figure 9.4).

### 4.7.2 Mitigation and compensation

Potential mitigation includes methods to facilitate the spread of species to their future climatic envelopes and changing management regimes at individual sites. A fundamental decision at individual sites will be the extent to which management aims to maintain existing assemblages of species, or cater for species yet to colonize. Given the huge uncertainty of predictions, the best approach will be to plan to minimize the risk of sites becoming impoverished by maintaining a range of options, rather than plan to cater for the effects of a more limited range of predictions that might prove incorrect.

**Fig. 4.10** Sea-level rise and freshwater and brackish habitats. This shingle bank was artificially maintained to protect the important saline lagoons and freshwater reedbed behind it (a), but breached 2 years later ((b) was taken from a similar position).

The incursion of seawater has created interesting and dynamic intertidal habitat, but has damaged the existing interest of the large freshwater reedbed inland of it. The shingle bank will not be re-instated: most of the shingle has washed away. The freshwater reedbed will continue to be protected from the sea in the short to medium term, but it will probably be impractical to do so indefinitely. Replacement freshwater habitat needs to be created now to provide sufficient time for it to be colonized by a representative range of species before this and other areas of coastal reedbed in this area are lost altogether (Walberswick, Suffolk, England).

Species will be better able to shift to their future climatic envelopes in areas where semi-natural habitat extends to higher altitudes, because they will have shorter distances to move to remain within their climatic envelope. Linking existing sites together will help with dispersal, although creation of sufficient links over the distances required will be unrealistic in most areas. A more practical solution will be to translocate species. Conservationists have been understandably wary of translocations. This will probably have to change. Why wait decades or more for the remote possibility that a small number of less mobile species might eventually colonize an existing or newly created wetland? Instead, you could bring tanker-loads of water and lorry-loads of soil and vegetation containing these less mobile wetland species from another wetland site. Many will probably not establish, but you will have little to lose, at least at newly created sites.

Maintenance of species already present might require different management to that currently undertaken. For example, some breeding waders/meadow birds on wet grassland require a relatively short and open sward in early spring. Higher winter temperatures might allow grass to grow throughout the winter, where formerly it stopped growing in late autumn, and make areas too tall for some breeding bird species the following spring. A solution would be to continue grazing during the winter. Where conditions threaten to become too hot and dry for some species, then management may aim to provide damper and more shady conditions for them to survive in.

Increases in the threat of wildfires due to higher temperatures and lower humidities might be reduced by increased use of burning to reduce fuel loads and the likelihood and intensity of wildfires. Another example of changing management involves the interaction between grazing and the hydrology of vernal pools (Figure 8.5). It will also be necessary to think a long time ahead when planning habitat management and creation, for example by not planting trees in areas where the climate is predicted to be unsuitable for them by the time they mature.

Changes in the hydrology of wetlands brought about by changes in precipitation patterns and higher evapotranspiration rates can to some extent be mitigated through management and design. If it is intended to maintain a similar hydrology, then the main challenges will occur if changes in climate reduce the volume of water entering the site by reducing precipitation and inflow and increase the volume lost through higher evapotranspiration. Dealing with excess water is less rarely a problem, since it can in most (but not all) cases be drained off the site into the usually drier surrounding areas. Lack of water at particular times of year can be addressed using the methods described in Section 8.1.

The alternative is to accept future changes in hydrology and manage or design the site accordingly. For example, if there is predicted to be an increased drawdown in water levels in spring and summer, then design sites to make use of this increased drawdown to provide valuable wetland habitats using so-called Blue-Border management (Figures 8.19 and 8.20).

Loss of intertidal areas can be compensated for by large-scale habitat creation behind existing sea walls, through managed re-alignment or regulated tidal exchange (Section 9.3). Suitable sites for creation of compensatory habitat can, however, be difficult to find.

# 5
# Dry grasslands

There are a number of terms used to describe areas dominated by grasses. Prairie refers to flat and rolling grassy plains in central North America. Steppe is used to describe semi-arid grasslands on plains in mid-latitudes of Europe and Asia. Grasslands that contain widely scattered trees and/or shrubs, typically in the tropics or subtropics, are called savannah. Rangeland is a term used mainly in North America for areas of grazed, near-natural vegetation comprising grasslands, shrublands, woodlands, and forests.

Wet grasslands are those with a high water table and/or which hold surface water. The suitability of wet grasslands for wildlife is highly influenced by their hydrology. They are discussed with other types of wetland in Chapter 8.

Dry grasslands are usually classified according to their:

- soil pH,
- extent of agricultural improvement,
- domination by so-called warm- or cool-season grasses,
- altitude.

There is a range of similar terms used to describe grasslands in terms of their soil pH. Grasslands with a pH of between about 4 and 6 are described as acid, acidic, or calcifugous. They occur over acidic/base-poor rocks, particularly on soils heavily leached by high rainfall, and on nutrient-poor, acidic sands. Grasslands with a pH between about 6 and 7.5 are called neutral or mesotrophic. Grasslands with a pH above about 7.5 are referred to as base-rich, calcareous, calcicolous, or alkaline. They usually support a more species-rich vascular plant flora than acid grasslands. As will be discussed, however, the botanical composition of all these grasslands can be heavily influenced by management.

Most grasslands have been created by people for agricultural use from other habitats, or are highly modified remnants of natural grasslands. The degree of modification for agriculture, usually termed agricultural improvement, or just improvement, greatly influences their existing and potential value for wildlife. Agricultural improvement of dry grasslands involves addition of fertilizers, mainly nitrogen and phosphorus, to increase plant growth and in many cases also re-seeding with a limited range of agriculturally productive grasses. The level of agricultural improvement is usually classified as improved, semi-improved, or unimproved.

Warm- and cool-season grasses differ in their carbon metabolism. Warm-season grasses have $C_4$ metabolism, which means they initially fix carbon dioxide into a four-carbon compound. Cool-season grasses initially fix carbon dioxide as a three-carbon compound. As their name suggests, warm-season grasses are typical of warmer climates. Where they occur together, warm-season grasses tend to be more productive in the warmth of summer and cool-season grasses in the cooler conditions of spring and autumn.

## 5.1 General principles of managing grasslands

Common aims of grassland management are to:

- provide the desired sward composition, the specific aim often being to maintain or increase plant species richness;
- provide the sward height and structure favoured by a desired range of animals;
- prevent colonization by scrub and trees.

As with most other 'dry' habitats, the main techniques used to achieve these are grazing, cutting, burning, and soil disturbance. Also, as in other habitats, these forms of vegetation removal and disturbance have the potential to damage the fauna in the short term. Hence a major consideration is the timing and spatial pattern of this management to minimize any short-term damage.

Grasslands are often associated with a number of other types of habitat and often managed together with them. These include small wetlands, areas dominated by tall forbs, scattered scrub, and limestone pavement (Figure 5.1). Upland and montane grasslands contain a range of features that are rare or absent from other types of grassland: rock cliffs and ledges, scree, flushes (areas with lateral movement of water), and snow fields.

### 5.1.1 Effects of management on vegetation

The most important management-related factors influencing vegetation composition of grassland are as follows.

- The intensity and selectivity of vegetation removal and, to a lesser extent, its timing. Both will influence competition between plant species.

General principles of managing grasslands | 89

**Fig. 5.1** Limestone pavement. These are areas of more-or-less horizontal limestone rock broken up by deep cracks known as grykes (a). They contain a relict flora of open-ground plants that has persisted since the last glaciation, because these pavements have never become completely afforested.

A very extensive and special example of limestone pavement is found in The Burren, County Clare, Ireland. The Burren contains a notable flora including relict populations of arctic-alpine plants at close to sea level, for example mountain avens, *Dryas octopetala* (b), uniquely mixed with small numbers of Lusitanian-Mediterranean species.

Traditional management in The Burren involves grazing the higher limestone pavement with cattle in winter, and moving these to lower pastures in summer. This seasonal altitudinal movement of stock is the opposite to that of most transhumance systems (see Fig. 5.9). The absence of grazing on the limestone pavement in summer allows plants to flower and set seed. In the complete absence of grazing, the limestone pavement and associated grasslands in the Burren tend to be colonized by hazel scrub, *Corylus avellana*.

- The extent to which management creates suitable gaps within the sward in which plants can germinate, and provides suitable conditions for their subsequent establishment.

The method, intensity, and timing of vegetation removal will also influence seed production and hence the ability of plants to regenerate.

The effect of vegetation removal on plant species richness will vary according to the type of grassland. The basic principle is that if the grassland is dominated by one or a limited range of large grasses or forbs in the absence of management, then removing vegetation will reduce the dominance of these larger species and allow less-competitive, smaller plants to survive. This will in most cases increase plant species richness (see review by Bakker 1998).

Most lichens in grasslands require open, sunlit conditions and bare and disturbed ground. They are generally unpalatable and hence benefit from selective grazing that reduces surrounding competition, but are damaged by heavy trampling.

The suitability of conditions for regeneration of plants will be important in influencing the grassland's long-term plant species composition. All plants need to reproduce via propagules at some stage. Propagules will only germinate and establish in patches of bare soil sufficiently free of competition from existing vegetation. These patches are called germination gaps, which are most commonly formed through trampling and other disturbance by grazing animals. A consistent supply of bare and disturbed ground will be necessary to maintain any ruderal (annual or biennial) flora within the grassland.

Increasing nutrient levels through inorganic fertilizer application will in virtually all cases damage the flora of existing, species-rich grasslands. There are, though, rare occasions when low inputs of organic fertilizer may be used to maintain high plant species richness through replacing nutrients removed as hay.

### 5.1.2 Effects of management on fungi

Fungi are an important component of many dry grasslands and perform an important role in nutrient recycling. A group considered to be of particularly high conservation value in Northern European grasslands are the waxcaps of the genus *Hygrocybe*. Diverse fungal assemblages are largely restricted to agriculturally unimproved and semi-improved grasslands, including some that are species-poor in vascular plants and consequently often considered of low conservation value. Increasing nutrients, particularly nitrogen through fertilizer application, is considered damaging to larger fungi. Leaving cuttings on site will also allow nutrients to accumulate (Griffith *et al.* 2004). Fungi usually fruit most prolifically amidst short turf. It is, though, not known to what extent suitability for fruiting influences the long-term persistence of fungi. Many types of fungi grow on the dung of large herbivores and hence benefit from grazing.

### 5.1.3 Effects of management on the grassland fauna

In the majority of grassland the most important factor influencing its suitability for birds and small mammals is its structure and the process that create it, rather than plant species composition *per se*. This is usually referred to as sward structure. In practice, plant species composition will affect structure to an extent and vice versa.

Sward structure refers to variation in the density and height of the vegetation. It is useful to distinguish between *fine-scale* variation (over tens of centimetres), often referred to as tussockiness, and larger, *coarse-scale* variation (over tens of metres or more).

For birds, sward structure and the availability of bare ground and abundance of litter will influence their ability to detect predators and feed efficiently and the suitability of the grassland for nesting (Vickery *et al*. 2001; Whittingham and Evans 2004). Bare ground probably increases access for birds to soil invertebrates (Perkins *et al*. 2000) as well as surface-living arthropod prey such as beetles. A dense litter layer probably makes it more difficult for most birds to access soil invertebrates. Vegetation composition will also directly influence food supply for birds, for example by providing suitable seeds or palatable grass species. Providing variation in sward conditions will increase the likelihood of there being at least some suitable conditions for nesting or feeding in a given area.

The grassland management itself can also affect populations of grassland birds by influencing nest and chick survival. Vegetation removal and other agricultural activities can destroy a high proportion of nests and young of birds. These losses can be minimized by altering the timing of management and by using other specific management techniques.

Scattered scrub and trees will increase the numbers of bird species using a grassland by providing nest sites and song posts for additional species. However, these will tend to be more generalist species, rather than grassland specialists. Trees and scrub may also provide nest sites and look-out posts for predatory birds and thereby reduce the breeding success or survival of grassland specialists (e.g. Green *et al*. 1990b). In North America trees and scrub will also encourage brown-headed cowbirds.

Densities and species richness of small mammals, such as voles, are higher in taller, ungrazed grassland with an abundant litter layer than in heavily grazed grassland (Rosenstock 1996; Tattershall *et al*. 2000; Evans *et al*. 2006). However, there is evidence that the highest densities often occur at moderate to high grazing intensities that provide a mixture of tall grass and cover and also fresh vegetation growth stimulated by livestock grazing (Schmidt *et al*. 2005b). Many slightly larger mammals that dig burrows and rely on detecting predators by sight, such as rabbits, prairie dogs, ground squirrels, and sousliks, *Spermophilus* spp., prefer (and maintain) short, open grassland.

A wide range of factors will be important in influencing conditions for invertebrates. In general, taller swards support a larger number of individuals and species of arthropods, although many species are restricted to short, open swards (Morris 2000). In cool regions bare and disturbed ground is particularly valuable, especially on well-drained and friable soils on sunny slopes. This provides a suitable microclimate for warmth-loving species towards the edge of their climatic range and open, compacted ground for solitary bees and wasps to excavate nests in. Disturbance also creates suitable conditions for annual plants, whose seeds can be important seed sources for invertebrates. Other features of particular value for invertebrates in grasslands are:

- the presence of specific food plants of suitable growth form;
- a continuity of suitable nectar sources for adult insects;
- litter for litter-dwelling species;
- seed heads, both as a source of seed for some species and to provide over-wintering sites;
- grass tussocks to provide over-wintering sites;
- dung and carrion for dung- and carrion-feeding specialists and their predators.

Other features of particular value for invertebrates often associated with grasslands are:

- quarries, eroding cliffs and faces, disturbed edges of paths, derelict land, and other disturbed areas that provide early-successional habitats;
- wet flushes, marshy areas, and temporary pools.

As in other habitats, a key principle when managing grasslands for insects is to maintain a range of suitable conditions in close proximity to help provide the range of microhabitats required for insect species to complete their annual life cycles.

## 5.2 Non-intervention

In the absence of vegetation removal through grazing, cutting, burning, or soil disturbance, most grasslands will eventually become dominated by a limited range of more competitive larger perennial grasses and forbs and accumulate a layer of dead plant litter. They will lack plants that need to regenerate from seed at frequent intervals, and be largely dominated by plants that spread vegetatively. Unmanaged grassland will provide suitable conditions for a range of plants, birds, invertebrates and small mammals that prefer tall vegetation and an abundant litter layer and require freedom from disturbance to complete their life cycles.

It is often assumed that if a grassland is left 'unmanaged', it will fairly quickly become colonized by scrub. This will not be the case, though, if there are no seeds of woody plants present or no gaps within the sward in which they can establish. Unmanaged grasslands can be surprisingly stable and resistant to scrub establishment (Figure 5.2). Woody plants commonly colonize grasslands where a period of heavy grazing or other disturbance that creates suitable conditions for their establishment is followed by a relaxation of management that allows the young seedlings to grow into bushes and trees. Establishment of trees and shrubs is discussed further in Sections 7.1.3 and 7.1.4.

**Fig. 5.2** Unmanaged grassland and scrub invasion. Grasslands that have been left unmanaged for a number of years, particularly those on more fertile soils, accumulate a thick layer of plant litter that smothers any gaps for scrub seedlings to establish in.

This is part of a large area of grassland that has been unmanaged for about 20 years. Despite this, the grassland contains virtually no scrub, instead comprising tall, rank grasses, common nettles, *Urtica dioica*, and tall umbellifers, Umbelliferacea, (Rainham Marshes, East London, England).

## 5.3 Differences between grazing, cutting, and burning

Grazing has a quite different effect on sward composition and structure to either cutting or burning. The differences are described below. For practical reasons, cutting can obviously only be undertaken on relatively flat sites. Cutting may be the only practical method of managing small grasslands which are impractical to graze or burn, such as those in urban areas. Upland and montane grasslands are usually impractical to manage by anything other than extensive grazing.

### 5.3.1 Sward composition and structure

Light to moderate grazing removes vegetation gradually and selectively and, in so doing, can maintain suitable conditions for animals living within the sward while this is taking place, so long as some of the vegetation is retained. By contrast, cutting close to ground level and burning are catastrophic events. Both suddenly remove virtually all the vegetation and make the grassland temporarily unsuitable for its inhabitants (see Morris 2000 for effects on invertebrates). They will be particularly damaging to species that are slow or unable to re-colonize as the vegetation becomes suitable for them again. These negative effects can be reduced by only cutting or burning a proportion of the grassland at any one time.

The selectivity of vegetation removal through grazing varies according to the type of grazing animal and grazing pressure. Grazing particularly benefit plants that either avoid being eaten, or recover well following defoliation. Plants avoid being eaten by:

- being unpalatable; that is, containing distasteful or poisonous substances;
- having physical defences such as spines;
- being low-growing and, in particular, forming rosettes that are difficult for animals to bite (Figure 5.3).

The group of plants best adapted to tolerate repeated defoliation are grasses and other monocotyledons. These grow from meristems situated at the base of the plant, which are out of the reach of grazing animals. Dicotyledons grow from apical meristems which, if protruding above ground, are vulnerable to being grazed.

Grazing creates coarse-scale variation in sward structure and composition by accentuating existing variations in plant composition due to differences in topography and former management. Animals also create variation in sward conditions by treading down vegetation and creating patches of bare ground by poaching and other forms of disturbance. Patchy deposition of dung and urine can further increase spatial variation in plant composition.

## Differences between grazing, cutting, and burning | 95

**Fig. 5.3** Benefiting from heavy grazing. Heavy grazing provides a competitive advantage for plants that are avoided by grazers.

The poisonous large Mediterranean spurge *Euphorbia characias* (a) has been avoided by livestock that have closely cropped all the surrounding vegetation.

Acanthus-leaved carline thistle, *Carlina acanthifolia* (b), is low-growing and difficult to bite, and also spiny.

The largely unselective vegetation removal by cutting and by burning tends to encourage greater uniformity in vegetation composition, height, and structure compared to grazing. Cutting and burning tend to encourage tall, bulky plant species that can out-grow smaller ones during the periods between the one-off cutting or burning events. These usually include larger, often more palatable plant species, which would otherwise be selectively removed by grazing animals (Figure 5.4).

Burning creates more bare ground than cutting, particularly with more intense fires that remove a greater proportion of litter. This favours the germination and growth of annual plants. Burning also often removes vegetation less uniformly than cutting due to variations in fire intensity. Fires often miss or only superficially burn some patches of vegetation. The flush of re-growth following burning attracts heavy grazing by livestock and wild herbivores.

Mowing and leaving the cuttings *in situ* smothers small plants and covers bare ground, including potential germination gaps, and so tends to reduce plant species richness. Management of meadows involves cutting and removing vegetation followed by a period of grazing (known as aftermath grazing). The aftermath grazing creates germination gaps.

## 96 | Dry grasslands

**Fig. 5.4** Flowery meadows. Agriculturally unimproved hay meadows typically contain a high diversity of plant species. These often contain an abundance of tall, bulky forbs and palatable forbs, which are scarce in, or absent from, more heavily grazed grasslands. This meadow was stunning in colour! (Lanslebourg-Mont-Cenis, Haute-Savoie, French Alps).

### 5.3.2 Scrub

Grazing, cutting, and burning differ in their effects on scrub. Grazing animals prevent scrub from establishing by eating their seedlings. They can weaken or kill more established scrub through browsing and bark removal, but this ability varies with livestock type, grazing pressure, timing of grazing, and availability of more preferred vegetation. Cutting prevents seedlings from growing above the height of the cutting blade. Burning removes fire-intolerant species, but leaves fire-tolerant species to continue growing with often reduced competition.

### 5.3.3 Soil nutrients

Grazing animals re-distribute a high proportion of the nutrients removed in the vegetation via their dung and urine, unless the livestock are folded (kept in a pen) on other areas at night to deposit dung on them. Patchy deposition of dung increases variation in vegetation composition and structure by increasing spatial variation in nutrients. Avoidance of dunging areas by grazing animals can further accentuate this variation. Folding was used widely to deposit nutrient-containing dung on arable land from animals grazed during the daytime on grassland or heathland.

Cutting and removal of vegetation removes nutrients from the grassland, and this can be useful when seeking to reverse the effects of artificially raised nutrient levels. Leaving cuttings *in situ* recycles them. Burning abruptly releases a proportion of nutrients in its ash.

### 5.3.4 Dung and carrion

Dung produced by grazing animals supports a diverse assemblage of invertebrates, particularly species of beetles and flies. These can provide prey for birds such as red-billed choughs, *Pyrrhocorax pyrrhocorax* (Roberts 1982), and bats. Carcasses of livestock can be important in sustaining scavenging birds and mammals, as well as supporting carrion-feeding invertebrates.

## 5.4 Management by grazing

The specific effects of grazing will depend primarily on:

- the type of grazing animals (and to a lesser extent their breed, age, experience, and herd structure);
- grazing pressure;
- the timing and duration of grazing.

The value of domestic animal's dung for invertebrates is reduced by the administration of anti-parasitic drugs (Section 4.4.3). A further consideration is the potential effects of domestic livestock on wild herbivores, either through reducing food for them, or by facilitating their use of the grassland by reducing sward height to within their preferred range.

### 5.4.1 Effects of different grazing animals

The most commonly used domestic grazing animals in habitat management are cattle, sheep, equines (horses, ponies, and donkeys), and goats. Wild herbivores can also be important grazers and beneficial creators of disturbance. Different types of livestock vary in their:

- spatial patterns and selectivity of vegetation removal;
- trampling effects;
- dung fauna;
- extent to which they browse woody vegetation.

Differences in sward structure between cattle, equines and sheep are illustrated in Figure 5.5. Goats browse woody vegetation far more than other livestock.

Before comparing the effects of different types of livestock, it is worth noting that their effects will vary with grazing pressure. In particular, variation in sward structure and differences in the effects of grazing animals will be greatest under medium grazing pressure (Section 5.4.3).

## Cattle

Cattle have a different method of feeding to equines or sheep and this produces quite different effects on vegetation composition and structure. Cattle feed by wrapping their large, rasping tongue around the sward and cutting it between their lower teeth and upper dental pad as they swing their head. This makes their feeding unselective at a very fine scale, since they cannot differentiate between different plant species within the tuft of vegetation they grab.

Cattle also differ from sheep and equines in the distribution and nature of their dung. They deposit most of their dung in a scattered manner, throughout the grassland, although it is sometimes concentrated in favoured lying-up areas. Cattle dung supports an especially diverse invertebrate fauna (Figure 4.8). Cattle avoid grazing within 10–20 cm of their pats, leaving scattered patches of fertilized taller vegetation (called avoidance mosaics). At medium grazing intensities, the combination of the removal of tufts of vegetation and avoidance of pats produces more tussocky vegetation than grazing by equines or sheep. Tussocky, cattle-grazed grassland is generally considered better for most ground-nesting birds than more homogenous swards maintained by sheep grazing. This is probably because the greater variation in cattle-grazed swards provides greater opportunities for concealment and camouflage of nests and young. European hares, *Lepus europaeus*, also prefer cattle-grazed swards to those produced by sheep (Smith *et al.* 2004b).

The hooves of cattle exert a greater trampling pressure than those of equines and a far higher pressure than those of sheep. Cattle are therefore better at squashing down and breaking up rank vegetation and litter and at creating bare ground by poaching. Adult cattle trample a lower proportion of birds' nests than sheep per quantity of vegetation removed, although frisky, young cattle trample a similar proportion (Figure 8.29). There is little information comparing nest-trampling rates of cattle and equines.

Cattle can reduce the vigour of woody plants by tearing off leaves and twigs. Horned cattle can inflict further stress on woody plants by breaking twigs and rubbing off bark with their horns.

## Management by grazing | 99

**Fig 5.5.** Grazing animals and sward structure. Cattle, equines, and sheep produce quite different sward structures at medium grazing pressures, as shown here. At high grazing pressures all create a short and uniform sward.

(a) Cattle usually produce the most tussocky vegetation. They also create more trampling and disturbance than either equines or sheep.

(b) Equines typically create a mosaic of short, heavily grazed lawns interspersed with patches of taller, lightly grazed or ungrazed vegetation.

(c) Sheep nibble the vegetation close to the ground. They tend to produce a less tussocky sward than cattle, although still avoid particular unpalatable plants, in this case small tussocks of mat-grass, *Nardus stricta*.

An important consideration when deciding which type of cattle to use is whether they will need to survive on nutrient-poor vegetation and in hostile weather conditions. This varies with age and breed (Section 4.4.2). Grasslands supporting relatively nutritious grasses can be grazed in summer by most modern, commercial breeds, including young cattle. Older breeds of more slowly maturing beef cattle are considered best for feeding on coarse, less nutritious vegetation. Cattle are better able to digest coarser vegetation once their rumen has fully developed at 18–24 months old. Milk production generally requires more nutritious forage and hence dairy cattle are generally unsuitable for grazing coarser vegetation. There is little suggestion of significant differences in grazing behaviour between cows and bulls.

If an aim of grazing is to reduce the abundance of more unpalatable vegetation, by forcing livestock to eat it through lack of alternative, more palatable forage, then more hardy or primitive breeds might be needed. This will especially be the case if it requires year-round grazing under severe weather conditions.

*Equines and sheep*

Equines and sheep feed by nibbling vegetation using their incisors. They are both highly selective grazers, being able to differentiate between different plants as they nibble. Sheep will preferentially feed on the flowers and buds of preferred forbs. They therefore have the potential to seriously deplete nectar sources for insects. Equines usually concentrate on eating palatable grasses and generally avoid forbs. The small size and nibbling behaviour of sheep make them ineffective at grazing down and breaking up tall swards. Cattle are much better at doing this (see above). Conversely, the selective nibbling and only limited trampling of sheep is better for lichens.

At medium stocking densities equines will accentuate existing large-scale variation in vegetation structure and composition, by heavily grazing some areas and avoiding others. This larger-scale variation in sward structure and composition can be further enhanced by equines' habit of concentrating dung in discrete areas. This results in localized nutrient-enrichment and often colonization by taller, more nutrient-demanding plants. At moderate grazing intensities grazing by equines produces a mixture of closely cropped lawns and patches of ranker, ungrazed vegetation. The amount of bare ground produced by equines is intermediate between that produced by cattle and sheep.

Sheep have a bad reputation for creating extremely uniform, closely cropped swards. They do tend to produce more uniform swards than equines and especially cattle, although this uniformity of composition and structure is to some extent due to the often high stocking densities at which they are kept. Sheep do not have specific dunging areas. They create less bare ground through poaching than either cattle or equines and cannot break up tall, rank vegetation.

Equines and sheep will both feed on coarser vegetation, including sedges and rushes, when there is little grass available, especially in winter. In these conditions equines will dig up rhizomes, browse woody plants and tear off and eat bark, often killing shrubs and saplings. Conventional sheep are poor at browsing scrub, even when there is little else to eat. However, primitive breeds of sheep browse woody vegetation far more than conventional breeds, and are useful at controlling young scrub (Section 4.4.2). Equines form strong social groups, including sub-groups of colts and fillies. In extensive and naturalistic grazing regimes the territorial nature of these different social groups can result in large-scale variations in grazing pressure and hence in vegetation composition and structure.

Virtually all equines used for conservation grazing of grasslands are ponies. Donkeys are rarely considered. They can, though, be useful for controlling thistles, *Cirsium* spp., rushes, and other coarse vegetation where these are a problem, such as on former arable land. Donkeys will preferentially graze thistles down to their basal rosettes and nibble down rushes, despite there being plenty of apparently more palatable grasses present. Donkeys are similar to sheep in being best suited to drier conditions and, unlike ponies, will not graze in shallow water.

As with cattle, a major consideration when deciding the type of sheep or equine to use is the level of hardiness required. Also, as with other livestock types, this will vary largely with age, and to some extent breed. Both very young and very old sheep and equines are less able to graze coarser vegetation. Hill breeds of sheep are generally better able to survive in more hostile weather conditions on poor-quality forage than upland breeds, which are in turn more hardy than lowland ones. Primitive sheep are extremely hardy. There is a suggestion that, in the case of at least some breeds, rams and wethers (castrated males) browse more and feed on coarser vegetation than ewes. There is no evidence in any marked difference in feeding behaviour between sexes in equines.

*Goats*

Goats are well-known for consuming a wide variety of vegetation. They will typically spend 50–75% of their time browsing the leaves of trees and shrubs, far more than cattle, sheep, or equines. Goats also strip the bark of woody plants, particularly in winter. This makes them very effective at controlling even established scrub. Goats also graze grass and forbs, but do not usually crop them as short as sheep do. The other important feature of goats is their agility. This allows them to exploit food sources out of the reach of other grazers. They can rear up on their back legs to reach vegetation, and climb bushes, small trees, and precipitous slopes. The disadvantage of this is they can destroy patches of grazing-sensitive vegetation that have previously survived on ledges and cliffs out of the reach of other grazing animals.

Goats form groups of females (nannies) and young. These can also include yearling males. Male goats (billies) may be solitary or form all-male groups. Billies and nannies are thought to have slightly different feeding preferences. In particular, billies strip bark more (Bullock and Oates 1998). The main problem with goats, though, is preventing them from escaping.

*Other grazing animals*

A number of wild grazing animals are especially important in influencing habitat conditions in semi-natural grasslands. Prairie dogs, pocket gophers, and American bison, *Bison bison*, are important grazers and creators of soil disturbance in North American prairies. European rabbits are important grazers and create valuable soil disturbance in many grasslands in Europe, especially those on loose, dry, especially sandy soils.

Prairie dogs, pocket gophers, and rabbits all cause soil disturbance by burrowing, and, in the case of rabbits, also by scraping the ground along the boundaries between warrens. Prairie dogs play a key role in providing suitable conditions for a wide range of other animals in short-grass prairie (Figure 5.6). Heavy rabbit grazing provides short vegetation and bare ground that, on suitable soils, can support vegetation characterized by lichens, cushion-forming mosses and winter annuals (Dolman and Sutherland 1992), and support warmth-loving invertebrates towards the edges of their climatic range. Very short, open, stony and disturbed ground produced by heavy rabbit grazing is favoured by Eurasian thick-knees, *Burhinus oedicnemus*, a rare and declining bird in much of Europe (Green and Taylor 1995; Beasley *et al.* 1999). The close, selective nibbling of rabbits and lack of heavy trampling by them favours lichens on grass heaths and grassy sand dunes. Lichens are otherwise easily trampled and destroyed by large grazing animals. Soil disturbance and local concentrations of dung produced by rabbits can provide suitable conditions for a range of plant species otherwise rare or absent within the grassland, including valuable nectaring plants. Rabbits are also important prey for a wide range of carnivores. They are the primary prey of 29 species of predators in their native Spain (Delibes and Hiraldo 1981).

Rabbits prefer short swards. They can be encouraged by using other livestock to keep the sward suitably short (Section 5.4.2) and by providing suitable cover and areas for them to establish warrens in (Figure 5.7).

American bison have similar feeding characteristics to cattle, but tend to concentrate on grasses rather than forbs to a greater extent and to browse slightly

**Fig. 5.6** Prairie dog towns. Colonies of prairie dogs, known as towns, are important in sustaining a range of other species in the short-grass prairies of North America. Their grazing provides short, open areas favoured by mountain plovers, *Charadrius montanus*, and their vacant burrows habitat for burrowing owls, *Athene cunicularia*, rabbits, hares, lizards, and rattlesnakes. The prairie dogs themselves are important prey for ferruginous hawks, *Buteo regalis*, swift foxes, *Vulpes velox*, and the endangered black-footed ferret, *Mustela nigripes*.

less (Plumb and Dodd 1993). Wallowing (dust-bathing) by American bison can create temporary pools. Deer will graze grasslands heavily close to the safety of woodlands. Wild boar and pigs can create valuable soil disturbance by rooting.

### 5.4.2 Using mixtures of grazing animals

Grasslands can be grazed using combinations of different livestock. The effects can be difficult to predict. Adding small numbers of cattle to otherwise sheep-grazed grassland, particularly in winter, can be used to add valuable bare and disturbed ground for invertebrates and regeneration gaps for plants. Combinations of cattle and sheep can be used to open up tall and rank swards. Cattle are first used to graze down and break up the tall sward. Sheep are unable to do this. Sheep are then introduced to nibble and pull out the remaining dense thatch from the short sward that cattle are unable to reach.

Combinations of livestock also have the potential to reduce the overall selectivity of vegetation removal, through one type of animals feeding on the plants

**Fig. 5.7** Encouraging rabbits. European rabbits can be important grazers and agents of soil disturbance in grasslands, but are reluctant to establish new warrens in areas far from cover.

Rabbits can be encouraged on to fields by depositing piles of brash (a). These provide safe cover for the rabbits while they excavate warrens in the soil beneath them. The brash piles can be removed once a new warren has been established (b; Minsmere, Suffolk, England).

A different approach is required on soils where rabbits have difficulty excavating burrows, such as on thin, calcareous soils overlying relatively hard chalk or limestone. In these areas mounds and banks of soil can be provided for them to excavate warrens in.

the other has avoided. This can reduce overall structural diversity, but increase plant species richness at the small to medium scale, by reducing the dominance of more competitive species that the other livestock type has selectively avoided. For example, horses avoid grazing their latrine areas, and this allows more competitive plant species to become dominant in these areas. Adding cattle reduces the dominance of these more competitive plant species in the horse latrines, and increases plant species richness in them (Loucougaray *et al.* 2004).

Grazing by one species can be used to facilitate use of the grassland by other grazers. Sheep can be used to graze down the sward to a suitable height for rabbits.

## 5.4.3 Grazing pressure

Grazing pressure will influence the structure and composition of the sward and thereby its associated fauna. At medium stocking levels animals will tend to increase variation in sward structure by selectively removing a greater proportion of the vegetation in some areas compared to others. The extent and pattern of this will vary with livestock type. Very high or very low levels of grazing will remove, respectively, virtually all or none of the vegetation. Neither will increase variation in structure. Higher levels of grazing will obviously create a shorter sward and tend to create more bare and disturbed ground, although the extent of this will again be influenced by livestock type, the timing of grazing, and soil wetness in the case of bare and disturbed ground. The short, open conditions and bare ground produced by heavy sheep, and especially rabbit, grazing are important in maintaining the characteristic species-rich assemblages of lichens that can occur on skeletal, especially calcareous soils (Lambley 2001).

Grazing pressure influences sward composition by selectively disadvantaging some plants more than others. At very low stocking levels animals will concentrate on the most palatable plants. As stocking levels increase, animals will remove an increasing proportion of bulky and palatable species, and thus increasingly benefit those plant species that avoid being eaten or recover well from it. It is only at very high stocking levels, where there is little else left to eat, that livestock will be forced to eat the least palatable plants. Plant species often become dominant on grasslands because they are relatively unpalatable and thereby gain an advantage over the rest of the vegetation. In many cases, though, increasing grazing intensities to levels high enough to reduce these less-palatable species may be damaging to the other grassland flora and fauna. It may also be unacceptable to commercial graziers.

Methods of estimating the approximate numbers of livestock to use are described in Table 5.1 and Figure 5.8. The effects of grazing on sward conditions will vary between years due to weather-related differences in plant growth. Although it is useful to set out desired stocking levels, these still have to be adjusted to achieve the desired sward conditions. In particular, the suitability of the sward for birds in spring can change rapidly in response to rapid grass growth, while periods of drought may mean that livestock numbers have to be reduced because there is insufficient food for them. Good conservation grazing management relies largely on the adjustment of stocking levels in a particular area through skill and judgement: knowing when to move them, and which areas to exclude livestock from at particular times of year. These may include areas with high densities of nesting birds or important food plants or nectar sources for

**Table 5.1** *Livestock units*

| Animal | Livestock unit for medium-sized animal | Range of livestock units for small to large animals |
| --- | --- | --- |
| **Sheep** | | |
| Ewes (including their lambs) and rams | 0.10 | 0.08–0.15 |
| Ewe followers and store lambs | 0.08 | 0.06–0.10 |
| **Cattle** | | |
| Dairy cows | 1.0 | 0.8–1.1 |
| Suckler cows (including their calves) | 0.9 | 0.7–1.1 |
| Beef and other cattle older than 2 years | 0.7 | 0.6–1.0 |
| Weaned beef and other cattle younger than 2 years | 0.6 | 0.5–0.7 |
| **Equines** | 1.0 | 0.8–1.2 |

These are the standard measure for describing grazing pressure. They show the relative quantity of vegetation removed by different grazing animals. From the figures shown here, it can be seen that one medium-sized dairy cow removes the same quantity of vegetation as 10 medium-sized ewes (including their lambs). Values of livestock units for other ruminants can be calculated by dividing their live weight (in kg) by 650.

*Source*: UK's Rural Development Service Technical Advice Note 33.

invertebrates. Livestock can be excluded temporarily from sensitive areas using electric fencing.

When introducing grazing to a formerly unmanaged sward it is prudent to only introduce it to a proportion of the area and only begin grazing at a low stocking level. Heavy grazing, as with burning and cutting, has the potential to cause the local extinction of existing populations of invertebrates. In particular, it might remove important larval foodplants, nectar sources, or over-wintering sites and thereby prevent populations of insects from completing their annual life cycle. Examples of where this has occurred are given by Waring (2001). Grazing pressure can, if need by, be increased later on based on the results of monitoring and any other experience gained. If the sward has been left unmanaged for a long time, there is unlikely to be any urgency in restoring it to a heavily grazed condition.

The practicality of achieving desired sward conditions depend to some extent on the size of the grazing units. Greater control can be achieved using smaller grazing units. Each of these can then be grazed at an optimum level. The ability to achieve desired stocking levels may also depend on the availability of alternative land of minimal conservation interest, on which livestock can be placed

**Fig. 5.8** Estimating how many stock to use. The standard measure of grazing pressure is the number of livestock unit days/ha per year. This provides an index of the approximate quantity of vegetation removed by grazing animals per hectare of land over 1 year. Grazing pressure can be calculated using the following formula.

Livestock unit days/ha per year = number of stock × livestock unit of stock type × number of days spent grazing per year/ha of habitat grazed

For example, six medium-sized beef cattle older than 2 years (=0.7 livestock units per animal; see Table 5.1) that graze for 180 days during the year on 5 ha of land will exert a grazing pressure of

6 × 0.7 livestock units × 180 days/5 ha

= 151 livestock unit days/ha per year.

These figures do, though, have to be treated with caution. In particular, the timing of grazing also influences the extent of vegetation re-growth and hence the grazing pressure required to remove it. They are, however, useful in estimating the approximate number of stock of a given type that are required at a site, although these numbers have to be adjusted depending on seasonal and between-year variation in vegetation growth. Ranges of grazing pressure used to manage different types of grassland are shown below.

| Type of grassland | Grazing pressure (livestock unit days/ha per year) |
|---|---|
| Dry, unimproved, acid and base-rich | 40–100 |
| Neutral semi-improved and improved | 100–400 |

when not required on areas of high conservation value. This is often referred to as sacrificial grazing.

Within very large grazing units the distribution of livestock, and hence variation in grazing intensities, was formerly commonly controlled by shepherding. This practice is rarely, if ever, economically practical in temperate areas nowadays.

In the uplands where flocks of sheep and herds of cows have been continually grazed in an area over a number of generations, they often form their own, defined home-ranges. These flocks and herds are known as hefted. Specific considerations when grazing upland and montane grasslands are highlighted in Figure 5.9.

**Fig. 5.9** Upland and montane grasslands. In upland areas livestock, like these sheep in the Spanish Picos de Europa (a), are traditionally moved to high pastures in summer and wintered in sheltered lowlands. This system of seasonal movement of people and their livestock in search of grazing is known as transhumance.

Cliffs and ledges inaccessible to grazing animals in upland areas can support relict populations of grazing-intolerant plants as on the ungrazed ledges (b) of the historically heavily grazed Ben Lawers range in Scotland (c). Introduction of goats can be especially damaging, since they can reach and graze out plants that have previously remained inaccessible to less nimble animals.

Care should also be taken in upland areas to ensure suitable levels of grazing in flushes, which can support very rich assemblages of plants.

### 5.4.4 Timing and duration of grazing

The timing and duration of grazing will influence:

- the suitability of the sward for its inhabitants, especially birds and invertebrates, at different times of year;
- the effect grazing has on individual plant species and hence the long-term sward composition;
- the damage caused to the fauna by the vegetation removal itself and trampling of livestock.

There are two main approaches to grazing: either maintaining a similar grazing regime each year or grazing different areas each year on rotation. Grazing can be carried out at any time of year, but in temperate areas grazing is typically carried out:

- from spring to autumn (i.e. during the growing season), often referred to as summer grazing;
- from autumn to spring (i.e. during the dormant season), often referred to as winter grazing;
- year-round.

*Summer grazing*

Most pastures are grazed in summer to maximize their agricultural output. The main aims of *conservation grazing* in summer are to maintain high plant species richness by preventing more competitive plant species from out-competing less competitive ones, and to provide a suitably short and open sward required by desired groups of breeding birds, wintering birds arriving after the growing season, and invertebrates. The disadvantages of grazing during the growing season are that it can:

- remove nectar sources for adult insects;
- prevent the development of seeds for seed-feeding invertebrates and birds, seed heads used by over-wintering invertebrates, and make conditions less suitable for larger and longer-lived invertebrates;
- trample birds' nests (Beintema and Müskens 1987; Green 1988).

Hence the optimal grazing regime will seek to provide the desired sward conditions while minimizing these negative effects. In less productive grasslands, especially those on thin soils where vegetation growth is limited by summer drought, it may be possible to achieve the desired open and varied sward conditions by grazing during winter and avoid these disadvantages of summer grazing.

Summer grazing will prevent larger, more vigorous plant species from outcompeting smaller, less competitive ones. The specific timing of grazing during the growing season will also have subtle effects on plant species composition. Plant species will vary in their ability to withstand defoliation during different periods of growth, and this will alter their relative competitive ability. In practice, though, the subtle effects of differences in the specific timing of grazing during summer are rarely an important consideration when deciding on grazing regimes. Instead, the timing of summer grazing is more often determined by agricultural objectives and the requirements of nesting birds.

Methods for minimizing trampling of bird's nest by livestock have been developed mainly specifically to benefit breeding waders/meadow birds on wet grasslands and are described in Section 8.12.3. These techniques can, though, also be applied to drier grasslands.

Grazing levels during summer and autumn will be important in creating short swards for birds to feed on larger soil invertebrates in winter (e.g. Buckingham *et al.* 2006), and where there is little or no vegetation growth in winter, sward conditions at the beginning of the following breeding season.

Grassland fungi possibly benefit from relatively heavy grazing in summer. Most species are thought to fruit best when the sward is short (less than about 10 cm) in late summer and autumn.

### Winter grazing

Winter grazing has traditionally taken place on grasslands on more free-draining soils in the lowlands that are less subject to poaching. Livestock grazed in upland areas in summer are often out-wintered in sheltered, lowland grasslands. Winter grazing has declined in many areas, with less hardy, modern commercial breeds usually now wintered indoors.

The presumption against grazing most types of grasslands in winter is based largely on its potential damage to the agricultural productivity of the sward. It is interesting to note that in original-natural grasslands there would undoubtedly have been a higher grazing pressure in winter *relative* to the quantity of food available. This is the opposite to most conventional agricultural systems. The higher grazing pressure in winter in natural systems might explain the large array of ruderal plants that require bare ground (created by poaching in winter), and the large number of forbs that are intolerant of heavy grazing during the growing season.

The main uses of conservation grazing in winter, or outside the main growing season, are to:

- provide suitable conditions for invertebrates in summer, without the deleterious effects of removing vegetation during summer (Figure 5.10);

**Fig. 5.10** Grazing outside the main growing season. This provides the benefits of reducing accumulation of litter and creating bare ground, but without removing nectar sources and food plants for insects during their main active periods. This forest-edge grassland is grazed for a period in early spring, but is left ungrazed during the middle of summer when the cattle are moved to higher ground. The mixture of bare ground and abundant flowers support a wealth of insects, including this superb cardinal butterfly *Argynnis pandora* (Arguébanes, Asturias, Spain).

- maintain specialized plant communities created and maintained by winter grazing.

Winter grazing may also supplement summer grazing in order to:

- help maintain a suitably short sward in spring in areas where vegetation continues to grow during winter;
- reduce the dominance of less palatable wintergreen plants that are largely ungrazed in summer.

Winter grazing can be used to provide suitable conditions for warmth-loving invertebrates on agriculturally unimproved grasslands on light, free-draining soils, especially over chalk and limestone. On these generally unproductive swards summer grazing is not necessarily needed to maintain the short, open conditions required by many invertebrates and less competitive plant species.

Winter grazing will maintain relatively open conditions by preventing the accumulation of litter and create bare and disturbed ground by poaching, but without the potentially damaging effects of summer grazing. Grazing pressure should be high enough to remove accumulated litter and create bare ground, but light enough to retain seed heads and tussocks, both of which can provide important over-wintering sites for invertebrates.

Where sward continues to grow during winter, additional light winter grazing may be beneficial in maintaining a short and open sward for wintering birds that feed on soil invertebrates, and for open-sward-loving species the following spring. Several species of birds that feed on soil invertebrates prefer fields with livestock in them during the winter (Perkins *et al.* 2000).

A common use of winter grazing is to reduce the vigour of unwanted, less palatable, wintergreen plants, such as rushes and some coarse grasses. These tend to be avoided by grazing animals in summer when more palatable herbage is available. However, wintergreen plants become the best forage available in winter once deciduous plants have transferred their food reserves below ground and any other more palatable vegetation has been consumed.

Although a degree of poaching caused by winter grazing is beneficial in grasslands, excessive poaching can damage species-rich grassland and result in colonization by some unwanted plants, such as thistles, ragwort *Senecio* spp., and rushes. Winter grazing needs constant attention to be successful. In winter a sward can rapidly change from being lightly poached and moderately vegetated to heavily poached and largely denuded of vegetation. This is particularly the case on wetter soils.

*Year-round grazing*

Year-round grazing using similar stocking levels is a feature of most extensive and naturalistic grazing regimes (Section 4.4.1). In original-natural conditions food supply is likely to have often limited large herbivore numbers in winter. Starvation of livestock is unacceptable under animal welfare legislation. This means that stocking levels for year-round grazing have to be based on the maximum *winter* carrying capacity of the area, which will be considerably lower than that in summer. Maximum stocking levels for year-round grazing are typically only a third, or even less, of those that a site is capable of supporting in summer.

Hence grazing with similar numbers of livestock throughout the year will usually make it difficult to maintain suitable conditions for species that require short, open conditions in late spring and summer. This will particularly be the case where vegetation growth has been artificially raised through previous fertilizer application.

Supplementary feeding can be used to maintain livestock in winter, but is generally disadvantageous because it will increase nutrient levels. However, supplementary feeding of livestock can be important in providing seed for some bird species in the late winter and early spring. Supplementary grazing should only be carried out on land of low conservation value.

## 5.5 Management by cutting with or without grazing

There are two types of cutting used on grasslands. These are mowing close to ground level to remove the bulk of the sward or cutting to remove specific patches of unwanted, taller vegetation. This is often referred to as topping. The main conservation benefits of mowing are its use in combination with grazing to maintain the conservation value of agriculturally unimproved hay meadows, and using it to provide a patchwork of different stages of re-growth on grasslands where vegetation removal by grazing is impractical. Agriculturally improved grasslands cut for silage have little wildlife interest, and have received little interest from conservationists.

### 5.5.1 Meadows, and silage and haylage fields

Meadows can be managed to produce hay (dried grass and forbs), silage (grass cut while still green and then fermented), or haylage (dry silage). Cutting is carried out close to ground level to maximize the off-take of herbage. Increasing the height of the cut can, though, in some cases increase overall grass yield. For example, in silage fields cut twice or more during the season, increasing the height of the first cut from 3 to 8 cm can increase the total annual grass yield (Binnie et al. 1980). The re-growth following cutting is then grazed. Common aims in conservation management of agriculturally unimproved hay meadows are to maintain their high plant species richness and high ratio of forbs to grasses while minimizing the loss of nests and chicks of ground-nesting birds during cutting. The main management-related factors that will influence the flora of grasslands managed by cutting are the timing of and number of annual cuts and the timing and pressure of subsequent grazing.

Hay meadows are traditionally cut once a year between June and August. The general timing of cutting varies between regions while its precise date in a given area will vary from year to year depending on weather conditions. The timing of cutting will affect at what stage of a plant's growth it is defoliated and, importantly, whether or not it is able to set seed before it is cut. Cutting earlier in the summer will reduce the number of plant species able to flower and set seed and, if carried out over may years, ultimately reduce the number of plant species persisting in the sward. Cutting very early in late spring or early summer will prevent more or less all plant species from setting seed. It will result in the sward becoming dominated by perennial grasses, which are able to spread vegetatively and persist without regularly setting seed. Cutting in late spring or early summer will also destroy bird's nests. Methods for minimizing losses of nests and chicks are discussed in Section 5.5.4.

Following cutting, hay is left on the field to dry, and turned to help expose it all to the drying sun. This management has the inadvertent effect of allowing ripened seed to fall back on to the field. The re-growth following cutting (the aftermath) is grazed in autumn, usually by cattle. Light grazing typically continues during winter, in some cases with the cattle replaced by sheep, until late winter or spring when the livestock are removed and the meadow 'shut up' to grow on for hay. This aftermath grazing is important in maintaining the high plant species richness of agriculturally unimproved meadows. Its poaching creates germination gaps and animal's hooves press seeds into contact with the soil, so helping them germinate. Grazing of the re-growth of more competitive plant species reduces their vigour and ability to out-compete smaller plants. Cattle produce more poaching and therefore more regeneration gaps than sheep, per quantity of vegetation removed. Too much poaching, though, can result in the establishment of some more competitive annuals and biennials such as docks, *Rumex* spp. Grazing animals also disperse seed between fields, either attached to their fur or deposited in their dung.

Cessation of aftermath grazing, or changes in its timing, will result in slow, but important, changes in plant species composition. For example, in hay meadows in northern England Smith and Rushton (1994) found that the high species richness of existing meadows was maintained by both autumn and spring grazing. Autumn grazing favoured autumn-germinating ruderals. Spring grazing disadvantaged plants that grow early in the season, and benefited a range of forbs over grasses that otherwise dominated in the absence of spring grazing.

Silage fields are heavily fertilized to increase their yield, and this enables them to be cut for silage two to three times in late spring and summer. The combination of fertilizer application and repeated cutting greatly impoverishes the flora. Most silage fields are in any cases re-seeded with a small number of agriculturally productive grasses, especially perennial rye-grass, *Lolium perenne*. This early and repeated cutting destroys many nests of birds that are attracted to nest in tall grass. Aftermath-grazed silage fields can, though, support a range birds in winter that feed on larger soil invertebrates. There is also potential for silage fields to provide food for seed-eating birds in winter (Figure 5.11).

**Fig. 5.11** Managing silage fields for wintering, seed-eating birds. Heavily fertilized silage fields are extremely species poor in both plants and animals, but comprise a high proportion of the grassland in some lowland areas. Research has recently found that they can at least be managed to benefit wintering seed-eating birds, especially buntings such as yellowhammers, *Emberiza citrinella* (shown here). This management involves not taking the final silage cut, to allow grasses to set seed, and leaving this standing grass crop over the winter to provide seeds for birds. This management regime slightly reduces the agricultural productivity of the sward the following year, but the grass seed can be used to restore the sward if managed correctly (Buckingham and Peach 2006; photograph by RSPB IMAGES).

### 5.5.2 Other grasslands

Where grass is not being cut and removed for agricultural reasons, the main considerations are whether the cuttings are removed and the:

- timing of cutting;
- height of cutting;
- frequency of cutting;
- size and spatial arrangement of cut blocks.

The effects of patch size and frequency of cutting are discussed in Section 5.7. There is unlikely to be any reason for cutting more than once a year on grasslands managed for conservation, because of the damaging effects of multiple cutting already described for silage fields.

It is important to remove dense cuttings to prevent them from smothering smaller plants and potential germination gaps. Leaving thick layers of cuttings is also thought to be damaging to waxcap and other larger fungi (Griffith *et al.* 2004). Removing cuttings will help deplete nutrients, which will invariably be beneficial in maintaining high plant species richness. Leaving cuttings on botanically uninteresting grasslands, though, might be beneficial in providing temporary cover for small mammals and some invertebrates until the vegetation has grown again.

The timing of cutting will influence plant species composition and the proportion of birds' nests destroyed during cutting, as described for meadows and silage and haylage fields. Even though grasslands managed by cutting are poor for invertebrates, flowers in them can still provide useful nectar sources for mobile, winged insects. Cutting small-scale patches can be used to add heterogeneity of vegetation structure and plant species composition. It can also be used to set back the flowering period of patches of flowers. This can be used to prolong their overall flowering season in a given area and the length of time that they are available as nectar sources for insects.

Increasing the height of cutting will probably benefit some sward-inhabiting invertebrates and small mammals, by at least retaining some cover. It will reduce direct damage to birds' nests. However, gathering of cut grass will destroy a large proportion of remaining nests and young. Any nests and young that do survive will usually be conspicuous and vulnerable to predators. Cutting to a higher level can, though, be used to reduce the height of the sward to within the range favoured by different groups of wintering birds, for example for wintering geese (Vickery *et al.* 1994). Regular cutting at a height of 15 cm is used to maintain a dense sward of 15–20 cm high to actually *discourage* feeding birds, such as flocks of common starlings, *Sturnus vulgaris*, gulls, corvids, and plovers, from grassland at airports to reduce bird strikes (Civil Aviation Authority 1998).

### 5.5.3 Topping

Cutting of specific patches of vegetation (topping) can be used to:

- reduce the vigour and prevent seeding of specific patches of tall, unwanted plants, which have been avoided by grazing animals;
- remove dead stems and seed heads to maintain very open conditions favoured by some open-ground bird species.

In both cases the benefits of removing this taller vegetation have to be carefully balanced against those of retaining it. Plants such as thistles are often the target of topping, but can be very valuable for insects, especially for nectar. Seed heads are valuable over-wintering sites for invertebrates. Topping prevents the development of seed for birds in winter and decimates populations of larger invertebrates.

In most cases, topping only produces small quantities of poor-quality vegetation that dry up and blow away, without smothering lower-growing plants. Even though annual topping reduces these plants' vigour, it is rarely effective at eradicating them.

### 5.5.4 Minimizing losses of birds' nests and chicks during cutting

A range of methods can be used to minimize losses of birds' nests and chicks. The most widespread technique, though, is delaying cutting until after birds have finished breeding. In temperate areas delaying cutting until early August will prevent loss of any nests. However, leaving hay cut until this late reduces its nutritional value and can thereby make management uneconomic. It can be difficult to dispose of poor-quality hay. Most hay fields managed for conservation in Atlantic temperate areas are cut between mid-July and early August.

An additional method used on drier grassland is to mow from the centre of the field outwards to reduce chick loss (Figure 5.12). Further techniques have been used specifically to reduce loss of nests and chicks of breeding waders/meadow birds. These are described in Section 8.12.3 on wet grasslands, but could equally be used on drier grasslands.

## 5.6  Managing by burning with or without grazing

Prescribed burning, with or without grazing, can be used to:

- maintain the characteristic species assemblages of fire-prone grasslands by removing fire-intolerant plants, including woody vegetation;
- provide a mosaic of different stages of re-growth (Figure 5.13).

Burning is also used agriculturally to increase re-growth of palatable grasses and remove woody species to maximize grazing. Herbivores are attracted to the lush re-growth following burning. This patchy grazing can enhance the heterogeneity of sward conditions created by burning different patches of grassland: the grassland consisting of areas of largely ungrazed, unburnt grassland, heavily grazed, recently burnt grassland, and areas intermediate between these. One-off burns are also used to remove litter when seeking to restore open conditions to rank grassland.

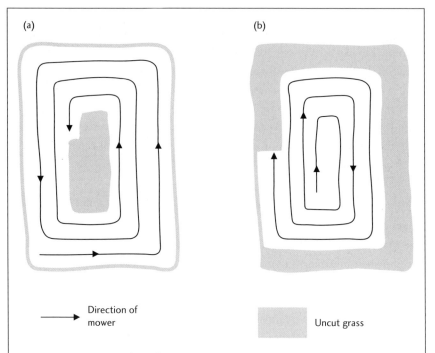

**Fig. 5.12** Corncrake-friendly mowing. Fields are conventionally mown from the outside of the field inwards (a). This concentrates any chicks in an ever-decreasing island of un-mown grassland until this is itself cut, killing the chicks.

Mowing from the centre of the field outwards (b) allows chicks to escape to surrounding fields, and has been used to increase productivity of corncrakes, *Crex crex*, in the UK (Tyler *et al.* 1998). Mowing from the inside outwards is more difficult, though.

The main considerations when burning are its method, season, and frequency, and the size of area burnt at any one time. The method, season, and frequency of burning will all influence the intensity of the fire. There are two main methods of burning. Fires can burn against the wind (back-fires) or in the same direction as the wind (head-fires). Burning against the wind means that the fire moves more slowly and burns with smaller flames than head-fires, and burns hotter at the ground surface. Back-burns are therefore safer and easier control and are the recommended method. Burning upslope is similar to burning with the wind, because in both types of burning the fire pre-heats the fuel in front of it, thereby increasing the rate of spread of the fire.

Season will influence fuel load by affecting the quantity and moisture content of the vegetation and conditions during burning. In practice, burning is usually

**Fig 5.13** Fire. Bare and litter-free ground produced by recent burning provides ideal conditions for some open-country birds, such as this black-winged lapwing, *Vanellus melanopterus* (Masai Mara National Reserve, Rift Valley Province, Kenya).

carried out using at times of year when the vegetation will burn sufficiently, but the fire easy to control; that is, at cool times of year. It is rarely, if ever, carried out during hot, dry periods when natural wildfires occur. This difference in season and fire intensity means that prescribed burning will have different effects to the natural wildfires that have maintained the grassland. As with other forms of vegetation removal, the timing of burns influences which plant species are most heavily negatively impacted by its defoliation, and this will influence subsequent species composition. For example, on North America prairies prescribed spring fires favour large, late-flowering warm-season grasses, while summer fires (similar to natural wildfires) favour a range of early-season grasses and forbs (Howe 1995; Copeland *et al*. 2002). Burning should be avoided immediately before periods of heavy rain that might cause unwanted erosion of recently exposed ground, especially on more erodible slopes.

Burning at less frequent intervals will allow a greater quantity of combustible material to accumulate and hence increase fire intensity, and also reduce the proportion of early stages of re-growth. The intensity of burns also affects conditions for germination. More intense burns remove a greater proportion of the litter and kill seeds close to the soil surface. It can either reduce invasion of alien/exotic species (e.g. Gillespie and Allen 2004) or increase it. Annual burning of large areas can result in dominance by a usually small number of fire-tolerant and ruderal plant species, sometimes including prolific, alien/exotic species (e.g. Trager *et al.* 2004).

Burning is subject to regulations and these should obviously be checked. Suitable precautions also obviously need to be taken. These include only burning when windspeeds are low, using trained personnel, creating firebreaks/fuel breaks, having water bowsers and a pressure hose or fogging machine present, and informing the local fire service. Burnt areas should be re-visited before nightfall to check for and extinguish any areas still smouldering.

Like other forms of catastrophic management, burning has short-term, damaging effects through removing most or all of the habitat. Hence, burning needs to be carried out on rotation, to provide refuges of un-burnt grassland for animals to survive in, as discussed in the following section.

## 5.7 Rotational management and the size of areas managed at any one time

In some situations management is necessary to maintain the desired conditions within the grassland in the long-term, through grazing, cutting, or burning, but the key interest of the grassland is associated with grassland that has not been recently managed. Examples include where the key interests are:

- small mammals associated with rank grassland containing abundant litter;
- invertebrates that require litter, tussocks for nesting (e.g. some bumblebees) or over-wintering in, or nectar sources;
- birds that require a deep layer of litter for nesting.

The best option is to only manage a proportion of the grassland at any one time (i.e. on rotation), to ensure there is always at least some of the grassland in the desired state for these groups at any one time. Rotational management is particularly important when undertaking catastrophic management (cutting, burning, and heavy grazing) that temporarily makes the grassland unsuitable for most of its fauna. Key decisions will be the:

- length of the management rotation;
- size and spatial arrangement of the blocks managed at different times.

The frequency at which individual areas are cut, burnt, or grazed heavily will depend on the specific requirements of the key interest of the grassland. For example, in dry prairies in Missouri, USA, cutting on a rotation of 1–2 years is considered best for grasshopper sparrows, *Ammodramus savannarum*, that require a light litter layer, whereas a rotation of 2 years or more is considered better for Henslow's sparrows, *Ammodramus henslowii*, that prefer deeper litter (Swengel and Swengel 2001). Light-to-moderate cattle grazing every 3 years or so is considered beneficial for maintaining high densities of small mammals by encouraging the growth of more succulent grasses among otherwise tall, dense grasses, while retaining the largely tall structure and dense litter layer also required by them. Cutting or burning on a rotation of more than 1 year can also be used to create coarse-scale variation in sward structure, by creating a mosaic of patches at different stages of re-growth. The best option for maintaining a diverse invertebrate fauna will be to cut different patches at varying times of year (i.e. on different rotations; Morris 2000). This will more closely mimic the effects of patchy grazing.

The optimal size and arrangement of areas managed at any one time is difficult to assess. While some bird species will require relatively large, continuous areas of suitable habitat, most invertebrates are likely to benefit from a small-scale patchwork, particularly for species with poor powers of dispersal. For a given area it will cost more to manage many separate small areas than fewer larger ones. Therefore, in practice the size of patches will be a compromise between a theoretical ideal and what is practical. For less mobile groups such as some invertebrates and small mammals, it will be best to arrange patches to ensure that they are always close to potential sources of re-colonists.

## 5.8 Soil disturbance

Valuable bare and disturbed ground can be created by:

- mechanical disturbance;
- human trampling, horse-riding, cycling, motorbikes, and other vehicles;
- creating erodable banks (Figures 6.13 and 11.8);
- the activities of herbivores (Section 5.4).

Disturbance of the soil below its surface (sub-soiling) is used to reduce soil compaction caused by heavy stocking levels and agricultural vehicles, including that created during former arable use at grassland reversion sites. The potential conservation benefits of improving soil structure by sub-soiling are poorly understood.

On flat areas and gentle slopes ploughing or rotovation can provide soil disturbance and reduce the organic content of the upper soil, thereby benefiting less-competitive ruderal plants and their associated fauna. On sandy soils periodic rotovation, typically once every 3–5 years, provides better conditions for usually scarcer, less competitive plants, especially lichens, than periodic or annual ploughing or annual rotovation. These beneficial effects of rotovation can still be apparent up to 17–20 years later (Dolman and Sutherland 1992, 1994). The effects of this periodic rotovation are similar to those of heavy rabbit grazing.

Soil disturbance by ploughing or rotovating can be undertaken at any time of year, but is most commonly carried out in late winter to:

- maximize the area of disturbed ground for invertebrates in the following spring and summer;
- provide suitable nesting conditions for birds requiring bare ground for nesting, such as Eurasian thick-knees and northern lapwings, *Vanellus vanellus*.

A novel use of soil disturbance is to diversify the structure of structurally uniform grassland using a chisel plough (Figure 5.14). A similar effect can be produced

**Fig. 5.14** Chisel ploughing. This can be used to break up compacted and structurally uniform, heavily sheep-grazed grassland to create small divots and hummocks for northern lapwings to nest among. The variation in sward structure created is similar to that of areas grazed and poached by cattle over a long period (Ynys-Hir, Powys, Wales; photograph by Ross Willis).

by sub-soiling with a conventional sub-soiler, but removing the packing roller that normally flattens the soil after it has been broken and lifted by the sub-soiler. On the heavier soils on which this technique has been used, the gaps created are sometimes colonized by unwanted agricultural weeds, especially thistles. The variation in topography produced using these methods can also make it slightly more difficult to carry out mechanical operations such as cutting.

Repeated trampling or the action of vehicles causes greater soil compaction than rotovation or ploughing. Compacted ground can remain unvegetated for a long period, especially if subject to continual trampling or vehicle action, and provide valuable habitat burrowing solitary bees and wasps (Figure 5.15). Land managers sometimes consider meandering, braided paths created by people as unsightly, and seek to confine people to more tightly defined pathways. While the centres of paths may in many cases be too regularly trampled to be of high value to invertebrates, reptiles, and ruderal plants, their intermediately trampled edges can be especially valuable for these groups. The ideal is to maximize the area of habitat disturbed through different levels of trampling and of different ages since the last regular disturbance event. This can be done by periodically

**Fig. 5.15** Soil disturbance and compaction. The compaction caused by vehicles in this overflow car park suppresses vegetation and provides a suitable level of soil compaction for nesting solitary bees and wasps. Their numerous excavations can be seen on the photograph to the right. These include burrows of the European bee wolf, *Philanthus triangulum* (shown here), a fascinating solitary wasp which provisions its nest chambers almost entirely with adult honey bees, *Apis mellifera* (Pulborough Brooks, Sussex, England).

moving paths, so that heavily tramped areas gradually become more vegetated, while new, early successional habitat is created elsewhere.

## 5.9  Use of fertilizers, lime, slurry, and farmyard manure

Application of inorganic fertilizers and slurry to grasslands to increase their agricultural productivity has greatly impoverished the value of grasslands for wildlife. The most commonly applied inorganic fertilizers are nitrogen (N), phosphorus (P), and potassium (K). High levels of plant-available phosphorus are generally associated with low plant species richness. In at least some types of grassland, the lowest plant species richness is associated with combinations of high levels of plant-available phosphorus, nitrogen, and potassium (Janssens *et al.* 1998; Critchley *et al.* 2002; Crawley *et al.* 2005). Typical levels of these nutrients in agriculturally improved and unimproved neutral grassland are shown in Table 5.2.

A number of specific mechanisms are thought to be involved in the reduction of the value of grasslands for birds caused by fertilizer application (Vickery *et al.* 2001):

- the rapid and uniform growth of grass in heavily fertilized swards in spring makes the vegetation too tall and dense for many nesting and foraging birds;
- increased fertilizer use and associated management probably decrease the abundance of large arthropod prey in the sward;

**Table 5.2** *Differences in key soil nutrients between agriculturally improved and unimproved grasslands*

|  | Agriculturally unimproved neutral, lowland hay meadows | Agriculturally improved perennial rye-grass-dominated grasslands |
|---|---|---|
| Extractable phosphorus (mg/l; Olsen extraction method) | 9.6 ± 4.4 | 18.5 ± 13.5 |
| Extractable phosphorus (mg/l; resin extraction method) | 18.2 ± 8.9 | 31.8 ± 21.7 |
| Extractable potassium (mg/l) | 135 ± 39 | 213 ± 120 |
| Total nitrogen (percentage of dry weight) | 0.67 ± 0.30 | 0.90 ± 0.44 |

This table shows the higher levels of phosphorus, potassium, and nitrogen found in the soils of agriculturally improved compared to agriculturally unimproved grassland. All three nutrients are commonly applied as fertilizer. Values are means ± one standard deviation. From Critchley *et al.* (2002).

- increased fertilizer use allows higher stocking densities and earlier mowing of fields, both of which can reduce nest survival of grassland birds;
- fertilizer-induced replacement of floristically diverse swards with uniform grassy ones reduces the quantity of suitable seed available for birds.

Lime has commonly been applied to neutralize unimproved or semi-improved acidic soils to increase nutrient availability to plants. This will obviously damage their characteristic acid grassland flora and its associated fauna.

Application of inorganic fertilizers can also has wider detrimental effects. Nitrate is water soluble and readily leached from grassland into watercourses, where it can cause problems of eutrophication of water bodies and wetlands (Section 8.2). Phosphorus is slow to be leached out of the soil, and residual high levels of phosphorus are a major constraint when seeking to recreate species-rich grassland.

For the reasons described, there will rarely be any justification for increasing nutrient levels in grasslands managed for conservation. The two main exceptions are:

- addition of small quantities of well-rotted farmyard manure to agriculturally unimproved meadows;
- increasing the productivity of already agriculturally improved swards to encourage use by feeding geese.

Small quantities of well-rotted farmyard manure (<12.5 t/ha per year) have traditionally been added to agriculturally unimproved hay meadows in spring, to offset the loss of nutrients removed in the hay. Replacing these nutrients is thought to help maintain the high plant species richness of such meadows. Addition of higher levels of fertilizer, particularly inorganic fertilizers, will be highly damaging.

Nitrogenous fertilizer can be applied to already agriculturally improved grassland to increase their use by feeding geese (Owen 1975, Vickery et al. 1994, Percival 1993). Management also needs to ensure that the sward is within their preferred height range (Vickery and Gill 1999).

Addition of low levels of farmyard manure has been found to increase use of grasslands in winter by birds that feed on soil invertebrates, probably because it increases the numbers of earthworms close to the soil surface (Tucker 1992). Any such benefit would have to be set against the previously mentioned damaging effects of increasing nutrient levels.

## 5.10 Diversifying botanically dull grasslands

Botanically rich swards are associated with relatively low nutrient levels (Section 5.9). Cessation of fertilizer application to already impoverished, agriculturally improved grasslands will result in a decline in nutrient levels, although it may be some time before they are low enough to support a species-rich sward.

The soil variable considered to most commonly limit plant species richness of grasslands is phosphorus, apart from where plant growth is limited by drought or water-logging. Where phosphorus limits plant growth, species rich swards are unlikely to develop on soils with a phosphorus content of much higher than 15 (Olsen extraction method) or 30 mg/l (resin extraction method). Successful diversification of swards can probably take place on soils with initially higher levels of phosphorus, if future management involves cutting and removal of vegetation, which will reduce nutrient levels over time.

The first option when seeking to diversify a sward is to see whether this can be achieved through a change in sward management. Reducing grazing levels in summer may allow formerly heavily grazed plants to flower and set seed, whereas creation of suitable germination gaps may allow seeds of desirable species to germinate. However, many plant species are unlikely to re-appear because their seeds are no longer present. Re-colonization by lost plants species may be more rapid if the grassland is adjacent to a more species-rich sward, especially if animals are allowed to graze both and disperse seeds attached to their fur and in their dung. Where this is not the case, the only realistic methods to restore the botanical interest of the sward are to encourage germination of any buried seedbank or to add seeds or young plants from elsewhere.

In all cases long-term sward management involving appropriate cutting dates and aftermath grazing to create germination gaps is necessary to increase and maintain high plant species richness (e.g. Smith *et al.* 2000b).

Methods of introducing seeds or plants into botanically dull swards have been the subject of a number of experimental studies, but the long-term effectiveness of these techniques has not yet been demonstrated. There are four possible techniques:

- over-sowing;
- addition of green hay;
- slot seeding;
- planting young pot-grown plants or seedling plugs.

An additional technique for diversifying swards, which can be used in combination with these methods, is by introducing yellow rattle, *Rhinanthus* spp., to reduce the sward's productivity (Figure 5.16).

When using any of these techniques it is important to first reduce the height of the existing sward to minimize competition with the introduced plants. This can be done by cutting and removing vegetation or by grazing. Seeding and planting

**Fig. 5.16** Using yellow rattle to increase plant diversity. Yellow rattle species are found throughout much of Europe and in North America and Asia. They are root hemi-parasites on a variety of plants, including a wide range of grasses and legumes. Addition of yellow rattle reduces dominance by grasses and enables less competitive plants to survive, at least in the short term (Davies *et al.* 1997; Pywell *et al.* 2004).

Yellow rattle is unusual among hay meadow plants, in being an annual and not surviving long in the seedbank. Therefore, for yellow rattle to persist, there must be suitable conditions for it to both set seed and germinate virtually every year.

is best undertaken in early autumn. The moist soil conditions during winter help plants to survive during the period when they have not yet established a deep root network. Spring-sowing or planting runs the risk of recently germinated or newly established plants dying from late spring or summer drought.

Addition of green hay will be the most practical method of diversifying the sward at many sites. It does not require specialized harvesting equipment and, providing there is a suitable source of green hay nearby, should be cheaper than other seeding or planting methods. Green hay will in most cases contain a wider range of suitable seed species of local provenance than commercially grown seed. However, it will not be suitable for establishing swards containing low-growing plants typical of heavily grazed grasslands, because it is difficult to harvest significant quantities of green hay from these. Planting young pot-grown plants or seedling plugs is far more expensive than other techniques per plant established. It can, though, be a useful supplementary method for introducing small numbers of plants of species that are difficult to establish from seed.

### 5.10.1 Obtaining suitable seed

Seed can be bought from commercial suppliers or harvested from a nearby donor site. It is important to check the provenance of commercially grown seed, to ensure it contains similar strains of plants to those found in the area. Introduction of seed from further away has the potential to swamp local strains, or it may be less suited to local conditions. Donor sites should have similar soil conditions to receptor sites, especially with regard to their pH and wetness, and be subject to similar management to provide a suitable range of plant species. Obviously, commercially grown seed should contain a suitable selection of plant species native to the area of the restoration site.

The best method of collecting seeds from donor sites is using a brush harvester. This uses a rotating brush to dislodge seeds from standing plants. There are a variety of models, including towed and hand-held ones. Seeds of different plant species ripen at different times of year. Harvesting on a range of dates will increase the number species of seeds collected. Grazing needs to be temporarily excluded from regularly grazed donor sites to allow plants to set seed prior to harvesting.

### 5.10.2 Oversowing and addition of green hay

Over-sowing involves spreading seed over the existing sward. Sowing rates of 5–10 kg/ha are recommended. Heavier seed can settle out and result in uneven sowing of different species. Seed can be spread more evenly by adding an insert material to bulk it up. Silver sand or fine sawdust are suitable.

Green hay should be cut just *before* the vegetation has dried, browned, and the majority of its seed ripened and fallen. Green hay contains a higher proportion of seeds of desirable forb species than grasses, compared to conventional dry hay (Pywell 2006). It needs to be collected and spread on the receptor site the same day as it is cut. This prevents it from shedding its seed back on to the donor site, and from heating up and possibly damaging the seeds within it. It is recommended to spread 1 ha's worth of green hay from the donor site over 3 ha of the receptor site.

Before over-sowing or adding green hay it is necessary to create germination gaps in addition to reducing the height of the sward. Germination gaps of about 10 cm diameter are recommended. These are best created using cattle. Where this is not practical, gaps can be created using a power-harrow or by light disking.

Livestock, ideally cattle, should be introduced following over-sowing or addition of green hay, to tread seed into the ground and help germination. Light rolling can be used if there are no livestock. Green hay should be left for between

1 and 3 weeks to allow it to shed its seed before any grazing is introduced. Some seeds may take time to germinate, so it is worth waiting several years before concluding whether the management has been successful.

### 5.10.3 Slot seeding

Slot seeding is a standard agricultural method for introducing seed into an existing sward. As with other techniques for diversifying grassland, it is important to cut or graze the sward short prior to sowing. The slot seeder cuts a shallow (approximately 15 mm) groove in the grass and drills seed into it. It is recommended to fit the slot seeder with a band sprayer. This applies a narrow strip of herbicide to kill the vegetation beside the grooves to reduce competition with establishing seedlings. If using a band sprayer, then the sward needs to be left a short while to green up following cutting, for any systemic/translocated herbicide (Section 4.5) to be effective. An advantage of slot seeding is that it results in a higher number of plants established per quantity of seed used compared to over-sowing. For slot seeding a sowing rate of 1–2 kg/ha is recommended. A disadvantage is that the introduced plants grow in lines. These lines may persist for a number of years.

Following slot seeding, the sward should be kept short to provide suitably open conditions for the establishment of seedlings. This can be done by mowing and removing the cuttings or by short periods of heavy grazing.

### 5.10.4 Planting young pot-grown plants or seedling plugs

Small pot-grown plants or seedling plugs are planted directly into gaps in the sward. As with other techniques, it is important to reduce the height of the sward prior to slot seeding or planting, to reduce competition from established vegetation.

### 5.10.5 Providing suitable aftercare

When using any method to introduce plants into the sward, it is important to provide suitable aftercare to help the introduced species establish. This should involve cutting and removing vegetation, or periods of heavy grazing, during the following spring and summer, to prevent existing grasses, and unwanted, competitive ruderal species, from out-competing the introduced plants. The timing of this management should take into account its potential damaging effects on breeding birds (Section 5.5.4). Continual, heavy grazing should be avoided, though, since this can prevent the fragile, small plants from establishing.

It may also be adding additional seed on successive occasions in different years to increase the chance of at least some individuals of different plant species becoming established. The benefits of successive seeding, though, have to be set against the potential damage that repeated soil disturbance might cause to establishing plants. Once a suitable range of species have been established, then ongoing maintenance management should be introduced, as described elsewhere in this chapter.

# 6
# Dwarf-shrub habitats and shrublands

These are vegetation types dominated by low (less than 3–4 m high), predominantly evergreen, shrubs, typically on soils with low nutrient availability. Characteristic dwarf-shrub vegetation and shrublands occur in the five regions of the world with Mediterranean-type climates of mild, wet winters and hot, dry summers:

- phrygana/garrigue and maquis in the Mediterranean;
- coastal sage scrub and chaparral in California;
- mattoral in Chile;
- fynbos in South Africa;
- mallee and heathland/kwongan in southern and western Australia.

In the Mediterranean garrigue/phrygana refers to open, dwarf-shrub-dominated vegetation typically a metre or less high, and maquis thickets of dense shrubs and small trees up to 4 m or so high. Similarly, in southern Australia, kwongan refers to dwarf-shrub/undershrub vegetation and mallee to taller shrubland.

A range of other types of dwarf-shrub vegetation occur in montane areas and in semi-arid environments where there is insufficient water for the growth of trees. An important type of semi-arid dwarf-shrub vegetation is sagebrush steppe, which covers large parts of the western USA. Shrubby vegetation in semi-arid and montane areas is rarely, if ever, *actively* managed for conservation. Apart from sagebrush steppe, it is not considered further in this chapter.

Another characteristic type of dwarf-shrub habitat is the European Atlantic cultural heathlands and moorlands. These are dominated by ericaceous shrubs on acidic substrates in the cool, wet conditions of north-west Europe. These can be divided into:

- lowland heathlands on nutrient-poor sands and gravels in the warmer and drier lowlands, including on acidic sand dunes;
- upland heaths and moorlands in cooler, wetter conditions, typically on peat and at higher altitude.

Management of these Atlantic heathlands differs from that of other dwarf-shrub-dominated vegetation and shrublands in that a variety of very specific management interventions is often used to maximize the conservation value, particularly in the case of lowland heathlands. Management of Atlantic lowland heathlands and upland heaths/moorland are discussed separately in Sections 6.5 and 6.6.

## 6.1 Key factors influencing the suitability of dwarf-shrub habitats for plants and animals

Most conservation management of dwarf-shrub vegetation and shrublands other than Atlantic heathlands aims primarily to maintain their intrinsic floral and faunal interest by perpetuating their characteristic vegetation types, rather than aiming to provide specific conditions for individual species or groups of species. All five regions with Mediterranean climates support characteristic and species-rich floras, containing high degrees of plant endemism and a range of endemic fauna. South African fynbos is especially botanically diverse (Figure 6.1). Shrubby vegetation outside of these regions is usually far less species-rich.

Conditions for plants and animals within dwarf-shrub habitats vary primarily in relation to the stage of re-growth following disturbance, although the structure and species composition of this re-growth can also be modified by grazing and browsing. The most common form of disturbance is fire. The early stages of re-growth following burning are open and typically support a diverse range of largely short-lived grasses and forbs, commonly referred to as fire ephemerals or fire-followers, together with associated open-ground invertebrates. Over time, this herbaceous layer is usually out-competed by dwarf shrubs, unless their re-growth is suppressed by heavy grazing. In some vegetation types, though, herbaceous vegetation persists in open ground between shrubs. The shrub fauna changes as the shrubs increase in height. For example, the bird fauna of Mediterranean shrublands changes from that of open, sparsely vegetated ground to species characteristic of dense shrubs, while taller shrubby vegetation supports species more typical of woodlands (e.g. Katsimanis *et al.* 2006).

## 6.2 General principles of managing dwarf-shrub vegetation and shrublands

Dwarf-shrub vegetation and shrublands contain assemblages of plants adapted to particular fire regimes. Many of theses contain small and thick-leaved, highly flammable, foliage and also require fire for germination. Dwarf-shrub

## General principles of management | 133

**Fig. 6.1** South African fynbos. This shrubby vegetation forms part of the Cape Floral Kingdom, the smallest and richest per unit area of the world's six floral kingdoms. Fynbos vegetation is highly diverse, containing the majority of the Cape's plant species (coastal fynbos at De Hoop Nature Reserve, Western Cape, Republic of South Africa; photograph by Graham Hirons).

vegetation and shrublands are thus prone to natural wildfires that periodically set back succession.

Fires in dwarf-shrub vegetation are effectively crown fires (Section 7.4.6) in that they consume all, or the majority, of above-ground vegetation. Prescribed burning is frequently used to remove entire stands of older shrubby vegetation with the aim of reducing the risk of catastrophic wildfires (hazard-reduction burning). This approach is slightly different to the use of surface fires in woodland to reduce fuel loads within *existing* stands of trees (Section 7.4.6). The effectiveness of hazard-reduction burning in shrublands is probably quite variable (Figure 6.2).

Large areas of shrubby vegetation have been managed to provide grazing for livestock. Trees have also been cut and removed from it for firewood or other uses. The shrubs themselves are often relatively unpalatable, at least compared to most grasses. Consequently, shrubby vegetation has typically been periodically burnt to remove unwanted, unpalatable woody plants and to increase nutritious re-growth, especially of grasses and palatable herbs. In South African fynbos burning is also used to maximize production of flowers for harvesting, particularly species of protea, *Protea* spp. However, heavy grazing of re-growth by livestock,

## Dwarf-shrub habitats and shrublands

**Fig. 6.2** Hazard-reduction burning. Californian chaparral, such as this, is especially prone to wildfires. Prescribed burning is often carried out with the aim of reducing fuel loads and the likelihood of large-scale, catastrophic fires. Research suggests that although these measures may be successful in suppressing wildfires under moderate weather conditions, the length of time since areas were last burnt has little or no effect in preventing large-scale fires driven by high winds (Keeley *et al.* 1999; Keeley and Fotheringham 2000; Moritz *et al.* 2004; Mount Tamalpais State Park, California, USA).

especially by very selective grazers such as sheep, and of more established shrubs by goats, has the potential to eliminate more grazing-intolerant plant species and thereby impoverish the flora. Combinations of frequent burning and heavy grazing of re-growth will eventually eliminate most, or all, dwarf shrubs to produce grassy vegetation. Frequent burning combined with heavy grazing (overgrazing) is considered to have severely degraded most sagebrush steppe and large areas of fynbos. In many areas management for grazing has involved the complete destruction and in some cases re-seeding with grasses, especially in the case of sagebrush steppe.

Prescribed burning for conservation also aims to provide a mosaic of different stages of re-growth to maintain a range of conditions suitable for species associated with these different stages, while, where necessary, grazing at suitable levels to sustain commercial grazing or maintain a diversity of open-ground vegetation and its associated fauna. Long periods of fire suppression can in some cases result in the loss of characteristic dwarf-shrub habitats through colonization by more competitive, fire-intolerant plant species.

The presence of alien/exotic plant species is an issue in many types of dwarf-shrub vegetation. Spread of alien/exotic annual grasses, especially from the Mediterranean region, is a particular problem in chaparral and sagebrush steppe in the western USA. Trees and shrubs, especially acacias, *Acacia* spp., pines, *Pinus* spp., and *Hakea* spp., are the most important invasive alien/exotic species in South African fynbos. Interestingly, Mediterranean vegetation appears to be relatively immune to significant invasion by alien/exotic plant species. Management therefore often aims to reduce or eradicate undesirable alien/exotic plant species.

Although the overall aim of management may be to maintain dominance by dwarf shrubs, in some types of vegetation much of the floral and faunal interest is associated with the herbaceous vegetation and open ground between the shrubs. Mediterranean vegetation in particular often consists of mixtures of open herbaceous vegetation, low-growing dwarf-shrub-dominated phrygana/garrigue, and taller, shrubby maquis. Much of the botanical and invertebrate interest in these mosaics is associated with open areas. Mosaics of vegetation types have in many cases been created by periodic cultivation that sets back succession (Figure 6.3).

## 6.3 Burning

There are three main factors to consider when deciding a prescribed fire regime for dwarf-shrub vegetation and shrublands. These are the:

- season of burning;
- frequency of burning;
- size of area burnt at any one time.

The season will influence fire intensity by affecting the moisture content of the fuel and the weather conditions during burning. Wildfires typically burn under hot, dry, windy conditions. Prescribed burning is virtually always carried out using back-fires (Section 5.6) under low windspeeds during the cooler, wetter periods of the year (cool-season-prescribed burning) when fires burn less intensively and are more easily managed. Precautions obviously need to be taken to minimize the risk of burns becoming out of control (Section 5.6). Thus, prescribed fires will burn both at a different season and less intensively than the majority of wildfires, to which the vegetation was originally adapted.

Burning should be avoided immediately before periods of heavy rain that might cause unwanted erosion of bare ground exposed by burning on more erodible

## 136 | Dwarf-shrub habitats and shrublands

**Fig. 6.3** Disturbance by periodic cultivation. Disturbance can be important in maintaining species-rich, early successional habitats within shrubby vegetation. This is especially the case in the Mediterranean region, where a large proportion of the endemic flora is associated with open and disturbed habitats.

Disturbance provided by periodic cultivation on these terraces supporting species-rich herbaceous vegetation containing a diverse (and confusing) variety of species and forms of *Ophrys* orchids, a highly diverse group in the Mediterranean. These include Bertoloni's orchid, *Ophrys bertolonii* (left), and early spider/Gargano orchid, *Ophrys sphegodes/garganica* (right). These orchids are absent from the often species-poor, dense stands of maquis at the top of the picture that often eventually develop in the absence of disturbance (near Mattinata, Gargano Peninsula, Puglia, Italy).

and steeper slopes. The frequency of burning will be determined primarily by the need to leave a long-enough interval between burns so that:

- the vegetation contains representative stages of the oldest desirable stages of re-growth;
- plants that recover from seed following fires (obligate seeders) have sufficient time to mature and produce seed before the area is re-burnt;

while not burning so infrequently that:

- there is an unacceptable risk of large wildfires that endanger people and property and run the risk of eliminating species within a given patch of habitat;
- the characteristic vegetation is out-competed by any potentially more vigorous, fire-intolerant plant species.

A further consideration in some cases is the likely effect that the frequency of burning has on the abundance of any unwanted, alien/exotic plant species. The ability of more palatable plants to persist will also be influenced by the grazing pressure on re-growth following burning.

Fire regimes in forests are often used to mimic what is considered to be their natural fire regime. However, unlike in forests, where fires leave scars on trees enabling past burning regimes to be re-constructed, burning usually removes entire stands of shrubs. Thus it is usually far more difficult to re-construct the frequency of natural fire regimes in shrubby vegetation.

The risk of large, catastrophic wildfires can be reduced by maintaining permanent bare or sparsely vegetated firebreaks/fuel breaks, to both help reduce the spread of wildfires and provide access for fire-fighters. These can be created by grubbing out vegetation and ploughing or bulldozing strips of ground. Firebreaks/fuel breaks are unlikely to prevent the spread of wildfires under very dry and windy conditions when they can cross even wide expanses of bare or sparsely vegetated ground. In some cases there is a danger of the open conditions along firebreaks/fuel breaks providing routes for colonization by unwanted, ruderal, or alien/exotic plant species (Merriam *et al.* 2006). Fuel loads can to some extent be reduced by grazing.

Different types of dwarf-shrub vegetation vary in the time they take to pass through their characteristic successional stages and thereby in the minimum length of burning rotation required to maintain the desired range of successional stages. Fynbos typically takes about 30 years for the longer-lived shrubs to attain maximum height, while following burning sagebrush steppe can take 35–120 years or more to attain a similar shrub cover to that of unburnt areas

(Baker 2006a). Mature Californian chaparral can remain relatively resilient to change in the absence of burning. Ancient, 150-year-old stands of chaparral differ in shrub species composition to younger stands, due mainly to loss of shorter-lived obligate seeder species, but can still recover almost as well as mature stands in terms of recovery of fire-followers/fire ephemerals (Keeley *et al.* 2005a).

The minimum desirable frequency of burning needs to be based on the length of time for the slowest-maturing, obligate seeders to mature and set sufficient seed to maintain their persistence within the vegetation. Shrubs can regenerate in two ways following canopy fires: by re-sprouting from underground rootstock (sprouters) and germinating from the seedbank. The extent to which individual species regenerate in either of these ways will to some degree depend on the temperature of the fire. However, some species are usually largely or totally dependent on regenerating from seed. These are called obligate seeders. If the area is repeatedly burnt at intervals shorter than the time it takes these plants to mature and set seed, it will denude the species' seedbank without replenishing it. This will eventually cause its disappearance from the vegetation. For fynbos vegetation it is recommended that at least 50% of the population of the slowest-maturing shrubs, typically protea species, have flowered for at least 3 years before being burnt again. In this case burning no more frequently than once every 10–15 years is considered sufficient to maintain the persistence of these slower-growing species (see Tainton 1999). Frequent burning is likely to favour sprouters over obligate seeders (Syphard *et al.* 2006), although very frequent burning may even reduce the capacity for sprouters to recover following burning.

The frequency of burning also has the potential to influence the abundance of unwanted, alien/exotic plant species. In shrublands in the western USA very frequent burning can favour alien/exotic annual grasses and other ruderal plants, while burning of ancient stands of shrub can open them up to colonization by alien/exotic plants (Keeley *et al.* 2005a, 2005b; Figure 6.4). Native shrubs in Californian chaparral are eliminated and replaced by alien/exotic, weedy grassland when areas are burned more frequently than once every 10–15 years. Alien/exotic trees in fynbos can be removed by combinations of felling, burning, and chemical treatment. The most successful methods vary between tree species (van Wilgen *et al.* 1994). A widely used method to control pines and *Hakea* species in fynbos is to fell them, leave them to drop their seeds and then burn the area to kill their seeds and any of their regenerating seedlings. However, there is evidence that the intense fires created by burning felled material possibly also prevents, or at least hinders, successful regeneration of native vegetation (Holmes *et al.* 2000).

**Fig. 6.4** Alien/exotic grasses in the western USA. Invasion of native vegetation by alien/exotic annual grasses is an issue in Californian chaparral and other fire-prone habitats in the Western USA. These annual grasses not only out-compete native herbaceous plants, but can also increase the fuel-load in the gaps between shrubs and allow fires to spread between shrubs even before they have grown large enough to coalesce. The increase in fire frequency further encourages growth of these alien/exotic annual grasses (Knick and Rotenberry 1997; Keeley 2005; Point Lobos State Reserve, California, USA).

As in all rotational management for conservation, it will generally be better to burn a larger number of smaller areas rather than a smaller number of large ones. This will maximize medium-scale variation in stages of re-growth and minimize the risk of local extinctions. However, this has to be set against the increased resources needed to burn many small areas compared to fewer larger ones. Hazard-reduction burning should focus on *strategically positioning* burns to protect *vulnerable areas*, and minimize the proportion of land elsewhere subjected to the ecologically damaging high frequencies of fire needed for successful wildfire reduction (Keeley *et al.* 1999; Keeley 2002).

Overall, it will be best to maintain a variety of burning rotations (especially frequency but to some extent also season) between the desirable upper and lower limits for that particular vegetation type. This will maximize variation in vegetation composition and structure.

## 6.4 Grazing and browsing

Most grazing and browsing of dwarf-shrub habitats and shrublands is by sheep, cattle, and goats. The main considerations are the effects of grazing on the:

- regeneration of shrubs following burning;
- potential effects of browsing on more established shrubs;
- composition of any associated herbaceous vegetation.

The effects of livestock on re-growth of shrubs will depend primarily on the:

- stocking levels;
- type of grazing animal;
- fire regime.

Grazing animals are attracted to palatable re-growth of grasses, forbs and regenerating shrubs following burning (e.g. Van Dyke and Darragh 2006). Grazing intensities in recently burnt areas will therefore be highest when overall stocking levels are high and only a small proportion of the habitat burnt at any one time. Very heavy grazing, especially immediately following burning, can thus result in areas of shrublands becoming dominated by grazing-tolerant grasses and unpalatable or otherwise grazing-tolerant shrubs (Figure 6.5). Therefore, in situations where grazing has the potential to damage re-growth, livestock should be kept off, or only grazed at low densities on, recently burnt land. This will obviously conflict with maximizing grazing income.

Sagebrush steppe is particularly sensitive to grazing. It has probably developed in the absence of significant grazing by large herbivores (Mack and Thompson 1982). The ability of many areas of heavily grazed sagebrush steppe to recover following reduction in grazing levels is questionable (Knick *et al.* 2003).

The feeding characteristics of different grazing animals are described in Section 5.4.1. Sheep are more selective grazers than cattle, and therefore have a greater potential to eradiate re-growth and young plants of more grazing-sensitive species. Cattle tend to reduce re-growth of different species more equally, thus not favouring less-palatable species to the same extent. For this reason, mixtures of cattle and sheep, rather than cattle alone, are recommended for grazing fynbos.

Goats feed more on woody plants than either sheep or cattle, and can influence the composition of more established shrub by selective browsing. Goats also climb up small trees to browse their lower branches. Heavy grazing by goats therefore has the potential to reduce the species richness of established larger shrubs, rather than just influence their abundance by affecting their early growth stages.

**Fig. 6.5** Heavy grazing and browsing of dwarf-shrub habitats. Heavy grazing and browsing can suppress the growth of dwarf shrubs, such as this broom, and provide a competitive advantage to more grazing-tolerant plant species, especially grasses (Castro Las Cerras, Asturias, Spain).

The effects of livestock on open herbaceous vegetation between shrubs are similar to those in other types of grassland (Section 5.4). The highest plant species richness, and probably in many cases invertebrate species richness, tend to occur at moderate grazing levels that produce a mosaic of moderately grazed herbaceous vegetation and dwarf shrubs (Verdú *et al.* 2000; Vulliamy *et al.* 2006). Very high levels of grazing will discourage large, palatable forbs, and favour lower-growing grasses that are tolerant of grazing, and plants that have chemical or physical defences against grazers.

Therefore, while grazing levels need to be relatively low to maintain dominance by shrubs, especially following burning or where the aim is to regenerate lost shrub cover, moderate-to-heavy grazing can also be important in maintaining high species-richness of associated open habitats.

## 6.5 European Atlantic lowland heathlands

Lowland heathland is a highly valued cultural landscape in north-west Europe. Most lowland heathland is thought to have been created and maintained through past human land use and now requires intervention management to prevent it from succeeding to woodland or becoming dominated by grasses or bracken, *Pteridium aquilinum*. Some areas of lowland heathland presumably existed prior to human influence in open woodland, glades, dunes, and exposed upland and coastal areas. Heathland on many coasts is maintained as climax vegetation through salt spray and exposure.

Lowland heathlands vary in wetness, ranging from dry heaths on free-draining acidic sands and gravels to humid heaths, wet heaths, and bogs on increasingly wetter and deeper, acidic peat. Management of bogs is discussed in Section 8.9. Heathland often also forms transitions and mosaics with these peatlands, acid grasslands, scrub, and woodland, and is usually managed in association with them.

This management has generally involved the following, with the specific details varying between regions:

- grazing;
- cutting of vegetation for fuel, fodder, and animal bedding;
- cutting of peat turfs from mires for animal bedding and fuel;
- periodic cultivation;
- periodic burning to promote lush re-growth for grazing and clear areas for cultivation.

Most of these types of management were used to transfer nutrients from the heathland to fertilize adjacent arable land. In continental north-west Europe livestock were grazed on the heathland during the day and brought inside at night. The dung and urine deposited at night was soaked up by vegetation and dried peat used for bedding. This, together with the remains of uneaten fodder and ash from burnt turfs, was spread on arable land. These practices created so-called plaggen soils. These contain a layer of dark, peaty material lying over the existing mineral soil. Burning and removal of vegetation also depleted nutrients. It is unclear the extent to which heathlands were used to fertilize arable land in the UK, although there is evidence that in at least some areas livestock were grazed on heathland during the day and folded (Section 5.3.3) on arable land at night to deposit nutrient-containing dung on them. More southerly heathlands in Brittany, Spain, and Portugal were also grazed, burnt, and periodically cultivated (Webb 1998).

### 6.5.1 Key factors influencing the suitability of Atlantic lowland heathlands for plants and animals

Dry heathland is invariably extremely species-poor for vascular plants, comprising mainly a small number of dwarf-shrub species. Heather, *Calluna vulgaris*, is the dominant structural component of virtually all types of drier lowland heath. More open areas can support a wider variety of mosses and lichens. The main vascular plant interest is associated with disturbed conditions, wet heath and associated mires, and seepage areas. Wetter areas can support a number of plants rare in the otherwise nutrient-rich lowlands.

An important concept when managing lowland (and upland) heathlands is that of heather structure. Heather has four recognized growth phases (Gimmingham 1972), which differ in their structure and associated fauna. These are described below. The time taken to reach these different phases will vary between sites according to growth rates.

- **pioneer**: young (0–5 years old) plants colonizing from seed or the early re-growth following burning or cutting of older plants;
- **building**: vigorously growing and dome-shaped plants (5–15 years old);
- **mature**: slower-growing plants with a more open canopy (15–25 years old);
- **degenerate**: plants that are starting to open out, collapse, and eventually die (25–> 40 years old).

Lowland heathlands are of particular value for a variety of southern bird, reptile, and especially invertebrate species that are largely confined to this and other similarly structured warm, sandy habitat towards the cooler and wetter edges of their climatic range. Lowland heathlands support low densities of breeding birds of only a limited range of species, but containing several species considered of high conservation value in north-west Europe. These include Dartford warbler, *Sylvia undata*, Eurasian nightjar, *Caprimulgus europaeus*, wood lark, *Lullula arborea*, and tawny pipit, *Anthus campestris*. Few bird species remain on heathlands in winter. The value of lowland heathlands for birds is mainly influenced by structure. Dartford warblers favour areas with mature gorse, *Ulex europaeus*, scrub, Eurasian nightjars require small patches of dry, bare or sparsely vegetated ground among vegetation in which to nest, whereas wood larks need disturbed ground, short grassland and scattered trees and tawny pipits sandy heathland and active dunes (Bibby 1979; Sitters *et al.* 1996; van den Berg *et al.* 2001). Extensive areas dominated by dwarf shrubs without trees and scrub are extremely poor for birds. The primary bird interest of wet heaths and associated mires is breeding waders/shorebirds.

Dry heathland supports virtually the entire, albeit small, reptile fauna of north-west Europe, including two highly-valued species, the sand lizard, *Lacerta agilis*, and smooth snake, *Coronella austriaca* (Figure 6.6). The acidic water bodies associated with heathlands provide breeding habitat for several species of

## 144 | Dwarf-shrub habitats and shrublands

**Fig. 6.6** Smooth snake. This is one of many southern, warmth-loving species that in the north west of its range in Europe is confined to the warm conditions provided by the free-draining soils of Atlantic lowland heathland. It is found in a wide variety of open habitats further south.

amphibians. Heathland is generally poor for mammals, only supporting low densities and lacking any characteristic species.

Lowland dry heathland supports a particularly rich variety of warmth-loving invertebrates. It is especially rich in solitary bees and wasps, spiders, and true bugs. The main features of importance for invertebrates are listed below.

- There is a mosaic of growth phases/structures of the dwarf shrubs, since different growth phases support different invertebrate assemblages.
- There are areas of bare, consolidated and disturbed sand intermixed with sparse and dense vegetation. These provide warm microclimates, open areas for hunting and a suitable substrate for invertebrates to burrow in. Areas of other bare ground are valuable, but probably less so than bare sand.
- There are steep slopes and banks, especially south-facing, bare or sparsely vegetated ones, that provide nesting habitat for solitary bees and wasps.
- There is a continuity of suitable nectar sources, including early-flowering shrubs and nectaring plants associated with disturbed and more nutrient-rich vegetation. Nectar sources can be otherwise scarce outside of the main

flowering periods of the dominant dwarf shrubs. Vegetation along verges and various composites and umbellifers are especially valuable.
- The scattered trees, especially birch *Betula* spp., are in a range of different growth stages, and there are blocks of structurally diverse gorse scrub, which each support different invertebrate assemblages.
- There are mires and pools.

As can be surmised from the above, large areas covered solely by dwarf shrubs are not very rich in invertebrate species.

Different successional stages of wet heaths and mires also support different assemblages of invertebrates. Acidic pools and mires associated with heathland can support important assemblages of invertebrates, probably due in part to the lack of fish. These pools and associated heathlands can be rich in dragonfly and damselfly species.

## 6.5.2 Overview of management

Agricultural use of heathland has become largely uneconomic. In the absence of vegetation removal through agriculture, many heathlands are susceptible to loss of dwarf-shrub vegetation through succession to woodland and grassland and expansion of bracken. Although bracken is a native component of many heathlands, dense stands of it support a very limited fauna. The main tree species that colonize lowland dry heathland are silver birch, *Betula pendula*, Scots pine, *Pinus sylvestris*, maritime pine, *Pinus pinaster*, lodgepole pine, *Pinus contorta*, and dwarf mountain-pine, *Pinus mugo*. Tree and scrub species commonly colonizing wet heath are alien/exotic rhododendron, *Rhododendron ponticum*, and native downy birch, *Betula pubescens*, and willows, *Salix* spp. Unmanaged heathland usually also lacks disturbed, early successional conditions (Figure 6.7).

Management often also aims to reduce nutrient levels. Not only does nutrient removal no longer take place through agricultural use, but heathlands are nowadays subject to deposition of anthropogenic atmospheric nitrogen as well as acidification from anthropogenic atmospheric sulphur. Increases in nitrogen levels encourage the growth of competitive grasses at the expense of dwarf shrubs (Heil and Diemont 1983). The main species involved are wavy hair-grass, *Deschampsia flexuosa*, and to a lesser extent sheep's-fescue, *Festuca ovina*, on dry heath and purple moor-grass, *Molinia caerulea*, on wet heath. Increases in nitrogen levels also encourage the growth of pleurocarpus (creeping, stringy) mosses. Replacement by grasses can also be initiated following killing of heather plants by heather beetles, *Lochmaea suturalis* (Heil and Diemont 1983; Berdowski and Zeilinga 1987; Bokdam 2001). The susceptibility to severe heather beetle

## 146 | Dwarf-shrub habitats and shrublands

**Fig. 6.7** Non-intervention and succession on Atlantic lowland heathlands. In the absence of management dwarf-shrub heath can be out-competed by other vegetation, such as this alien/exotic dwarf mountain-pine. This unmanaged dune heath also contains relatively uniformly structured heather with little or no bare ground: see Fig. 6.8 for a comparison (west coast of Jutland, south of Hanstholm, Denmark).

defoliation is also thought to increase with nitrogen levels (Brunsting and Heil 1985; Power *et al.* 1998). Acidification from sulphur deposition can decrease overall plant species richness in wet heaths and mires (Roem *et al.* 2002). Thus, the over-riding aims of managing dry heathland for conservation usually involve:

- maintaining dominance by dwarf shrubs and providing the desired mixes of growth phases by combinations of burning, cutting and removing vegetation and by grazing (Figure 6.8);
- preventing dominance by trees and shrubs by removing individual trees and areas of scrub, while *retaining* suitable densities and distribution of scattered pines, birches, and blocks of dense and structurally diverse gorse scrub for their value for birds and invertebrates;
- eradicating other alien/exotic, invasive plant species such as rhododendron and the dwarf shrub shallon, *Gaultheria shallon*;
- preventing expansion, and in some cases decreasing the area, of dense bracken;

**Fig. 6.8** Good heather structure. A key management aim in most Atlantic lowland heathlands it to provide good (i.e. varied) heather structure comprising mixtures of structure interspersed with bare and disturbed ground to maximize the range of suitable conditions for invertebrates and reptiles. Compare these photographs with the more uniform heather structure in Fig. 6.7 (Kalmthoutse Heide, Flanders, Belgium).

(a) Young, pioneer-phase heather interspersed with bare ground and grassland.

(b) Patches of bare and disturbed ground and grassland scattered among building and mature heather.

(c) Degenerate heather containing some regenerating, pioneer-phase heather and grasses among it.

- preventing expansion, and in some cases decreasing the area, of purple moor-grass and wavy hair-grass by grazing, cutting, and sod cutting/turf stripping;
- providing the desired structure of any associated grassland, primarily by grazing;

- maintaining suitable bare and disturbed ground, especially on southerly facing slopes for invertebrates, ruderal plants, and reptiles.

The key for most groups, especially invertebrates, is to provide a mixture of conditions. For example, sand lizards prefer areas of varied topography supporting structurally diverse vegetation and bare ground (House and Spellerberg 1983). It is important to recognize the value of mixtures of habitats when re-creating habitat. Whereas it may be tempting to convert an entire area of former arable land to heathland, there may be greater benefits in providing some areas of flower-rich neutral grassland as well to provide nectar sources for heathland insects.

Management of wet heaths and mires usually also aims to maintain a:

- near-natural hydrology by blocking any artificial drainage;
- variety of successional stages and vegetation structures from open pools to *Sphagnum*-dominated areas, wet dwarf shrubs, sedge, and grass-dominated areas to scrub, by arresting or reversing succession to rank, grass-dominated vegetation and scrub.

Succession in wet heaths and mires can be set back or retarded by sod cutting/turf stripping, removal of scrub and trees and by grazing. Cutting of dwarf shrubs and sedge and grass-dominated areas can be impractical in very wet areas and burning these areas is often contentious. Grazing is usually considered the best form of management. Raising water levels too rapidly runs the risk of flooding out existing important wet heath fauna and flora (WallisDeVries 2002).

There is evidence that anthropogenic acidification reduces the plant species-richness and abundance of characteristic wet heath species, and may reduce potential beneficial effects of sod cutting (e.g. Sansen and Koedam 1996). This has prompted attempts to increase the pH of acidified wet heaths by spreading lime to restore their characteristic plant assemblages, in some cases in combination with sod cutting (Beltman *et al.* 2001; Dorland *et al.* 2005). Liming has only so far only been used on a small scale in restorative management.

### 6.5.3 Cutting and burning lowland heathland vegetation

Cutting and burning dry heathland vegetation can both be used to:

- maintain dominance by dwarf shrubs, by preventing heather from reaching its degenerate phase, whose open conditions can be vulnerable to colonization by other potentially dominant plants;
- prevent shrubs and trees from becoming dominant;

- remove patches of above-ground growth of dwarf shrubs and thereby diversify the structure of uniform stands;
- remove nutrients.

Burning can remove more of the accumulated litter and expose more bare ground than cutting, although the specific effects will depend on the intensity of the burn. Prescribed burning and mowing typically remove similar quantities of nutrients. The quantities removed during typical cutting or burning rotations are, though, low compared to those removed by sod cutting/turf stripping. For example, on heathland in north-west Germany prescribed winter-burning and mowing and removal of vegetation both only removed the equivalent of 5 years of atmospheric nitrogen deposition, compared to 89 years' worth for sod cutting/turf stripping (Härdtle *et al.* 2006).

Both cutting and burning return heather to its pioneer phase, but differ in the origins of this re-growth. Burning stimulates germination of heather seed. Regeneration following prescribed burning is usually from both the underground rootstock of burnt plants and from seed. In some cases regeneration is virtually entirely from seed (Sedlakova and Chytry 1999; Nilsen *et al.* 2005). Most regeneration following cutting is by re-sprouting from underground rootstock. Cutting only retards growth of small seedlings among the dwarf shrubs. Burning has the potential to kill more established trees and scrub. Again, the specific effects vary according to the intensity of the fire.

The major practical difference is that fires have the potential to get out of control, with obvious risks to people, property, and of burning unacceptably large areas of habitat. This should not be a problem, though, providing it is undertaken at appropriate times of year and with suitable precautions. It will obviously be unacceptable to burn heathland close to habitation. Cutting is impractical on steep slopes, in rocky or otherwise bumpy terrain, and in very wet areas.

The need to manage dwarf-shrub vegetation to maintain its dominance will vary according to the potential for colonization by grasses. This will be higher in areas with high nitrogen levels and a nearby source of grass seed (Britton *et al.* 2000a; Barker *et al.* 2004). Conversely, the open conditions produced by burning in particular can themselves provide opportunities for tree seedlings and bracken to establish (Bullock and Webb 1995). Re-growth of heather following both burning and cutting can be suppressed and vegetation composition altered by grazing (Vandvik *et al.* 2005).

Although both cutting and burning can be used to increase vegetation structure at a large scale through providing areas of differently aged re-growth, it is important to realize that the growth phase and vegetation structure within any area cut or burnt at the same time will itself be relatively uniform. More small-scale variation in heather structure can often be produced by allowing heather to pass through its growth phases and regenerate naturally, providing it does not become out-competed by other plants, and by judicious use of grazing.

*Burning*

The effects of fire will also vary according to its time of year and weather conditions immediately before and during burning. As in other habitats, wildfires will occur more often during hot, dry conditions in summer and burn at high temperatures. In contrast, prescribed burning is carried out in late winter, when the vegetation is dry enough to burn but the soil wet enough to prevent the fire from becoming too intense and difficult to control. Wildfires are thus, compared to prescribed burns, more likely to kill larger trees and scrub, remove a greater proportion of organic matter in the soil, create more bare ground, kill dwarf shrubs by burning their rootstock as well as above-ground vegetation, and kill seeds near the soil surface. Most regeneration following intense fires is from seed. However, if fires burn deeply and hot enough to destroy most of the seedbank, then regeneration of dwarf shrubs may be poor and burnt areas instead colonized mainly by widely dispersing mosses and birch (Clément and Touffet 1990; Gloaguen 1993; Bullock and Webb 1995). Regeneration following prescribed burning typically results in an increase in plant diversity between 2 and 4 years afterwards, typically comprising grasses, forbs, and bryophytes (Vandvik *et al.* 2005).

Prescribed fires should be carried out by back-burning (Section 5.6), since back-fires are easier to control than head-fires and more effective at removing above-ground vegetation and litter. The same precautions need to be taken as when burning other types of vegetation (see also Section 5.6). Firebreaks/fuel breaks on heathland consist of bare or only sparsely vegetated ground that can be created by cutting and removing vegetation and rotovating strips. These can provide a valuable source of bare and disturbed ground. Firebreaks/fuel breaks need to be 5 m or more wide (Symes and Day 2003), be consolidated enough to prevent fire appliances from becoming stuck and include turning circles for them. Ponds are sometimes excavated in heaths to provide water for fire-fighting as well as open water habitat. This destroys potentially highly valuable existing habitat, and should only be undertaken after very thorough consideration.

Burning of humid, wet heath and mires is contentious, since it can damage the moss layer if carried out in dry conditions and also encourage dominance by species-poor stands of purple moor-grass (Brys *et al.* 2005). However, Bullock and Webb (1995) found that humid and wet heath had returned to close to its assumed pre-burn condition by 11 years after intense summer fires during drought conditions. Burning purple moor-grass-dominated wet heaths and mires when the moss layer and peat is too wet to burn is, though, considered by some to be an acceptable method of opening up and removing rank vegetation to encourage less competitive plant species (Symes and Day 2003).

Burning should not be carried out in areas known to support important concentrations of reptiles and lichen-rich heath.

### Cutting

The best method for cutting and removing heathland vegetation is by using a double-chop forage-harvester. This cuts the vegetation with knives, resulting in better regeneration than when stems are shattered using a flail cutter. Older, degenerate heather, though, may not regenerate from rootstock. Cutting is usually undertaken in autumn and winter, to minimize any damaging effects on nesting birds and active reptiles and invertebrates. Cut material can be used to provide a source of heather seed for heathland restoration. Where this is the case, the heather should be cut between about mid-October and early December to maximize the quantity of ripe seed harvested before it is shed.

It can be difficult to cut wet and humid heaths without causing unacceptable soil damage, although some rutting will create beneficial disturbance. The best option is to use low-ground-pressure tyres and cut during the drier conditions of early autumn or when the ground is frozen.

### Frequency of cutting and burning

The frequency of cutting and burning will depend on the desired proportions of different growth phases, more frequent rotations being required where the aim is to maintain a high proportion of early growth phases. Species that prefer short, recently cut or burnt (or heavily grazed) areas include silver-studded blue butterflies, *Plebejus argus*, especially towards the edge of their climatic range and wood larks and red-billed choughs (on maritime heath). Dartford warblers prefer taller, older heather. Prescribed burning of heathland for conservation is typically carried out on a rotation of 15–30 years. It is, though, also important to leave some areas unmanaged to also provide older and degenerate Heather for its associated species and, if regeneration is successful, provide areas with greater small-scale vegetation structure than can be achieved by burning or cutting.

### Sizes of areas cut and burnt

Cutting or burning only small patches of heathland at any one time increases spatial diversity in vegetation structure, and decreases the likelihood of inadvertently damaging localized populations. However, this has to be set against the greater time needed to cut or burn many small areas. Burning patches between 0.25 and 1.0 ha is a good compromise. A commonly used technique when cutting is to mow long, sinuous, one cut-wide strips.

### 6.5.4 Grazing and browsing

The effects of grazing on heathland are difficult to predict, and vary widely between sites and grazing regimes (Bullock and Pakeman 1996). Its main potential benefits are to:

- provide structural variation within areas of dwarf shrubs by arresting or delaying their aging process, including providing areas of short, heavily grazed heather that is structurally similar to its pioneer phase;
- prevent taller grasses, especially purple moor-grass and wavy hair-grass, from out-competing heather;
- prevent or reduce establishment of scrub;
- maintain open, trampled, and grazed conditions on wet heath, especially to benefit less-competitive plant species found on damp, bare acidic ground;
- maintain areas of associated short and open grassland;
- increase overall plant species richness (Bullock and Pakeman 1996; Bokdam and Gleichman 2000).

Grazing is generally considered better for invertebrates on heathland than cutting or burning, because it can create more small-scale variation in vegetation structure, and provides more soil disturbance and a source of dung (Kirby 1992b). The value of different types of dung for invertebrates and information on the use of anti-parasitic drugs in livestock are given in Section 4.4.3.

Nowadays, conservation grazing of heathland invariably involves allowing livestock to roam and graze within relatively large grazing units, instead of being shepherded and removed indoors at night or on to adjacent land. This type of grazing regime will not deplete nutrients from the heathland as a whole, but can re-distribute nutrients within it by selective deposition of dung.

As in other habitats, livestock require access to shade and water. The distribution of these will affect grazing patterns. Supplementary feeding should be avoided because it will increase nutrient levels. The availability of nearby alternative (sacrificial) grazing to and from which livestock can be moved can be important in enabling suitable grazing levels to be achieved on the heathland itself.

The basic grazing characteristics of herbivores used in conservation grazing have been described in Section 5.4.1. Additional information regarding their use on heathlands is given below.

*Cattle*

Cattle create more tussocky vegetation than either sheep or ponies. They also produce far more poaching and trampling per quantity of vegetation removed than sheep, and slightly more than ponies. Cattle can concentrate their dung at habitual resting sites, but do not have specific latrines like ponies. Cattle are similar to ponies in preferring feeding on grasslands, particularly more nutrient-rich ones, and wet heath/mire to dry heathland. As with both ponies and sheep, cattle only take significant quantities of dwarf shrubs in winter when there is little grass available. They can, though, differ slightly from ponies in their seasonal use of habitats as described above.

Cattle have the potential to achieve all the potential benefits of grazing listed at the beginning of this section, with the exception of controlling established scrub. They are, though, less suitable for grazing areas dominated by short, acid grassland than ponies and especially sheep, because they do not nibble vegetation close to the ground. Cattle graze back and create variation in structure of taller grasses, especially purple moor-grass, although often less so than ponies, and graze and trample wet areas. Cattle do not browse scrub as much as ponies and primitive sheep breeds but can still reduce establishment of seedlings of trees and shrubs by removing them among mouthfuls of other vegetation.

The most favoured hardy cattle for conservation grazing of heathlands and purple moor-grass dominated grasslands in continental north-west Europe are Galloways.

*Ponies*

Ponies have a similar nibbling action to sheep, are highly selective and can create short, closely cropped grass swards and carpets of ericaceous shrubs. Their level of poaching and other disturbance is intermediate between that of sheep and cattle. Ponies prefer grassland, especially more nutrient-rich types, to dry heathland, and again only feed to a significant extent on dwarf shrubs in winter when there is little grass available. They differ from sheep in grazing wet heath and mire, and from conventional sheep in browsing trees and shrubs, especially the wintergreen gorse. Browsing is greatest in winter when other forage is scarce. In the New Forest in England (Figure 4.5) the diet of ponies varies seasonally more than that of cattle, feeding more on the deciduous purple moor-grass in

**Fig. 6.9** Browsing trees and scrub. A common aim of using livestock on heathlands is to reduce or prevent scrub encroachment.

Primitive sheep, such as these Hebrideans, are intermediate in their behaviour between conventional sheep and goats. They spend a larger proportion of time browsing trees and shrubs than conventional sheep. Hebridean sheep are commonly used to control birch scrub on heathland. Goats are even better at browsing, but also much better at escaping (Pembury Walks, Kent, England).

wet areas in summer, and making increased use of browse in winter (Pratt *et al.* 1986; Putman *et al.* 1987).

Ponies can concentrate their dung in limited latrine areas, resulting in localized soil enrichment and more nutrient-rich vegetation.

*Sheep*

Many areas of heathland, particularly areas of dry heath and acid grassland, have traditionally been grazed by sheep, often by specialist heathland breeds. Conventional, as opposed to primitive, sheep prefer grasses to dwarf shrubs, and only take significant quantities of the latter in winter when there is little or no grass available (Bakker *et al.* 1983). Heavy grazing by sheep can create short carpets of closely cropped heather. Sheep do not concentrate their dung in limited areas, as cattle and particularly ponies do. Conventional breeds of sheep are rather poor at achieving the potential benefits of grazing described above. They nibble down shorter acid grassland to produce a tight, often relatively species-rich but uniform sward. They nibble off flowers, which can be important nectar sources for heathland invertebrates at critical times of year. Grazing down the sward short using sheep can be used to encourage rabbits (Section 5.4.2). Sheep also have the disadvantages, especially compared to cattle, of:

- being less good at breaking down tall grasses where these are competing with dwarf shrubs, and producing less variation in structure of these tall grasses;
- producing less poaching and consequently less bare ground;
- generally avoiding grazing wet heath and mires.

Many heathlands are open to public access where sheep are vulnerable to attack from dogs, thus requiring restrictions on dog walking. Conventional sheep browse the leaves of scrub very little, but primitive sheep breeds are excellent at browsing (Figure 6.9). They are particularly valuable at controlling birch regeneration.

### Goats

As in other habitats, goats spend a greater proportion of their time browsing and feeding on other woody vegetation than other livestock. They can therefore be effective at controlling young scrub and small trees and grazing dwarf shrubs. They are similar to sheep in avoiding wet areas and being vulnerable to attacks by dogs. As in all other habitats, though, the main issue with using goats is the difficulty of containing them.

### Rabbits

European rabbits can be important grazers and creators of disturbance on heathlands and associated acid grasslands. They can maintain short, open heather, bare and disturbed ground, and provide valuable dung and carrion. Methods for encouraging European rabbits are described in Section 5.4.2.

### Deer

Browsing by red deer can be effective at controlling pine saplings. Browsing by deer is most intense close to woodland, where most tree regeneration is likely to take place, and in areas less heavily disturbed by people.

## Grazing pressure and timing of grazing

There are just two main options for the timing of grazing on heathlands and its associated habitats: summer grazing (i.e. spring to autumn) and year-round. Introducing, or re-introducing, livestock to formerly ungrazed heathland has the potential to damage existing valuable features. As when introducing grazing to other habitats, it is therefore prudent to begin at a low grazing pressure and, if necessary, increase livestock numbers over time in response to the results of monitoring and the experience gained. Fencing can be used to exclude grazing from sensitive areas, for example taller, mature heather containing localized populations of sand lizards. In mixtures of heathland and other habitats, grazing pressure within different habitats will be a product of:

- overall grazing pressure;
- the relative proportions of these different habitats;
- the relative preferences of livestock for these different habitats.

Preferences of livestock vary during the year. The effects of grazing on the composition and structure of dwarf shrubs will therefore be greater where overall grazing pressure is high and there is little or no alternative forage. Thus, sheep only eat significant quantities of dwarf shrubs if there is little alternative dry grass to eat, and ponies and cattle only eat significant quantities of dwarf shrubs if there is little or no dry grassland, wet heath or mire to feed on. In particular, all types of livestock usually eat more dry dwarf shrubs in winter when they have eaten all or most of the remaining grass and any deciduous purple moor-grass has died down (Pratt *et al.* 1986; Putman *et al.* 1987).

Grazing animals influence the structure of dry heathland by grazing shoots and trampling plants. Moderate levels of grazing can benefit dwarf shrubs by reducing competition from more vigorous grasses, although the effects of grazing in benefiting dwarf shrubs will vary. Grasses are more competitive relative to heather on more nutrient-rich soils (Bokdam and Gleichman 2000). Heather is usually grazed preferentially to cross-leaved heath, *Erica tetralix*, often the dominant species in humid and wet heath, and slightly more than bell heather, *Erica cinerea*. Hence grazing will tend to decrease the abundance of heather relative to these species. On wet heath and mires grazing and trampling can provide open conditions for a range of scarce plants (Figure 6.10).

Higher grazing pressure, especially from sheep, ponies, and European rabbits, produces short lawns of dwarf shrubs. Very heavy grazing leads to replacement of the grazing-intolerant heather with grazing-tolerant grasses and forbs. Heather is particularly susceptible to damage by grazing in autumn. Very heavy grazing

Fig. 6.10 Heavy grazing and trampling. Heavy grazing produces pioneer-type heather structure (a), and can eventually result in its replacement by grazing-tolerant grasses.

Heavy trampling on wet heath, such as along this track created by ponies and people (b), provides suitable open conditions for a range of less-competitive plants. The margins of this track support abundant marsh clubmoss, *Lycopodiella inundata*, which is rare and declining throughout most of its European range, and the scarce brown beak-sedge, *Rhynchospora fusca*. Both are absent from the denser vegetation in untrampled areas (near Matley Wood, New Forest, Hampshire, England).

and churning up of large areas of wet peat by livestock or deer will be damaging. In particular this can be the case where there is a small area of wet heath/mire on which animals concentrate their activities, set within a larger area of dry heathland and grassland.

Grazing mixtures of dry heathland and grassland in winter will be a balance between reducing the abundance of unwanted wintergreen plants that can out-compete dwarf shrubs, without reducing the abundance of the dwarf shrubs themselves. Winter grazing can be useful in reducing the vigour of wavy hair-grass in areas where it produces a flush of growth in late autumn following autumn rains, and begins re-growing early the following year. Winter grazing by cattle can also be used to reduce the vigour of the tall grasses wood small-reed, *Calamagrostis epigejos*, and Yorkshire-fog, *Holcus lanatus*.

Grazing densities of heathlands are typically between 20 and 70 livestock unit days/ha per year (see Table 5.1 and Fig. 5.8 for an explanation of how to estimate grazing pressure). As a rule of thumb, stocking levels need in winter to be about a third of those used in summer.

Livestock can affect regeneration of trees and shrubs by eating their seedlings when feeding on grass and other low vegetation, and by specifically browsing the leaves of trees and in some cases stripping their bark. Most livestock only browse trees and shrubs to any extent when there is little other food, especially in winter and under high stocking levels. Often, though, grazing levels need to be so high to suppress tree and scrub regeneration that this conflicts with other objectives. Hence even though grazing can reduce encroachment by trees and shrubs, additional cutting and removal is often required (Bokdam and Gleichman 2000; Piessens *et al.* 2006).

Periodic grazing can be also be used to graze down areas over a short period, and temporary or permanent fencing can be used to exclude grazing from specific areas. For example, grazing is often temporarily excluded to prevent livestock from eating the flowers of marsh gentian, Gentiana pneumonanthe, on wet heath used by the larvae of the Alcon blue butterfly, *Maculinea alcon*. Grazing is typically excluded from these key areas between the end of June and mid-September. This allows sufficient time for the adult butterfly to lay their eggs at the base of the gentian's flowers, the eggs to hatch, the caterpillar to eat the gentians seeds, and finally fall to the ground and be adopted by ants.

Burning, by producing a flush of palatable re-growth, can also be used in combination with grazing. Livestock preferentially graze the re-growth and keep previously burnt areas short.

### 6.5.5 Removing individual trees and patches of scrub

Removing individual trees and patches of scrub can present a dilemma. Removing all of them will make the area less interesting for most birds and many invertebrates, whereas leaving them will provide a source of seeds for further establishment of trees and shrubs (Manning *et al.* 2004). A compromise is to leave scattered single and clumps of trees (Figure 6.11).

The value of heathland for birds and invertebrates can also be enhanced by maximizing the length and structural variation of its interface with woodland. Sheltered hollows facing towards the sun can be created to benefit warmth-loving invertebrates and woodland edge diversified to provide feeding edge for Eurasian nightjars. Principles are similar to those described for enhancing the edges of rides and margins of woodlands in general (Section 7.4.1).

The method used to remove trees and patches of scrub will depend on their size. Saplings can be pulled up by hand but is a laborious process. Larger trees can be cut using chainsaws. Dense stands of trees and scrub can be removed using

**Fig. 6.11** Scattered trees and scrub. Trees and scrub can out-compete lowland heathland vegetation, but their presence also enhances the value of heathland for birds and many invertebrates. A compromise is to remove the majority of the trees and scrub, but retain scattered individuals and patches. Densities of three or four trees per hectare are generally recommended (Symes and Day 2003), with their locations and distribution also taking account of aesthetic considerations (Grange Heath, Dorset, England).

forestry mulchers. Subsequent treatment with herbicide is usually necessary to prevent re-growth of deciduous trees where this is not controlled by browsing. As with scrub removal in general (Section 7.3.1), it is best to concentrate on areas most recently colonized by trees, since these will be easier to restore to heathland vegetation. Older, leggy, gorse can be cut to ground level (coppiced) and allowed to re-grow or regenerate from seed to provide denser stands favoured by Dartford warblers.

Disturbance of the ground layer during tree removal will probably be sufficient to expose any buried heathland seed and provide suitable conditions for its germination. Removal of accumulated tree litter and humic material to *just above* the mineral layer helps in re-establishing heathland vegetation beneath dense stands of trees by maximizing removal of nutrients, while still retaining a smear of organic matter above the mineral soil containing buried seeds (Allison and Ausden 2006). Heathland invaded by Scots pine is usually easier to return to heathland than area colonized by birch and bracken, although there is much variation in the success of management between sites (e.g. Mitchell *et al.* 1999). Birch is more invasive on phosphorus-rich soils (Manning *et al.* 2004).

### 6.5.6 Controlling bracken

Bracken often forms large, monospecific stands of minimal conservation value, although it can be beneficial in providing nest sites for Eurasian nightjars and wood larks. There is often a desire to eradicate large areas of dense bracken. However, as with controlling other invasive plants (Section 4.5), any decision should take account of whether it is expanding and the likelihood of successfully reducing it and establishing more valuable habitat in its place.

There are four methods of controlling bracken: spraying with the herbicide Asulam, cutting, rolling, and bulldozing. The most effective method is spaying with Asulam. This can be applied to dense stands using a boom-sprayer or weed-wiper, or by spraying from a helicopter. Scattered bracken can be treated by spot-spraying. Follow-up spraying is always necessary to prevent surviving bracken from re-expanding. Other techniques can only be used on large expanses of bracken with little other conservation interest. Cutting reduces the abundance of fronds but will not kill bracken. However, 2 years of cutting mature fronds in mid-summer and replacement fronds as they reach maturity in late summer should significantly reduce its density, although annual cutting will subsequently be needed to prevent it from re-expanding. Squashing bracken using a roller with crimping edges has a similar effect to cutting and needs to be undertaken at the same frequency to achieve similar results. Bulldozing the fronds, rhizomes, and litter is also very effective (Mitchell *et al.* 1999).

Bracken produces a dense litter that inhibits the growth of most other plants. Thick layers of litter beneath long-established stands need to be removed, or at least scarified so that most breaks up and blows away, to allow other vegetation to establish following control.

### 6.5.7 Sod cutting/turf stripping to reduce nutrient levels and other methods of creating bare and disturbed ground

Sod cutting/turf stripping is the most effective method for reducing nutrients, especially nitrogen levels, within the vegetation, litter, and organic layer to favour dwarf shrubs at the expense of competitive grasses (Diemont and Linthorst Homan 1989; Britton *et al.* 2000b; Härdtle *et al.* 2006). Sod cutting/turf stripping will also create bare ground. Stripping down to as far as the mineral layer removes a higher proportion of nutrients, but delays the re-establishment of

**Fig. 6.12** Sod-cutting/turf-stripping. Anthropogenic atmospheric nitrogen deposition increases nitrogen levels in the vegetation and soil. This favours grasses at the expense of heather.

(a) Sod-cutting/turf-stripping can be used to remove this accumulated nitrogen, thus favouring heather, which can be seen establishing in sod-cut areas. Uncut areas are still dominated by grasses.

(b, c) An extreme form of sod-cutting/turf-stripping has been carried out in the National Park de Hoge Veluwe, Gelderland, in the Netherlands. Here, 65 ha of drifting, wind-blown sand has been re-created by felling pine woodland and removing the topsoil. The sparsely vegetated sand supports a number of species that are rare in the Netherlands, including the spectacular ladybird spider, *Eresus cinnaberinus*, blue-winged grasshopper, *Oedipoda caerulescens*, and sand lizard. The mobility of the dune can be appreciated from the bottom photograph which shows a cut stump now left standing high above the sand.

dwarf-shrub vegetation, because it removes any existing seedbank and probably provide less favourable conditions for germination of dwarf shrubs (Diemont and Linthorst Homan 1989; Allison and Ausden 2006). The best option is to remove material to just above the mineral layer as described in Section 6.5.5.

It can be difficult to dispose of topsoil removed by sod cutting/turf stripping, making it impractical to undertake over very large areas. There has been some success in marketing the material as acidic garden mulch, but in some areas it contains levels of heavy metals that are too high for this. Stripping of large areas has been used to expose sand to create mobile sand dunes (Figure 6.12). Sod cutting/turf stripping can also be used to set back succession in wet heath and mires, by lowering the surface of the peat relative to water level and exposing the buried seedbank.

Small-scale sod cutting/turf stripping can be used to provide suitable conditions for a range of scare plants (e.g. Jacquemart *et al.* 2003; Jansen *et al.* 2004), using the same methods and principles as described for fens (Section 8.8.4).

Other methods for exposing bare ground, other than through trampling by livestock and the activities of European rabbits, include:

- cutting and removing the vegetation and then rotovating the soil;
- scraping away the vegetation and topsoil using a bulldozer or angled blade attached to a tractor;
- creating vertical faces and eroding slopes;
- disturbance and compaction caused by human trampling, horse-riding, cycling, motorbikes, and other vehicles.

Where a thick layer of organic matter is present, sod cutting/turf stripping will be better than rotovation at exposing more bare sand and any buried seedbank. Rotovation partially re-buries the organic matter within the upper layer soil. Rotovating, or even ploughing, can be used at less frequent intervals on formerly cultivated heathland soils to set back succession (Degn 2001). When creating bare sand using a bulldozer or tractor-mounted blade, the scraped soil and vegetation should be mounded on its northern side to prevent it shading the bare ground created.

A continuity of disturbed conditions can be created by constructing vertical sand faces and erodable slopes (Figure 6.13). These can provide important nesting sites for solitary bees and wasps, and basking and warm, dry, over-wintering sites for reptiles. They should ideally face towards the sun to provide the warm conditions favoured by the majority of species associated with bare and disturbed ground. The area of suitable face and bank can be maximized by excavating a sand face and using the material removed to build an additional bank. These

**Fig. 6.13** Banks and slopes. Artificially created banks can provide a continuity of bare and disturbed ground through erosion. In this case, periodic clearance of small areas is also used to maintain open conditions, especially on flatter areas. Trampling by sheep also helps maintain open conditions. The areas of bare ground sheltered by overhanging vegetation here support larvae of the antlion, *Euroleon nostras*, which in the UK is very localized and confined to heathland (Aldringham Walks, Suffolk, England).

areas may require periodic small-scale vegetation clearance to keep them open, although they are often kept open by trampling and rubbing by livestock.

Trampling along paths and disturbance by vehicles is sometimes considered a nuisance on heaths as in other habitats. However, as in free-draining grasslands, it can produce a variety of valuable microhabitats for invertebrates, particularly consolidated and sparsely vegetated sand on the path edges. Areas *continually* churned up by vehicles and horses will be less valuable.

As with all forms of disturbance, it is important to carry it out on rotation. Disturbance should ideally be created successively in adjoining plots to increase the chances of species colonizing newly created plots from those that are becoming unsuitable. As a rule of thumb, only create disturbance over a proportion, for example between a tenth and a third, of the habitat at any one time, to minimize the risk of destroying all the suitable habitat for a species at once. It is common practice to periodically rotovate entire areas, especially firebreaks/fuel breaks. A better option is to maintain a far wider firebreak/fuel break, but only rotovate a proportion, for example a third to a half its width each year. This will maintain a variety of successional stages in close proximity to one another, while still maintaining its function (Kirby 1992b).

## 6.6 Atlantic upland heaths and moorlands

Dry and wet dwarf-shrub upland heaths and moorlands dominated by heather are restricted to upland areas of north-west Europe, mainly in the British Isles. Most areas have been created by deforestation and been subsequently maintained by grazing and burning. Trees are usually absent. Burning (known as muirburning in Scotland) has been used to maintain dominance by heather and encourage a flush of palatable grasses and heather re-growth for grazing by sheep and red deer, or to provide a small-scale mosaic of young and old heather for red grouse, *Lagopus lagopus scoticus*, to maximize numbers for shooting. Areas of open upland managed for sheep are known as sheep-walk and for deer known as deer forest. Peat cutting for fuel has also been undertaken in wetter areas.

Upland moorlands invariably occur, and are managed, together with areas of blanket bog, wet flushes, acid grassland, and bracken. Upland heath and moorland is also farmed in association with a fringe of enclosed upland grassland at lower altitude known as in-bye.

The dwarf-shrub vegetation of drier upland heath and moorland differs from those on lowland heaths in the presence of several montane dwarf shrubs. Upland heaths and moorland also lack many of the rarer plants associated with disturbed ground that are found on lowland heath. Wetter areas are, as on lowland wet heath, typically dominated by mixtures of heather, cross-leaved heath, and purple moor-grass. These grade into waterlogged blanket bog and bog pools dominated by mixtures of heather, hare's-tail cottongrass, *Eriophorum vaginatum*, and *Sphagnum* mosses. Wet flushes, particularly base-rich ones, can be botanically rich and contain many plant species with localized distributions.

Despite the similarities in dominant plant species, the fauna of upland heaths and moorland is quite different to that of lowland heaths, although there is some overlap of species, especially at lower elevations. Whereas lowland heaths support a range of mainly southerly species, the cool, wet, and cloudy upland heaths and moorlands and associated habitats support a range of mainly arctic-alpine, alpine, and boreal invertebrates, and a number of arctic and boreal bird species.

Areas consisting largely or entirely of dry upland heath and moorland support only a very limited avifauna, with only one species, red grouse, confined to this habitat. However, a wider range of birds are associated with mixtures of

**Fig. 6.14** Black grouse. Like many birds found on Atlantic upland heaths and moorlands, black grouse actually require a mixture of habitats. They often nest in tall heather, but the hens require protein-rich food prior to laying, such as flowers of cottongrass in bogs and buds of other plants. In summer chicks feed on invertebrates, especially in wet flushes. In winter black grouse feed on dwarf shrubs and, especially when these are covered in snow, on the twigs and needles of various trees (photograph by RSPB IMAGES).

upland heath and moorland and other upland habitats. These include northern harrier, *Circus cyaneus*, merlin, *Falco columbarius*, black grouse, *Tetrao tetrix* (Figure 6.14), Eurasian golden-plover, *Pluvialis apricaria*, short-eared owl, *Asio flammeus*, ring ouzel, *Turdus torquatus*, common stonechat, *Saxicola torquata*, and twite, *Carduelis flavirostris*. Vegetation structure is important in influencing habitat use by many of these upland birds, with heterogenously structured vegetation probably supporting the widest range of species (Pearce-Higgins and Grant 2006). Scattered trees and woodland will increase bird species richness and favour species such as black grouse and scrub-associated songbirds, but decrease suitability for most species typical of open habitats, particularly breeding waders/shorebirds.

The invertebrate fauna varies with the growth phase of heather following cutting or burning (Usher 1992), as on lowland heaths, but lacks the suites of species of high conservation value associated with early successional habitat.

## 166 | Dwarf-shrub habitats and shrublands

The invertebrate fauna of upland heaths and moorland is rarely, if ever, specifically taken into consideration during management.

### 6.6.1 Overview of management

Most upland dry heath and moorland is managed by periodic burning to provide grazing for sheep and red deer and maximize red grouse numbers. Judicious burning also maintains dominance by heather by setting back its growth stage, thereby encouraging the vigorous, building, and mature phases of heather, which are less vulnerable to colonization by grasses. Management for red grouse also involves control of its predators and parasites.

A large proportion of upland heath and moorland and associated blanket bog, wet flushes, and acid grassland is considered to be in an unfavourable conservation condition due to the effects of heavy grazing (over-grazing) by sheep (Fuller and Gough 1999) and too frequent burning. High levels of grazing favour grasses over heather-dominated vegetation, and frequent burning, especially in wetter areas, can result in the replacement of heather-dominated vegetation by purple moor-grass. Replacement of heather by grasses is exacerbated by anthropogenic atmospheric nitrogen deposition (Hartley and Mitchell 2005). Heavy grazing by sheep has been encouraged by agricultural subsidies. Large areas of dense bracken are often controlled to increase the area available for grazing. Frequent burning of associated bogs (Section 8.9) is considered to have damaged its flora and, together with heavy grazing, can cause unwanted erosion. Wetter areas have been damaged by drainage.

Thus, although upland heaths and moors are again relatively similar to lowland heaths in terms of their dominant plant species, the conservation value of upland heath and moorland has usually been damaged by over-exploitation, and that of lowland heathland usually by lack of recent management that has resulted in its loss through succession.

Conservation management of upland heaths and moorland and associated habitats invariably has to take account of its commercial management for livestock, red deer or red grouse, and therefore usually aims to:

- provide a mosaic of heather growth phases by periodic burning and in some cases cutting to benefit red grouse or provide sufficient re-growth for livestock and red deer, and to maintain heather mainly in its building phase and thereby minimize the risk of its replacement by grasses;
- maintain, or restore, wet heath, blanket bog, and wet flush vegetation by leaving it ungrazed or only grazing at low levels, preventing burning and blocking any artificial drainage (grip-blocking; Section 8.9);

- provide an overall mosaic of suitably lightly or moderately grazed wet habitats — blanket bog, species-rich flushes, and wet grassland and heterogenous dry upland heath and moorland — to maintain a diverse range, and high densities of the previously mentioned bird species.

Whereas there is a general consensus that heavy grazing has in most cases been damaging the conservation value of these upland habitats, there is less certainty that these damaging effects can be reversed simply by reducing grazing levels, at least in the short term.

Management may also aim to encourage re-afforestation by native woodland by reducing levels of grazing and browsing by livestock and, in Scotland, by red deer (Figure 7.2).

## 6.6.2 Burning and cutting

The basic principles of burning and cutting are similar to those on lowland heathland. They differ in that heather in wetter areas of upland can perpetuate itself by rooting from prostrate stems. Layering prevents the heather from reaching its degenerate phase, and thus does not require periodic cutting of burning to maintain its vigour. The bird nesting season is later in the uplands, so burning can take place slightly later into spring than on lowland heaths. In the wetter climate of the uplands there are usually even fewer days suitable for burning in winter and early spring.

The large size and remote nature of most areas of upland heath and moorland means there is less of an issue of fires getting out of control and threatening people and property, than on most areas of lowland heathland. Burning is therefore used far more widely. Cutting is in any case often difficult in the more remote terrain.

Management for sheep and red deer usually involves burning relatively large areas to provide re-growth of grasses and heather for them to feed on. Management for red grouse involves burning narrow strips (muirburning in Scotland; Figure 6.15). Densities of breeding northern lapwings, Eurasian golden-plovers and Eurasian curlews *Numenius arquata* tend to be higher on moors managed for red grouse compared to on other heather-dominated moorland with similar vegetation, but it is difficult to differentiate between differences due to vegetation management and those caused by control of predators. Densities of most songbirds tend to be lower on managed grouse moors (Tharme *et al.* 2001). Eurasian golden-plovers, though, benefit from heather burning because they favour the short re-growth for nesting and chick-rearing, providing this is close to earthworm-rich grassland for adults to feed on (Whittingham *et al.*

# 168 | Dwarf-shrub habitats and shrublands

**Fig. 6.15** Muirburning. Management to increase numbers of red grouse for shooting involves burning narrow (typically less than 30 m wide) strips. This provides a small-scale mosaic of nutritious heather re-growth for feeding intermixed with taller heather for cover and nesting. The management aims to prevent heather from becoming more than about 30 cm high, with burning rotations typically in the range of 10–25 years (near Kingussie, Highland, Scotland).

2000). Areas that should not be included within the heather-burning rotation include:

- exposed ridges, summits, and slopes where dwarf-shrub heath is maintained by exposure and where burning only results in slow regeneration and has the potential to initiate erosion;
- wet heath and blanket bog, which can be damaged by burning, and where there is a risk of severe damage to the underlying peat;
- wet flushes, whose vegetation will be damaged by burning;
- damp slopes and gullies supporting important assemblages of bryophytes that can be damaged by burning;
- grass/heather mosaics subject to high grazing pressure are therefore vulnerable to replacement of the heather by grasses;
- scattered trees and scrubs, which are important in their own right and which would be damaged by burning;

- areas of taller, older heather important for nesting northern harriers and merlins (Redpath *et al.* 1998) and those that have well-developed layering.

As with lowland heaths, grazing animals concentrate on the nutritious re-growth following burning. Therefore, burning a small proportion of areas supporting high stocking levels, can greatly suppress or even prevent successful regeneration of dwarf shrubs. Bracken can expand following burning, as on lowland heathlands.

Blanket bog should not be burnt, since frequent burning, particularly in combination with heavy grazing and increased atmospheric nitrogen deposition, can lead to loss of peat-building *Sphagnum* and virtually complete dominance by hare's-tail cottongrass. As in lowland areas, burning upland wet heath can encourage species-poor stands of purple moor-grass at the expense of heather (Ross *et al.* 2003).

### 6.6.3 Grazing

The large size and remote nature of most areas of upland moorland means that grazing units are invariably large and encompass a range of habitats. Grazing pressure therefore has to be set at appropriate levels to maintain the whole suite of habitats within the grazing units within the desired state. Virtually all commercial grazing on upland heath and moorland is by sheep. However, cattle are generally preferred to sheep for conservation grazing for similar reasons to those described for heathlands and grasslands. Flocks of sheep and herds of cattle in upland areas are often hefted (Section 5.4.3). Ponies and goats are rarely used.

Heather can tolerate 40% of its current season's shoots being removed by grazing for a few years, but removal of 80% or more of shoots results causes it to die back, leaving it vulnerable to being out-competed and replaced by grasses. Heather is most vulnerable to defoliation in autumn (Grant *et al.* 1978, 1982). As on lowland heathlands, autumn grazing is most damaging. Modelling suggests that levels of utilization need to be less than 21–27% of potential maximum shoot production over the longer term to maintain dominance by heather (Read *et al.* 2002). The precise effect will depend on the age and vigour of heather and whether or not grass is already in the area. Tell-tale signs of grazing beginning to reduce the dominance of heather are when its pioneer-phase forms a low carpet and more mature heather begins to resemble topiary, and eventually a drumstick-like form, comprising woody stems with shoots reduced to heavily-grazed tufts at their tops.

As on lowland heaths, sheep, cattle, ponies, and also red deer prefer feeding on grasses and most other monocotyledons to heather and most other dwarf shrubs.

Goats take a larger proportion of dwarf shrubs and other woody plants (Bullock 1985; Grant *et al.* 1987; Gordon 1989; Jewell *et al.* 2005). Therefore, as on lowland heath, heavy grazing of dwarf shrubs will only occur at high stocking levels, and when there is little or no alternative forage, especially in winter when there is little grass left.

To maintain dominance by existing young and vigorous heather, overall stocking levels (including those of wild red deer) need to be kept below approximately 70 livestock unit days/ha per year. Stocking levels need to be less than half these to maintain heather dominance on poor soils and at high altitude, and no higher than 20 livestock units days/ha per year is recommended to maintain dominance by old heather and where it is in competition with purple moor-grass or hare's-tail cottongrass (Thompson *et al.* 1995). Supplementary feeding will influence the distribution of sheep, with high grazing pressures and die-back of heather often occurring close to feeding blocks.

Recovery of previously over-grazed heather on upland dry and wet heath has been found to occur under year-round grazing at, respectively, 40 and 24–27 livestock unit days/ha per year (see Table 5.1 and Figure 5.8 for an explanation of how to calculate grazing pressure). In both cases removing grazing completely resulted in an even more rapid recovery of heather (Hulme *et al.* 2002; Pakeman *et al.* 2003).

Heather can re-grow well from small, heavily grazed plants following a relaxation of grazing pressure, but can be difficult to re-establish once lost. The main effect of reducing grazing levels on grass-dominated former upland heath and moorland can therefore be to increase grass height and cause of the decline of lower growing plants, but with little or no (re)-establishment of other plant species (Hill *et al.* 1992; Hope *et al.* 1996). Disturbance to create germination gaps and expose any existing seed, or spreading of seed, will probably be necessary to re-establish it.

The effects of changes in grazing levels to benefit upland birds are poorly understood. Creation of a mosaic of vegetation heights and structures by reducing stocking levels over only a proportion of sites has resulted in increases in numbers of black grouse (Calladine *et al.* 2002). Conversely, heavy grazing that results in partial replacement of heather by grassland will increase numbers of field voles, *Microtus agrestis*, and meadow pipits, *Anthus pratensis*, thereby increasing prey for breeding short-eared owls (field voles) and northern harriers (both field voles and meadow pipits). High densities of breeding northern harriers can reduce the harvestable surplus of red grouse, thereby causing potential conflicts on commercially managed grouse moors. The best option for maximizing numbers of red grouse will be to restore suitably managed heather moorland lost through over-grazing. This will increase the area of suitable habitat for red grouse, while reducing numbers of nesting harriers in these areas by reducing densities of their main prey, field voles and meadow pipits (Redpath and Thirgood 1997, 1999; Thirgood *et al.* 1999; Smith *et al.* 2000a). Within the grassland itself, densities of field voles will be highest in ungrazed or only lightly grazed grassland (Evans *et al.* 2006). Heavily grazed, degraded hare's-tail cottongrass-dominated blanket bog can, though, provide suitably short and open conditions for breeding of Eurasian golden-plovers (Whittingham *et al.* 2000; Pearce-Higgins and Yalden 2004).

### 6.6.4 Controlling bracken

Principles of controlling bracken are similar to those on lowland heathland. However, the larger and more remote nature of most upland heath and moorland mean than bracken control is usually carried out on a far larger scale, often involving spraying from helicopter. As on lowland heathland, while extensive areas of dense bracken are of extremely limited value for wildlife, bracken-dominated areas containing scattered trees and shrubs can support a diverse range of breeding songbirds (Fuller *et al.* 2006). Open bracken is important for the larvae of the pearl-bordered fritillary butterfly, *Boloria euphrosyne* (Feber *et al.* 2001).

# 7
# Forests, woodlands, and scrub

Forest and woodlands are used to describe land dominated by trees. Scrub consists of small trees and bushes. Management of areas of dense, medium-sized, predominantly evergreen shrubs found in Mediterranean climates is discussed in Chapter 6. Management of forest, woodland, and scrub on land with a high water level is described in Sections 8.10 and 8.11.

The dominant structural components of forests and woodlands, their trees, are longer-lived than the dominant structural components of other habitats. This means that any management that takes place in woodlands and forests has to be viewed over a far longer timescale. Changes in the dominant tree species will take tens or even hundreds of years. The dominant tree species present now may have established under quite different conditions to those currently. Because of climate change, any trees that are currently establishing are likely to reach maturity under quite different conditions to those at present.

Because of the length of time for seedlings to grow into mature trees, there is a far greater presumption for maintaining the existing dominant structural components of the habitat—its trees—rather than seeking to radically change their composition, as may sometimes be the case in other habitats. Because of this, the approach taken when managing forests and woodlands will vary greatly depending on their existing tree-species composition, structure, and history.

## 7.1 Important features of forests and woodlands for wildlife

The main features of forests and woodlands that will influence their value for wildlife are the:

- dominant tree-species composition;
- continuity of forest or woodland on the site;
- age and structure of stands (a stand is a term for a growth of similar plants in a particular area; it is commonly used to describe a group of trees of similar age and species composition);

- quantity and types of dead wood;
- presence of forest edge and other associated habitats;
- variation in soils, topography, and drainage.

### 7.1.1 Tree-species composition

Tree-species composition, in particular whether the dominant tree species are broad-leaved or coniferous, is important in influencing the biodiversity of a forest or woodland. Areas dominated by broad-leaved trees support a distinctly different avifauna from those dominated by conifers, while mixtures of the two support one containing species from both. Different tree species vary in their fauna of plant-eating insects. Individual tree species also differ in their suitability for foraging birds (Peck 1989) and nest sites (Hågvar *et al.* 1990). The presence of different tree species will strongly influence the mycorrhizal fauna.

### 7.1.2 Growth stage and structure of stands

The growth stage and structure of different areas will particularly influence their bird, invertebrate, and amphibian fauna and assemblage of herbaceous plants. There are a number of terms used to describe aspects of the structure of forests and woodlands. The canopy comprises the crowns of the largest trees. The understorey describes the shrubs and herbaceous vegetation beneath the canopy. Field layer refers to just the herbaceous vegetation.

The structure of a given stand changes in relation to its stage of growth since establishment or catastrophic disturbance, although this structure can be significantly modified by management. There are two main theories of woodland establishment or regeneration following catastrophic disturbance. These are described in the following two sections.

### 7.1.3 Theory of woodland regeneration

The theory of woodland regeneration (Oliver and Larsen 1990; Peterken 1996) proposes the following four phases of development (Figure 7.1).

*Stand initiation or regeneration*

Where woodland is establishing, then *stand initiation* will involve the establishment of tree and shrub species from seed, mainly dispersed by wind or birds (Figures 7.1a and 7.2). Establishment may be slow, because of lack of seed. The shrubs and trees that establish will be of light-demanding species. These are often called intolerant species (intolerant of shade), as opposed to tolerant species

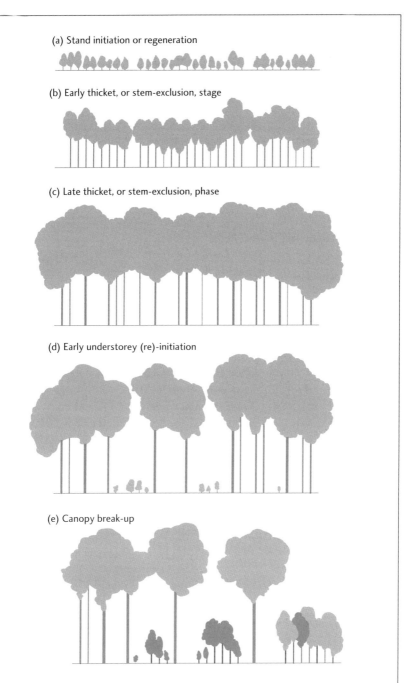

**Fig. 7.1** The theory of woodland regeneration (from Oliver and Larsen 1990; Peterken 1996). Tree species intolerant of shade are shown in light grey and those tolerant of shade are in dark grey. See Section 7.1.3 for details.

**Fig. 7.2** Deer and woodland regeneration. Red deer are important grazers and browsers in parts of temperate Europe, Asia, and North America. They can profoundly influence regeneration of woodland and scrub.

In this area of Atlantic moorland numbers of red deer have been reduced to encourage woodland regeneration. Note the patchy distribution of establishing trees. Alders, *Alnus glutinosa*, are restricted to areas beside streams and downy birch to those close to the seed source of mature trees. Only the bird-dispersed, berry-producing rowan, *Sorbus aucuparia*, is at all widespread in more open areas (Creagh Meagaidh, Inverness-shire, Scotland).

(tolerant of shade). In practice, these terms represent the extremes of behaviour, with tree species exhibiting a range of responses to different light intensities.

*Thicket, or stem-exclusion, phase*

Following stand initiation or regeneration, the canopies of the more vigorous saplings and shrubs eventually coalesce and shade out weaker shrubs and saplings (Figures 7.1a and 7.1b). This closing of the canopy creates a simple, uniform structure, reduces levels of light reaching the ground, and creates relatively uniform levels of light and humidity at ground level. This has a profound effect on the field layer, invertebrate fauna, and avifauna.

During the thicket or stem-exclusion phase seed is produced by the maturing trees, but any seedlings that do establish have little opportunity to develop into saplings because of the low light levels beneath the closed canopy (Figure 7.1c).

*Understorey (re)-initiation*

Eventually the canopy starts to open up, as trees fall, enabling a new cohort of saplings to grow up and (re)-initiate an understorey beneath it (Figure 7.1d). There is still, though, relatively little light beneath the canopy. These saplings will therefore generally be of tolerant tree species. The development of this understorey starts to increase structural complexity.

*Canopy break-up*

As the stand continues to age, more gaps are created in the canopy as trees fall over (Figure 7.1e). The regeneration that takes place within these gaps will depend largely on the size of the gap, the extent of any re-growth from fallen trees, and whether saplings of tolerant tree species are already present in the gap. Any grazing and browsing also influences regeneration. If the gap is very small, then it will probably be filled by expansion of the crowns of surrounding canopy trees. If the gap is larger, then it will eventually be filled by newly establishing trees, unless filled by re-growth from fallen ones or if browsing prevents re-growth. Tolerant tree species are likely to compete best and eventually form new canopy trees in smaller and more shady gaps, such as those created by a single falling tree. Intolerant tree species, and other light-demanding forbs and grasses, are only likely to establish in larger, more open and sunny gaps formed, for example, by large-scale windthrow or widespread death of trees caused by insect herbivory.

Canopy break-up greatly increases structural complexity by increasing small-scale variation in age structure. If the succession is allowed to continue unhindered, then trees will age, eventually achieving the characteristics of old-growth (Section 7.2.1).

### 7.1.4 Cyclical succession of woodland, scrub, and grassland mediated through grazing (from Olff *et al.* 1999; Vera 2000)

Starting with open grassland (Figure 7.3a), the proponents of this controversial theory suggest that grazing and browsing by high densities of large herbivores prevent the establishment of trees and shrubs. During periods when densities of herbivores are low, for example following disease of severe winters, thorny shrubs establish, particularly within the protection of other unpalatable forbs, such as thistles, (Figure 7.3b). These unpalatable forbs are themselves encouraged by soil disturbance by animals and deposition of dung. As the thorny scrub grows, it provides protection from grazing and browsing to palatable tree species that establish within it (Figure 7.3c). Colonization by woody species is aided by

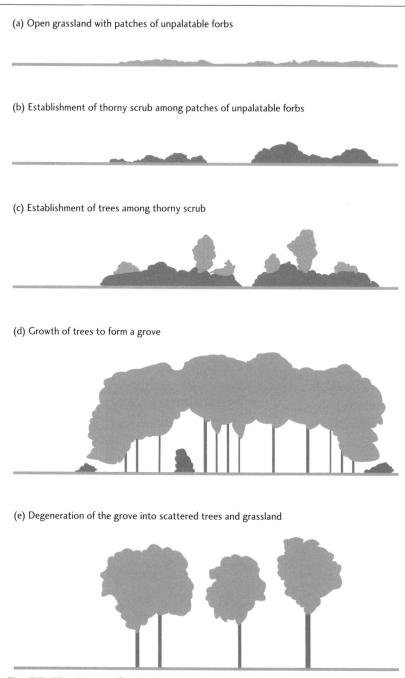

**Fig. 7.3** The theory of cyclical succession of woodland, scrub, and grassland mediated through grazing (from Olff *et al.* 1999; Vera 2000). Tree foliage is shown in light grey and scrub foliage in dark grey. See Section 7.1.4 for details.

spreading of seeds by birds, for example by caching of acorns by Eurasian jays, *Garrulus glandarius*. As these trees grow, they eventually shade out this scrub to produce a woodland grove (Figure 7.3d). Even if gaps appear in these groves, regeneration of trees is prevented by grazing and browsing by large herbivores. Exceptions may occur where seedlings are sheltered from large herbivores by the branches of fallen trees, allowing regeneration to take place. Otherwise, though, the degeneration of the canopy results in an open parkland landscape with scattered individual and groups of trees interspersed with grassland (Figure 7.3e). The process is re-started when thorny scrub re-establishes in the grassland.

The cultural landscape thought to most closely mimic this cyclical succession is wood-pasture (Figure 4.5 and Section 7.4.4).

### 7.1.5 Changes in biodiversity in relation to growth stage and structure

The age and structure of stands profound affect their fauna. In general, the richness of the flora and fauna of woodlands and forests increases with their structural complexity—there are more niches for different species to occupy in a more structurally complex habitat. Several groups show fairly consistent patterns of change. Changes in bird assemblages have been particularly well studied.

Species richness of breeding and wintering birds usually increases with age of stand and measures of structural and vegetation diversity (e.g. Buffington *et al*. 1997; Donald *et al*. 1997, 1998; Manuwal and Huff 1997; Laiolo *et al*. 2004). The proportion of tropical migrants also changes with stand age, but follows different patterns in Europe and eastern North America. In Europe the proportion of tropical migrants (mainly Old World warblers Sylviidae) is highest in early successional forest and scrub, particularly in vegetation 1–4 m high, with this proportion declining during the thicket/stem-exclusion phase but then increasing again in more mature forest. In eastern North America the proportion of tropical migrants increases with vegetation height, and is greatest in vegetation taller than 10 m (Mönkkönen and Helle 1989; Donald *et al*. 1998). The proportion of cavity-nesting birds also tends to increase with age of stand, in response to the increase in numbers of suitable large and decaying trees. Large stands of old-growth typically support a range of species associated with ancient large trees and decaying wood that are rare in or absent from younger woodlands. Many bird species are associated with scrubby, edge habitat, while forest-interior species require large blocks of old-growth or other suitable high forest.

### 7.1.6 Quantities and types of dead wood

Dead wood is particularly valuable and supports an exceptionally diverse assemblage of invertebrates, particularly beetles and flies, as well as fungi, mosses, liverworts, and lichens. It is the process of decay that provides the conditions required by this diverse range of species.

Larger-diameter dead wood on the ground is often termed coarse woody debris (CWD). There is no strict definition of the minimum diameter of CWD, with definitions usually varying between greater than 10 cm and greater than 15 cm diameter. Invertebrates that are dependent on dead wood during at least part of their life cycle, wood-inhabiting fungi and other species associated with this habitat, are termed saproxylic. Saproxylic invertebrates are also commonly referred to as dead-wood invertebrates.

Dead wood is important for woodpeckers and other birds that feed on invertebrates in it. Snags (standing dead trees) and other standing dead wood provide habitat for cavity-nesting birds. The cavities excavated by woodpeckers provide nest sites for other birds and small mammals. Holes in dead trees provide roost sites for bats.

There is a wide range of forms of dead wood and associated microhabitats, each supporting its own characteristic invertebrate fauna (Hilszczanski *et al.* 2005). These include:

- heart rot—decay that occurs primarily in the heartwood of living trees,
- dead wood on living trees,
- birds' nest cavities,
- fungus-infected bark,
- fine dead branches and twigs,
- fallen dead wood,
- rot holes,
- bracket fungi, particularly long-lived fungal fruiting bodies,
- sap runs,
- snags,
- stumps,
- burnt wood,
- wet fallen wood,
- roots.

Invertebrate species also differ in their preference for dead wood in full sunlight, semi-shade, or complete shade (e.g. Jonsell *et al.* 1998; Lindhe *et al.* 2005).

Trees can rot in two ways: from the outside (sapwood decay) or the inside (heart-rot decay). Heart-rot decay (rotting from the inside of the tree) is especially valuable. It supports a more specialized invertebrate fauna than sapwood and is used by woodpeckers to excavate nest holes. In general, large-diameter wood is most valuable and tends to be less common than smaller-diameter wood.

Different types of dead wood also support different fungal assemblages. Factors explaining variation in fungal diversity include the species of tree and the volume, diameter, age and stage of decomposition of the dead wood, and its degree of contact with the soil (Heilmann-Clausen and Christensen 2003; Norden *et al.* 2004; Heilmann-Clausen *et al.* 2005; Kuffer and Senn-Irlet 2005). Species composition of dead-wood-inhabiting mosses and liverworts appears to be less influenced by tree species, and more by the suitability of the surrounding microclimate. Many species of dead-wood-inhabiting mosses and lichens require relatively stable, humid conditions (Heilmann-Clausen *et al.* 2005).

Dead wood is of such high value for wildlife that there can be no justification for removing it from areas managed for nature conservation. All dead wood should be left to undergo its natural decay process, unless it poses unacceptable safety risks. Where it does endanger life, then the first option to consider is re-routing public access so that it avoids the dangerous tree. If this is not possible, then removal of tree limbs by competent trees surgeons will be necessary. Any tree surgery should remove the minimum quantity of wood necessary to make the tree safe, and should minimize further cutting of the removed branches. Removed timber should, wherever possible, be left where it falls. It is surprising, though, how often timber removed from nature reserves is cut into short lengths and neatly stacked. Worse still is leaving it on site so that it becomes colonized by the larvae of saproxylic invertebrates, and then removing it. If it is necessary to remove the timber, for example if it falls across a track, then placing it in dappled shade will probably provide the widest range of conditions for saproxylic invertebrates (Kirby 1992b; Lindhe *et al.* 2005). Always think of the best way to mimic natural processes.

In commercially managed woodlands the quantity of dead wood retained will be a compromise between the needs of timber production and conservation.

## 7.1.7 Presence of gaps, glades, forest edge, and associated habitats

A large proportion of the species found in large blocks of forest and woodland are associated with patches of non-forest habitat, particularly grasslands and wetlands, and the edges of wooded and open ground. Many insect species require a mixture of woodland and open areas.

## 182 | Forests, woodlands, and scrub

### 7.1.8 Variation in soils, topography, and drainage

Variation in soils, topography and drainage will increase the variety of ground conditions within the woodland and thereby increase the range of plants and animals, especially invertebrates, within it. Small-scale variations in topography are caused by falling over of shallow-rooted trees (Figure 7.4).

### 7.2 Types of forest and woodland

A number of terms are used to describe forests and woodlands in terms of their age and history. Virgin or primary are used to describe forest and woodland

**Fig. 7.4** Pits and mounds. Exposed root plates, such as on this fallen Norway spruce, *Picea abies*, create valuable diversity in soil conditions and topography. The root plate provides nest sites for birds and eventually breaks down to form a mound. The pit formed by pulling up of the root plate exposes the mineral soil beneath the litter, increasing the diversity of soil types and conditions for plant growth. Pits that fill with water form valuable temporary and permanent ponds.

Pits and mounds created by fallen trees are characteristic of older-growth forests. Root plates should always be left intact to provide this diversity of conditions (Yremossen, Västergötland, Sweden).

undisturbed, or virtually undisturbed, by human activity. They can be applied to areas of both very young, or very ancient, forest or woodland. Old-growth is a useful term for old forest (more than about 150 years old) containing large trees that have been subject to minimal human influence. Ancient woodland is a specific term used to define woodlands that have had a continuous cover of native trees since at least AD 1600 in England and Wales and AD 1750 in Scotland. We will use the term recent secondary woodland and forest for areas of trees that have established recently on un-forested land and have not yet attained the characteristics of old-growth. Types of woodlands and forest are also defined in terms of their management, as will be described below.

## 7.2.1 Old-growth

Most old-growth is coniferous and confined to high altitudes and high latitudes. There are few remaining areas of old-growth broad-leaved forest. They are particularly valuable as reference points against which to compare the features of managed high and secondary forests. In particular, old-growth contains:

- the majority, if not all, of the forest tree and other plant species native to that given area, and in more similar proportions to those that would occur in a present-natural (Section 2.3) situation than in managed forests and woodlands;
- a greater number of large, mature, and over-mature trees compared to managed forests, and far higher abundance of standing and fallen dead wood (Figure 7.5);
- relatively complex vertical structure and variation in horizontal spatial structure due to a long history of tree falls and other canopy loss at different times in different areas of the forest, resulting in different stages of re-growth; in many cases, though, they may contain large areas of relatively uniform structured re-growth following large-scale disturbance;
- natural variation in soils, topography, and drainage; a long history of large tree falls will have helped to increase the smaller-scale variation in topography through the pulling up of root plates of shallow-rooted trees.

Old-growth often supports a range of species that are rare or absent in more recently and more intensively managed forest. These include birds that require tree cavities or large-crowned trees for nesting and mature trees, snags, dead limbs, and branches for feeding (e.g. Newton 1998; Imbeau *et al.* 2000; Poulsen 2002). Old-growth is especially important for its fungi, lichens, mosses, liverworts, and dead-wood invertebrates. The quantity of dead wood

**Fig. 7.5** Old-growth. An often striking feature when entering old-growth forest is the massive quantity of standing and fallen dead wood and the huge size of the trees.

This mixed, old-growth forest in Białowieża National Park, Podlaskie, Poland, contains many enormous, relatively straight-trunked deciduous trees reaching 35–42 m in height. These are quite different in size and form to trees in managed forests and in parkland (Fig. 7.12). The bottom photograph illustrates the large quantities of fallen trunks and branches, here outlined by snow.

varies enormously between different types of old-growth, for example between 20–60 m³/ha in Scot's pine- and Norway spruce-dominated boreal forests near to the timberline in Lapland (Sippola *et al.* 1998) and 476–1189 m³/ha in Douglas fir–hemlock, *Pseudotsuga–Tsuga*, forest in the north-west USA (Harmon *et al.* 1986).

### 7.2.2 Recent secondary woodland and scrub

Scrub and secondary woodland establish following abandonment, or a reduction in intensity, of management of agricultural land, grassland, and dwarf-shrub habitats, or as the result of succession in wetlands. Scrub is also an important component of wetlands, grasslands, and dwarf-shrub habitats.

More recent secondary woodlands will in most cases have been subject to little if any traditional management. They will thus lack the attributes of cultural habitat, but neither have attained the valuable attributes of old-growth. As they are only recently established, many will still be in the thicket/stem-exclusion or understorey-initiation phase and hence be relatively structurally uniform and uninteresting for biodiversity. Their trees and shrubs will usually comprise mainly shade-intolerant, vigorous-growing, and widely dispersive species. The lack of structural and tree-species diversity means they usually support only an impoverished woodland flora and fauna, in many cases exacerbated by isolation from potential colonists.

### 7.2.3 Managed high forest

These are forests containing relatively straight, single-stemmed trees that are, or have been, managed primarily to provide straight, unblemished timber. They are usually made up of only a limited range of favoured tree species, often alien/exotic ones. Managed high forest will have little variation in vertical and horizontal structure and usually lacks a well-developed understorey and field layer. The extent of this impoverishment, though, depends on the type of forestry management (see Section 7.4.7). Managed high forest can be derived from old-growth, secondary woodland established on un-forested land, or have been specifically planted for timber production.

### 7.2.4 Coppice

Coppicing involves cutting broad-leaved trees close to the ground to produce harvestable re-growth of straight poles from their stumps (stools). Coppicing was formerly widespread in much of Europe, where it was undertaken in areas of long-established woodland, in many cases representing a continuation of forest cover from original primary forest. Most areas of **coppice woodland** contain scattered, more mature trees, known as standards. These are allowed to grow tall and are harvested periodically for timber. As in other woodland-management systems, tree-species composition has usually been highly modified through selection for species that produce the best coppice products and timber from standards. The structure of coppice-with-standards woodland is most similar to that of dense scrub containing scattered, taller trees. An important feature of coppice woodlands is the presence of open rides, along which timber is extracted.

Most **short-rotation coppice** has been developed relatively recently on former arable land in Europe and North America to provide wood for power-generation, so-called bioenergy. Short-rotation coppice also includes osier, *Salix viminalis*, beds grown in wet areas to produce flexible willow stems for basket-making and other uses. Short-rotation coppice managed for production of bioenergy comprises single-species stands of fast-growing willow cultivars, (usually based on osier), or poplar cultivars, *Populus* spp. Management of short-rotation coppice for bioenergy differs from that of coppice woodland in its extremely short coppice rotation, use of fertilizer to increase coppice growth, control of so-called weeds to reduce competition with the coppice, and lack of standard trees.

### 7.2.5 Managed open woodland systems

There is a range of open woodland-management systems that have been used for combinations of grazing, cultivation, and harvesting of tree products such as cork, olives, fruit, timber, and wood for charcoal production.

**Wood-pasture** (also known as pasture-woodland) is open woodland used for grazing and browsing by cattle, sheep, horses, ponies and deer, pasturing of pigs in autumn to feed on acorns and beechnuts, and pollarding or shredding of trees. Pollarding involves cutting branches of broad-leaved trees to a height above that of livestock and deer, to prevent the re-growth from being browsed. The branches removed are used for timber and fuel and their leaves for forage. Pollarding has also traditionally been carried out on trees to mark boundaries and on willows in open habitats. Shredding involves cutting the side branches of the tree to leave one main branch (the leader) to form a tall pole.

**Wood-meadow** is an ancient and now rare system in Europe. It is similar to wood-pasture in consisting of usually widely spaced, pollarded, or shredded trees. It differs in containing scattered, usually coppiced bushes and small trees, and because the grassland beneath and between the fields was cut for hay and aftermath grazed. Branchwood was also burnt and the ashes used to fertilize the grassland.

Other systems include **orchards, olive groves** (Figure 7.6), and the **wooded dehesas** (Spanish)/**montados** (Portugese) of the western Mediterranean (Figure 1.3).

**Fig. 7.6** Olive groves. Ancient olive, *Olea europaea*, groves and other low-intensity silvicultural systems can be of high cultural and aesthetic value as well as providing good wildlife habitat.

The olive groves of Puglia in southern Italy contain many beautiful, ancient olive trees. The periodic cultivation used to prevent establishment of woody plants maintains a rich ruderal plant fauna, here supporting numerous field marigolds, *Calendula arvensis*, and star of Bethlehem, *Ornithogalum umbellatum* (Madonna Incoronata organic olive farm, Mattinata, Puglia, Italy).

## 7.3 Managing scrub

Scrub can be made up of a diverse range of shrub species, particularly on base-rich soils. It is especially valuable where it:

- is open and structurally diverse;
- has plenty of edge;
- forms transitions from open habitat to scrub and through to woodland;
- is scattered within open habitats.

These types of scrub can all support high densities of breeding birds, although usually comprising mainly relatively widespread species. Mosaics of habitats provided by scattered scrub, glades, and edges provide a wide range of vegetation types and microclimates for invertebrates. Scrub can protect grassland plants from grazing, benefiting grazing-intolerant species, especially some bulky forbs, and providing areas of taller vegetation, litter, nectar, and seed sources that are scarce in more heavily grazed habitat. Scattered scrub in open habitat can provide nest sites, foraging, song, and look-out posts for birds that otherwise exploit open grassland and dwarf-shrub habitats, including perch-hunters such as shrikes Laniidae and raptors. Scattered scrub may also have a negative effect by deterring species that require extensive areas of open habitat, such as many open-ground nesting birds.

As described in Section 7.1.2, a marked change takes place in the structure, flora, and fauna of scrub following coalescence of the canopies of individual shrubs. Canopy closure reduces levels of light reaching the ground, restricting the field layer to a small number of shade-tolerant plants. It is also associated with a marked reduction in bird species richness and loss of invertebrates associated with open, sunny conditions. Few, if any, bird or invertebrate species prefer large blocks of uniformly aged, closed-canopy scrub. Closed-canopy scrub can be surprisingly stable and long-lived, especially where there is no seed source of tolerant tree species to grow up beneath it.

Where scrub occurs in association with open habitats, the primary decision will be its desired area, distribution, and age structure. This will depend on:

- the conservation value of the scrub compared to that of the open habitats it is replacing;
- the extent to which scattered scrub enhances or decreases the value of these open habitats.

Options for management include:

- removing the scrub;
- diversifying the structure of uniform, closed-canopy scrub;
- increasing the extent of scrub.

In practice, patchy removal of scrub can be used to also diversify its structure. The other consideration when scrub is out-competing more valuable, open vegetation, is whether the management of these other vegetation types can be altered to prevent the scrub from establishing. For example, grazing pressure can be lowered to reduce the availability of germination gaps for scrub seedlings, or livestock such as primitive sheep introduced that are effective at controlling young scrub (Section 4.4.2).

Scrub and associated open habitats can also be important components of otherwise dense woodlands and forests. Scrub and open edge are often in limited supply in many types of woodland. Where scrub occurs within otherwise closed-canopy forest, the main options will be to:

- maintain it as scrub, by preventing its succession to woodland;
- allow the scrub to eventually develop into woodland through non-intervention.

## Managing scrub | 189

If the woodland or forest lacks scrubby and open habitats, then it may be a priority to increase the extent of scrub (Figure 7.7).

### 7.3.1 Scrub removal

In some cases the aim may be to remove the entire area of scrub. If it is only intended to remove a proportion of it, it is usually best to remove more recently developed scrub, since this will usually be easiest to revert back to the habitat it has encroached on. Older scrub will have accumulated a deeper litter layer, making it more difficult to revert to nutrient-poor, species-rich grassland or

**Fig. 7.7** Soft and hard edges. Many patches of woodland, especially in lowland areas, are surrounded by often intensively managed cropland or grassland of little value to woodland-edge species. Even if the woodland is bounded by semi-natural habitat, then this boundary is often hard and abrupt, and lacks a soft, gradual transition from grassland to scrub and woodland required by many species ((a) Goor-Asbroek, Antwerpen, Flanders, Belgium).

It can be tempting to expand woodlands by planting gaps and adjacent areas with trees. In many cases, though, there will be greater conservation benefits in providing grassy areas and soft, scrubby edges to the woodland instead.

(b) shows a former hop, *Humulus lupulus*, field, which has been left to revert naturally for about 10 years. The mixture of flowery grassland, low brambles, *Rubus* spp., and silver birch scrub that has developed provides superb woodland-edge habitat. These open areas also provide summer nectar sources for insects, which are in short supply in the surrounding, dense woodland (Tudeley Woods, Kent, England).

dwarf-shrub heath. Older scrub is also likely to have developed on land generally more conducive to scrub establishment in the first place. Leguminous scrub fixes nitrogen, which can favour the growth of more competitive plants following its removal. If the scrub is of very low conservation value, though, and removed using large-scale machinery such as a forestry mulcher, then it may be more efficient to remove large blocks in one go.

The main method for removing scrub is by cutting it close to ground level and treating the cut stumps or re-growth with herbicide. Herbicide can be poured into holes made in cut stumps or by spraying the re-growing foliage. Livestock and deer can prevent re-growth of more palatable scrub.

Removing scrub from the margins of large blocks provides opportunities to diversify its edge structure (Figure 7.8). This can be done by removing more widespread and uninteresting shrub species, while retaining those of greater intrinsic value and which provide more valuable habitat for other species, such as valuable food plants and sources of nectar.

**Fig. 7.8** Sensitive scrub management. Scrub is often removed to prevent it from out-competing more species-rich herbaceous vegetation. However, scattered and structurally diverse scrub can be a valuable habitat in its own right, and its presence can enhance the value of these other habitats.

Small patches of scrub have been removed sensitively from the front and centre-left of this photograph to prevent undue encroachment onto the grassland, while still retaining a variety of different types and age of scrub elsewhere (Thompson Common, Norfolk, England).

### 7.3.2 Diversifying the structure of closed-canopy scrub and preventing scrub from succeeding to woodland

The structure of closed-canopy scrub can be diversified by cutting small patches to ground level to produce areas of younger re-growth. Cut scrub may need to be fenced to prevent browsing by livestock and deer.

Periodically cutting and removing scrub prevents it from succeeding to woodland. Alternatively, relatively stable scrub can be created within woodland by cutting down young trees and treating their stumps with herbicide, to allow shrubs and creepers to remain dominant. This technique has been used to prevent trees from touching power-lines running through forests and has benefited a number of scrub bird species (Askins 1994). In the USA birds nesting in these corridors of scrub through woodland have been found to suffer lower rates of brown-headed cowbird brood parasitism and nest predation than those breeding in scrub close to open habitats (Yahner 1995).

### 7.3.3 Increasing the extent of scrub

Trees and shrubs establish best when there has been a period of heavy grazing or other disturbance that has created suitable germination gaps for seedlings to establish in, followed by relaxation of grazing of cessation of other disturbance that then allows these seedlings to grow. Alternatively, scrub can be planted and protected from grazing by wild grazing animals and livestock using tree tubes.

## 7.4 Managing woodlands and forests for conservation

The range of options available for managing woodlands and forests for conservation depends on the type of woodland or forest, its history, and current state.

Old-growth does not usually require management to maintain its conservation value. The only exceptions may be removal of alien/exotic plants and the re-introduction and management of previously exterminated large herbivores and their predators to facilitate more natural forest dynamics. Because of the rarity and immensely high conservation value of old-growth, any further intervention will be damaging.

The various forms of managed high forest can be regarded as impoverished forms of old-growth. Stands of trees in managed high forest usually lack the following valuable attributes found in old-growth:

- a diversity of tree species;
- structural diversity;
- a well-developed field layer;

- large, mature trees with thick stems and branches;
- standing and fallen dead wood;
- pits and mounds on the forest floor caused by windblown trees;
- variation in hydrology;
- surrounding open habitats of high conservation value.

Quantities of dead wood in high forest managed by clear-felling or types of selective felling are typically between only 3 and 30% of those in comparable unmanaged old-growth (Kirby *et al.* 1991; Christensen *et al.* 2005; Gibb *et al.* 2005; Marage and Lemperiere 2005).

The overall structure of managed high forest can be more diverse, because individual stands are cut for timber at different times, creating a range of different age classes and structures within the forest as a whole.

The main priorities when managing forests impoverished by past management for timber production will be to increase their structural diversity, the quality of their edge habitat and increase quantities of dead wood. Lack of management is an issue in the eastern USA, where there is a lack of early successional, scrubby habitat in many woodlands.

Recent secondary woodland also lacks structural variation, gaps, old trees, and dead wood, in this case because these features have not yet had time to develop. The main management priority in these recent secondary woodlands will be to enhance these features where possible.

Woodlands managed as coppice, wood-pasture, and open woodlands, such as wooded dehesas/montados, orchards, and olive groves, lack most of the features of old-growth, but instead comprise distinctive and highly valued cultural habitats that support characteristic assemblages of species, many of high conservation value. The main conservation objective for these cultural habitats will be to maintain their existing interest through continuation or re-instatement of their former management. The exception to this involves the management of abandoned coppice woodlands that have lost their assemblage of open coppice species. Here, an option will be to allow the former coppice to develop into high forest. Most short-rotation coppice is relatively recent and the main conservation aim is usually to maximize its potential for breeding birds.

**Grazing and browsing** by domestic and wild herbivores has important effects on tree and shrub regeneration, the field layer and structure of the understorey in a range of types of forest and woodlands. Fire-prone types of woodland maintained can be managed by controlled **burning** to maintain their characteristic species-composition and structure, and to decrease fuel loads to reduce the risk of larger, more catastrophic wildfires. In all types of forest and woodland there will be a presumption to **remove alien/exotic plant species**.

Removal of the litter, humic and grassy layers in woodlands, usually referred to as **sod cutting/litter stripping** has been used on a small scale to improve conditions for germination and growth of seedlings, reduce nutrient levels, raise soil pH, and increase species richness of ectomycorrhizal fungi (Devries *et al.* 1995; Baar and Kuyper 1998; Dzwonko and Gawroński 2002). This management mimics the effects of litter removal that formerly took place as part of traditional woodland management in some parts of Europe. It is thought to counteract the effects of nutrient enrichment and increased litter production caused by inputs of anthropogenic atmospheric nitrogen deposition. However, this technique has not been applied widely in the conservation management of woodlands and forests and is not discussed further.

Regeneration of trees is necessary for woodlands and forests to persist, and the extent of regeneration will be a consideration in all types of woodland management. However, because of the long lifespan of most tree species, significant regeneration usually only needs to take place at very infrequent intervals to replace trees lost. Patchy regeneration will maximize the variation in growth stage and structure.

## 7.4.1 Enhancing formerly managed high forest and recent secondary woodland

The main techniques for enhancing these types of forest and woodland for wildlife are by:

- thinning to increase structural diversity and creating and maintaining gaps, glades and rides;
- providing dead wood;
- restoring natural variation in hydrology by blocking any artificial drainage.

Grazing and browsing by large wild and domesticated herbivores can also affect the structure and composition of the understorey and, by affecting tree regeneration, also influence long-term tree-species composition and structure. The effects of grazing and browsing are discussed in Section 7.4.5. Improving the value of dull forests and woodland and wildlife may also involve felling of undesirable alien/exotic trees and shrubs. In some types of woodland and forest burning can be used in combination with these techniques to provide the desired structure and tree species-composition (Section 7.4.6).

### Thinning and creating gaps, glades, and rides

Structural complexity within a structurally uniform woodland will increase without intervention as trees out-compete one another (self-thinning), and individual and groups of trees fall over and create gaps in the canopy. However, these natural processes can take a long time. The rate of increase in structural diversity can be accelerated by felling individual, or groups of, trees or by pulling trees over using a winch. Thinning can be carried out selectively to modify tree-species composition.

Pulling trees over using a winch not only creates gaps but also provides additional benefits by exposing the root plate (Figure 7.4). Felling or winching single or small groups of trees is only safe to carry out in relatively open woodlands, or along existing edges, where the felled or winched trees will fall to the ground, rather than lean un-fallen against other trees. Felled areas can either be left to regenerate into scrub or woodland, or if larger, maintained as more open habitat by subsequent grazing or cutting. Care is needed to avoid prejudicing the long-term future of the woodland for short-term gain. Medium-aged trees are potential ancient/veteran trees of the future.

The ideal when creating gaps in forest for regeneration of trees and shrubs is to leave the fallen trees *in situ*, thus providing a source of dead wood and cover for small birds and mammals, and more closely mimicking the process of natural windthrow. Gaps produced by felling groups of trees will provide suitable conditions for a range of clearing specialists, particularly insects that feed on flowers, the foliage of young saplings, fallen tree crowns, and the regenerating understorey. Retaining the dead wood will provide suitable conditions for a different range of dead-wood invertebrates to those in more closed canopy forest, including warmth-loving species and those whose adults visit flowers, such as many longhorn Cerambycidae and jewel beetles Buprestidae (Bouget and Duelli 2004; Bouget 2005). Tangles of branches from fallen trees can protect establishing seedlings and saplings from grazing animals. Dead trunks and branches can themselves provide important micro-habitats on which tree seedlings can then establish.

In practice, the cost of creating gaps in the canopy is usually met by the sale of the timber removed. A compromise is to cover the costs of the management by sale of a proportion of the timber, but to leave the remainder where it falls.

Most creation and enhancement of open areas within woodlands takes place along existing rides, where it is logistically easier to undertake. Much of this work has been focused on widening existing rides and creating new rides and open glades to provide open, sunny conditions, primarily to benefit woodland-edge butterflies and other woodland-edge invertebrates (Warren 1985; Greatorex-Davies *et al.* 1992; Figure 7.9). This enhancement may also involve creating soft edges around the margins of existing woodland or forest (Section 7.3).

## Managing for conservation | 195

**Fig. 7.9** Ride-widening and box junctions. Usually the most efficient way to increase woodland-edge habitat is to widen existing narrow rides and create glades at their junctions, so-called box-junctions. The connectivity of rides also facilitates dispersal of woodland-edge species.

(a) Narrow rides

(b) Widened rides and box-junction

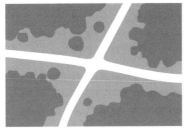

Trees | Herbaceous vegetation

The photograph shows a sunny ride that has been widened to benefit butterflies and other insects. It contains a range of types and structures of herbaceous vegetation across and along it, containing a variety of food plants and nectar sources, together with scattered scrub. The edges have been scalloped to further increase the length of edge and to provide sunny bays sheltered from the wind (Oakens Wood, Surrey, England).

The width of ride necessary to maintain sunny conditions for warmth-loving invertebrates will depend on the ride's orientation and the height of trees on its sunward side. Wide (typically about 30–40 m in high forest) east–west running rides are considered best, since they receive sunlight from the morning through to evening. North–south running rides are in partial or whole shade in the early morning when insects need warming up. The presence of flowering forbs and shrubs along rides benefits insects whose larvae utilize the woodland interior, but whose adults require nectar sources which can otherwise be rare in many areas of high forest and secondary woodland.

Rides need continual management to prevent them from becoming colonized by scrub or woodland, in the absence of high densities of deer or other herbivores. Grazing by domestic livestock is usually impractical, so open conditions are normally maintained by periodic cutting. Cutting regimes should aim to provide a variety of types and structures of grassland, dwarf-shrub vegetation, and scrub along rides to maximize the variety of food plants, nectar sources, variation in sward structure, and shelter for invertebrates. As with all such management, managing short stretches on rotation is best to help maintain a continuity of suitable conditions in close proximity to one another.

### Providing dead wood

All dead wood should be maintained through non-intervention in woodlands and forests managed for conservation, unless it poses unacceptable safety risks. It is also possible to increase the quantity of dead wood, although this is rarely done in practice.

There is no simple answer to how much dead wood to create. In old-growth not subject to recent catastrophic disturbance by wind or crown-fire, approximately 10% of all standing trunks are dead. This figure is remarkably consistent across a range of forest types in Europe and North America. The percentage of dead trunks in old-growth tends to be highest among the largest size classes (greater than 50 cm diameter at breast height; Nilsson *et al.* 2003), because these trees remain standing for longer. However, it is important to recognize that the quantities of dead wood present at any one time are the product of ongoing processes. Large quantities of dead wood can be due to a small number of large, dead trees that have stood for many years. Targets for creating dead wood should be more modest and the long-term impacts on stand dynamics should be considered. It is also unclear whether artificially created dead wood functions as well as natural dead wood. Some of the key fungi involved in decay need to establish in the dying tree.

There are two approaches to creating dead wood. The first is to inflict damage on parts of trees to initiate the decay process and thus provide a continuity of dead wood in a variety of stages of decay. This method is primarily aimed at benefiting dead wood invertebrates. Fungal infection and decay can be induced by drilling holes in parts of trees to allow water to enter, particularly in axils between large branches. In practice, this method can be logistically difficult and dangerous and is rarely undertaken. Damage inflicted on trees during thinning and gap, glade, and ride creation can also provide a route of entry for fungi. Any damage created during such management should be left, rather than cleaned up to prevent fungal attack.

The second approach is to kill whole trees to produce snags, or to kill large branches. Snag creation has generally been used to provide habitat for cavity-nesting birds. The following methods can be used:

- killing parts of or whole trees by girdling (making a continuous cut around the trunk);
- injecting with herbicide;
- blowing off the crown of conifers using dynamite.

These methods can also be accompanied by inoculation of fungi to speed up decay. These techniques should only be used on younger trees, preferably healthy ones, and not on older or partly dead trees that are already of high value. It is preferable to damage only part of the tree, since whole dead trees fall over more quickly and thereby lose their value to cavity-nesting birds. However, narrow-diameter snags will be unsuitable for species that require larger cavities. Summers (2004) suggests that snags created from Scots pine trees in Scotland should ideally be at least 40 cm diameter at breast height to provide suitable cavity nest sites for birds and habitat for other wildlife. Brandeis *et al.* (2002) found no difference in decay characteristics and woodpecker activity in Douglas firs, *Pseudotsuga menziesii*, killed by girdling, herbicide injection, and cutting off of the base of the live crown with or without inoculation of fungi.

### 7.4.2 Managing coppice woodland

Coppice woodlands are important cultural habitats in Europe and, where subject to a long history of coppice management, usually support a characteristic flora and fauna whose persistence is largely dependent on continuation of this form of management. In particular, coppice woodlands can be important for their:

- spring-flowering, woodland forbs;
- invertebrates associated with the early stages of coppice re-growth and rides;
- songbirds found in the earlier stages of coppice re-growth.

In general, managed coppice contains little dead wood, and is therefore unlikely to support specialized saproxylic species. For example, Kirby (1992a) found quantities of dead wood in 0–30-year-old mixed coppice of between 1 and 7 m$^3$/ha. The repeated cutting damage to old coppice stools can, though, sometimes create a reasonable quantity of dead wood in old stools, especially if the stumps have been cut relatively high.

The main considerations when managing coppice woodland are the:

- length of the coppice rotation;
- size of individual areas cut, known as 'coupes';
- spatial arrangement of coppice coupes and their linkage together by rides;
- level of grazing and browsing by deer.

Soil type will influence the woodland flora and its response to coppicing.

### Length of the coppice rotation

Conditions for wildlife change rapidly during the coppice cycle (Figure 7.10). Harvesting of coppice re-growth suddenly enables light to reach the ground. The coppice stools re-grow quickly, in a similar way to establishing, or re-establishing, scrub during the stand-initiation or regeneration phase of woodlands (Section 7.1.3). This is usually referred to as the establishment phase. The re-growth from adjacent stools then coalesces to form a closed-canopy, this period being known as canopy closure. The time until canopy-closure depends on the rate of coppice re-growth and density of stools, but is typically 4–10 years. The prevention of light reaching the ground marks a significant change in conditions for most groups. This process is akin to the thicket or stem-exclusion phase of establishing woodlands. The canopy then remains closed, this period being known as the maturation phase, and is then re-harvested.

The field layer of coppice consists of two main groups of plants:

- shade-tolerant, woodland forbs;
- shade-avoiding annuals, biennials, and short-lived perennials with a long-lived seedbank.

The flora of coppice is usually distinct from that of even sensitively managed, selection, or group-selection high-forest (Section 7.4.7), in supporting a greater proportion of spring-flowering, shade-tolerant, perennial forbs, especially species that propagate from bulbs, tubers, and corms (e.g. Decocq *et al.* 2004).

The persistence of the characteristic coppice field layer is dependent on alternate periods of open conditions following harvesting of coppice poles, and shady conditions following canopy closure. The shady conditions are necessary

**Fig. 7.10** Flowers in coppice woodland flowers. Conditions for flowers in coppice woodlands changes dramatically during the coppice cycle.

The increase in light following coppicing (b) causes prolific growth and flowering of vernal plants, such as this oxlip, *Primula elatior* (a).

As the coppice re-grows the canopy closes, allowing only more shade-tolerant plants such as ramsons, *Allium ursinum*, to persist (c).

A long period of canopy closure eventually limits growth and flowering of most of the field layer altogether (d). At this stage the coppice is usually re-cut, thus re-starting the cycle.

This woodland has been under continuous coppice management since 1252 (Bradfield Woods, Suffolk, England).

to prevent the flora becoming dominated by more competitive light-demanding species, typically grasses, similar to the flora found along continually open, sunlit rides. The light conditions following harvesting of the coppice are important in allowing woodland plants to flower in profusion and in providing suitable conditions for any characteristic open-ground coppice invertebrate fauna. Flowering of the field layer typically peaks in the second spring following harvesting.

The main conservation interest of coppice woodland for invertebrates is that of the small, but distinctive, range of species associated with the open conditions during the short period between harvesting and canopy closure, especially several species of fritillary butterfly. The larvae of these fritillaries feed on the prolific growth of woodland flowers. Both adults and larvae require the warm, light conditions provided by these open conditions. Numbers of butterflies can rapidly increase in numbers following coppicing, mirroring the growth of their food plants. However, these fritillary butterflies are rapidly lost as the canopy closes, typically after 3–4 years following cutting.

There is little information on the value of the coppice for invertebrates following canopy closure. It is assumed that there is only limited invertebrate interest associated with post-canopy-closure stages, which are too shaded to open-ground species, but which also lack the valuable features of older-growth forest for invertebrates.

The birdlife of coppice woodland is dominated by songbirds typical of scrub, and changes rapidly in relation to the height of re-growth. Total densities of songbirds tend to increase up to about the time of canopy closure and decline thereafter. However, there are differences in changes in density in relation to age of re-growth between species, and in particular between densities of migrant and resident birds. The proportion of migrants is highest in younger re-growth (e.g. Fuller and Moreton 1987; Fuller and Henderson 1992). Birds associated with later stages of forest growth are invariably scarce or absent.

Small mammals also show a succession in relation to the stage of coppice re-growth. For example, in Italy white-toothed shrews, *Crocidura* spp., and wood mice, *Apodemus sylvaticus*, are associated with young coppice re-growth, and yellow-necked mice, *A. flavicollis*, bank voles, *Clethrionomys glareolus*, and black rats with older coppice re-growth (Capizzi and Luiselli 1996).

The length of the coppice rotation in areas managed for conservation will depend on the desired proportions of different stages of coppice re-growth. Depending on the individual woodland, the length of the rotation is usually chosen to maintain a high proportion of:

- establishment-phase re-growth to increase the flowering of aesthetically pleasing woodland flowers and suitable conditions for any characteristic

open-ground invertebrate fauna, while also maintaining a suitable period of maturation phase to provide the shady conditions necessary to maintain this characteristic coppice field layer;
- establishment to canopy-closure phases to maintain high densities of songbirds, particularly of migrant species.

The length of the rotation required to maintain a given proportion of these different growth phases depends on the rate of coppice re-growth and density of coppice stools. A shorter rotation will be needed if coppice re-growth is rapid and the stools closely spaced. Rotations are usually between 10 and 25 years.

## Deer grazing and browsing

A common issue in coppice management is heavy grazing and browsing by deer. This can:

- prevent re-growth from coppice stools;
- kill regenerating coppice stools;
- reduce the abundance of desired species of woodland flowers;
- remove the dense understorey required by some birds;
- modify other aspects of habitat structure.

Browsing of recently cut stools coppice stools can be reduced to some extent by fencing coupes (Cooke and Lakhani 1996) and protecting individual stools by covering them with brash, similar to the protection afforded to saplings growing among fallen branches. However, these techniques are labour-intensive, whereas fencing is only successful if constructed to a high standard and checked regularly for damage. The only long-term solution at most sites is to reduce deer numbers.

## Size of coppice coupe

Individual coupes need to be large enough to support viable populations of the key species associated with them. For breeding birds individual coupes, or areas of adjacent, similar-aged coppice, need to be large enough to support at least one territory of the given species to maximize their benefits. Territories of most breeding songbirds in coppice woodland are larger than 0.5 ha (Fuller 1992). For less mobile open-ground invertebrates, the coupes need to be large enough to support a viable population in a given year, and to ensure that enough individuals disperse successfully to new areas before the existing coupes becomes too shady for them. However, because the flora can often vary greatly between individual coupes, there is also an argument for cutting a larger number of different small areas of coppice, to maximize the likelihood of there being a continuity of that at least some suitable habitat present. Plots cut to maintain woodland butterfly populations are usually 0.5–2.0 ha.

### Spatial arrangement and linkage of coupes

As with all rotational vegetation management, species will be more likely to re-colonize suitable areas of re-growth if they are closer to existing sources of colonists.

Management can be arranged to minimize the distance between early stages of re-growth. In coppice woodlands, there is the potential to link areas of early re-growth by sunlit rides. The adults of at least some species of fritillary butterflies use rides to disperse between newly cut coppice blocks. Suitably managed rides also provide habitat for other woodland-edge species. However, for the reasons mentioned earlier, these permanently open rides tend to become dominated by more competitive, light-demanding competitive grasses and forbs, and so not contain the high densities of woodland food plants and in some cases open ground required by the larvae of fritillary butterflies (Greatorex-Davies *et al.* 1992).

A specific type of linkage of coupes is suggested for hazel dormice, *Muscardinus avellanarius*. These prefer the middle stages of coppice re-growth and are reluctant to cross wide areas of open ground. Corridors of medium-aged re-growth can be left to facilitate movement of dormice between suitable stands, and particularly over wide, open rides.

### Density of standards

The main value of standard trees is in providing nest sites for cavity-nesting birds and high song posts, which are otherwise absent from coppice re-growth, and additional foraging habitat including dead wood. The main factors influencing the suitability of standard trees for birds are likely to be their density, age (and in particular state of decay), and to some extent type, although there is little qualitative information on the importance of these factors.

The shade cast by a high density of standard trees suppresses coppice re-growth and prevents the development of open conditions during the first few years following coppicing. Again, most of the quantified information on management is from studies of woodland fritillary butterflies. For these, a cover of standard trees of less than 25% is considered best, with characteristic coppice species generally absent from areas with a cover of exceeding 60% (Warren *et al.* 1984; Warren and Thomas 1992). As standard trees mature they increase their canopy cover, and so thinning might be needed to maintain suitably open conditions.

*Tree-species composition*

Some stands of coppice are dominated by a tree species of particularly low value for wildlife. A notable example of this is sweet-chestnut coppice, which usually has a poorly developed field layer, although it can be valuable for fungi.

Any attempts to change the composition of the coppice stools will obviously be a long-term process, and unlikely to be a high priority unless the existing coppice tree species are of exceptionally poor value for wildlife. Tree-species composition of coppice can be altered by killing existing stumps by spraying the first year's re-growth following coppicing with a herbicide and planting replacement trees. The benefits of this, though, have to be evaluated against the destruction of old coppice stools, which may support interesting fungal assemblages and a continual, albeit usually limited, resource of dead wood.

*Managing abandoned coppice*

Coppice management has ceased in many areas, because it is no longer profitable. When faced with abandoned coppice, a decision needs to be made whether to re-instate coppice management.

Coppicing can be re-instated to trees that have not been cut for at least 50 years or more, with little difference in structure of the restored coppice compared to that of continuously managed coppice. The small differences that do exist can be related to differences in canopy cover, restored coppice having a slightly higher canopy cover that result in slightly delayed coppice re-growth (Joys *et al*. 2004). When re-instating coppice management, it is worth considering establishing appropriate standard trees if none are present. The field layer of coppice abandoned has been found to vary little, if at all, during the period up to about 35 years beyond a normal coppicing cycle of 15 years, at least in situations where there is little potential competition from more shade-tolerant woodland forbs (Petersen 2002). However, the characteristic open-ground coppice invertebrate fauna is unlikely to survive within abandoned coppice. Furthermore, many of these species have limited powers of dispersal and, unless the abandoned coppice is connected to continually managed coppice, many of these species are unlikely to re-colonize naturally.

An alternative is to allow this abandoned coppice to develop into high forest. A common strategy is to continue coppicing a core area of the wood that has had a continuity of coppice management, while leaving areas of older coppice to revert to high forest. This will maximize variation in the age and structure of stands within the woodland as a whole.

Abandoned coppice will, like other structurally uniform woodland and forest, increase in structural complexity over time through non-intervention, as it passes through the thicket/stem-exclusion, understorey (re)-initiation, and canopy break-up phases described in Section 7.1.3 (Figure 7.11).

These increases in structural and plant species composition following the thicket/stem-exclusion phase will increase the abundance, species richness, and diversity of woodland birds (e.g. Laiolo *et al.* 2004). The time taken to reach the more structurally diverse canopy-break-up phase will vary between different types of woodland and the degree of natural disturbance events, particularly windfall. Changes in structure are fastest on steeper slopes, where gap creation through tree-fall is more frequent, due to the thinner soils and asymmetrical growth of trees.

A method to increase the value of abandoned coppice for birds is to thin dense, multi-stemmed, abandoned coppice to produce single-stemmed trees (singling). This probably increases shrub cover and overall densities of birds, particularly

**Fig. 7.11** Abandoned coppice. This area of former coppice has not been managed since 1944. It is developing into high forest as its trees grow large, and is developing a more patchy structure and accumulating dead wood as trees and branches fall.

The wood's transition to high forest, though, is being affected by heavy deer grazing, which is currently converting it towards wood-pasture (Lady Park Wood, on the English/Welsh border in Monmouthshire, Gloucestershire, and Herefordshire).

Old World warblers, and probably also densities of hole-nesting species as singled trees mature (Fuller and Green 1998).

### 7.4.3 Managing short-rotation coppice

The management, and wildlife interest, of short-rotation coppice for production of bioenergy differs substantially to that of coppice woodland. Short-rotation coppice grows very rapidly and is usually harvested after just 3–5 years. Management during establishment of the coppice usually involves control of ruderal plants using broad-spectrum herbicides and application of fertilizer to increase coppice growth. Rides to allow access by machinery will probably become a common feature of large-scale, short-rotation coppice plantations.

The field layer of short-rotation coppice is initially dominated by ruderal plants, which are then out-competed by widespread perennial grasses and forbs. Short-rotation willow is planted in wetter conditions than other types of coppice, and this is reflected in the composition of the field layer. Hence short-rotation coppice does not support the specialized flora and invertebrate fauna found in many areas of long-established coppice woodland. The grassy nature of the field layer in short-rotation coppice supports a small-mammal fauna more typical of agricultural land than of woodland (Christian *et al.* 1997). Any open areas and rides within the coppice are usually dominated by taller and more vigorous plant species and so have a flora and invertebrate fauna more similar to that of rank grassland. Short-rotation willow coppice can, however, provide a useful early-season nectar source for bees (Reddersen 2001).

The birdlife of short-rotation coppice differs from that of traditional coppice woodland in a number of ways, probably in part due to the intrinsic qualities of the coppice, but also undoubtedly due largely to its location within agricultural land rather than existing woodland. As well as supporting typical scrub birds, short-rotation coppice can also support the following groups of species (from Christian *et al.* 1997; Berg 2002a; Anderson *et al.* 2004; Sage *et al.* 2006):

- breeding wetland songbirds, particularly in willow coppice, such as sedge warblers, *Acrocephalus schoenobaenus*, marsh warblers, *Acrocephalus palustris*, and reed buntings, *Emberiza schoeniclus*, in north-west Europe;
- breeding and wintering open-country birds present during the establishment of new areas of coppice, including breeding Eurasian sky larks, *Alauda arvensis*, and northern lapwings in north-west Europe;
- breeding birds typical of grassy and other open habitats that use trees on the edges of plantation as song posts and open areas of failed tree growth within the coppice blocks, such as eastern kingbirds, *Tyrannus tyrannus*, clay-colored

sparrows, *Spizella pallida*, field sparrows, *Spizella pusilla* and savannah sparrows, in North America.

As in coppice woodlands, total densities of songbird species tends to increase with the age and height of the coppice, up to the time of canopy closure and harvest. Willow short-rotation coppice tends to hold a higher proportion of summer migrants, higher overall densities of birds, and higher bird species-richness than poplar short-rotation coppice (Sage and Robertson 1996).

Short-rotation coppice does not generally support types of farmland birds considered of high conservation value in Europe, or grassland species or neotropical migrants considered of high conservation value in North America. However, it does contain a higher density of birds and a more diverse avifauna compared to its main alternative land uses: agriculturally improved grassland or intensively managed arable land (Christian *et al.* 1997; Anderson *et al.* 2004; Sage *et al.* 2006). Short-rotation coppice could increase habitat diversity within polarized arable and grassland systems and benefit some farmland bird species, such as sparrows and finches, by increasing their range of foraging habitats.

So far, most of the research on the value of short-rotation coppice for wildlife has been on relatively small, trial so-called pre-commercial stands. Future, fully commercial short-rotation coppice will probably comprise much larger, mono-specific stands and be managed more intensively. Both are likely to affect their value for wildlife. Because of this, there is little information on the options for managing commercial short-rotation coppice to maximize its benefit for wildlife. In terms of birds, the group that has been best studied and for which short-rotation coppice has probably the greatest potential to benefit, the key factors affecting its value for wildlife are likely to be the:

- tree-species composition, with willows supporting a richer avifauna than poplars;
- extent of weed control during the establishment phase;
- variation in age classes of different stands; because the avifauna varies with age of re-growth, areas containing smaller blocks of a wider range of age classes should support a wider range of bird species than large, even-aged stands;
- size of area of short-rotation coppice and particularly the extent of edge habitat; this will affect the suitability of the coppice for species utilizing its margins in adjacent grassy, herbaceous, and arable habitat; the margins of short-rotation coppice plots tend to support higher densities of breeding birds than their interiors (Sage *et al.* 2006), and therefore creating smaller

stands of short-rotation coppice and interspersing the interior of large blocks of coppice with open rides is likely to be beneficial;
- timing of harvesting; most harvesting is likely to take place in winter, when woody material is driest and the leaves have fallen. However, in some cases the risk of damage to waterlogged soils in winter, or a need for a more continuous supply of material, might result in harvesting in summer. This would have negative effects on any birds nesting in the coppice.

Provision of conservation headlands adjacent to short-rotation coppice is also likely to increase the densities and range of bird species using them, just as it benefits farmland birds using adjacent hedgerows (Section 10.2).

The location of short-rotation coppice will also affect its value for wildlife relative to that of alternative land uses. Positioning short-rotation coppice in otherwise open landscapes should maximize its value for species that use the edges of scrub and woodland and scattered trees, but reduce the value of surrounding habitat for species that prefer very open habitats. There is a strong likelihood that short-rotation coppice will preferentially be placed on land of marginal agricultural value first; that is, the same land that currently has the greatest biodiversity value on farmland, and that would otherwise be most likely to be entered into agri-environment schemes. Loss of set-aside to short-rotation coppice is a major concern for farmland birds, as agri-environment schemes have yet to provide the same nesting and food resources to those found on set-aside.

Short-rotation coppice appears of limited value for most other groups. As in other wooded habitats, the presence of wide rides will benefit open-ground butterflies and other invertebrates (Section 7.4.1).

### 7.4.4 Managing open woodland systems

Many grazed, or in some cases also periodically cultivated, open woodland-management systems comprise important cultural habitats, as well as being of value for wildlife. These open woodlands are thought by some to be the closest modern analogues to what they consider to be open, original natural woodland maintained by the grazing of wild, large herbivores. Two types of open woodland system are of exceptional value for wildlife: ancient wood-pasture (Figure 7.12) and wooded dehesas/montados.

The feature of wood-pasture of greatest value for wildlife is its veteran/ancient broad-leaved trees and their associated fauna and flora. There are no strict definitions of ancient or veteran trees. They are simply trees that are very old for their particular species and which contain characteristics of old-growth, particularly

**Fig. 7.12** Wood-pasture and veteran/ancient trees. Many areas of wood-pasture are notable for their ancient (veteran) trees, such as this pedunculate oak, *Quercus robur*. The continuity of dead wood provided by large groups of ancient trees can support exceptionally important assemblages of saproxylic invertebrates, fungi, and epiphytic lichens.

This site, Windsor Great Park in Berkshire, England, is a royal park originally enclosed for hunting in the thirteenth century. It supports probably the most important saproxylic beetle fauna in the UK (Fowles *et al.* 1999).

valuable dead wood features. Areas supporting large numbers of ancient trees and a long-term continuity of dead wood can support important assemblages of saproxylic invertebrates and probably also be important for fungi. Ancient trees can also support characteristic and valuable assemblages of lichens (Kirby *et al.* 1995). Cavities in ancient trees can be important roost sites for bats.

Extensive, wooded dehesas/montados are especially important for their birdlife. The grasslands in wooded dehesas/montados can be very botanically species-rich. Wooded dehesas/montados usually support more diverse assemblages of songbirds and butterflies than neighbouring more densely wooded habitats or grasslands (see review by Diáz *et al.* 1997). Olive groves support high densities of wintering songbirds.

Overall, the main conservation priority in these open woodland systems will be to maintain their characteristic species assemblages by continuing existing management. This will involve continuing low-input grazing, without over-grazing or agricultural improvement, and continuing low-input periodic cultivation where this disturbance helps maintain high plant species-richness. It may also be a priority to minimize or avoid treating livestock with anti-parasitic drugs, especially in areas important for bats (Section 4.4.3).

The most important features of these grazed systems are the trees. The priority should be to maintain ancient trees through pollarding, where this is integral to the maintenance of the cultural habitat such as in wooded dehesas and montados, and to ensure that replacement trees are planted and protected from grazing. It will be difficult to predict which species of trees will be best suited to future climatic conditions.

Veteran/ancient trees in wood-pasture have often been managed by pollarding. This is thought in some cases to increase their life of a tree by preventing them from becoming top-heavy and falling over. Pollarding also helps create dead wood, by the cutting of branches leaving openings for the entry of fungi into the heartwood. Re-pollarding trees that have not been pollarded for a long period (lapsed pollards) can kill the tree (Green 1996). It is probably usually better to leave lapsed pollards alone, the only requirement for management being if limbs or trees become so unstable that they present a health and safety risk. If so, the options outlined in Section 7.1.6 should be followed including, of course, leaving any cut or fallen wood *in situ*.

Many dead-wood invertebrates feed on nectar during their adult stage, so in ancient wood-pasture it is important to retain or provide additional, suitable nectar sources. In Western Europe a large number of dead-wood invertebrates emerge as adults in early summer, when hawthorn, *Crataegus monogyna*, can be a particularly important nectaring shrub in wood-pasture. Various umbellifers also provide valuable nectar sources later in the summer. Stocking levels should be low enough to allow development of these areas of scrub and to allow flowering of suitable nectar-providing plants within the sward.

## 7.4.5 Grazing and browsing

Grazing and browsing by domestic livestock and wild deer influence the:

- structure and species composition of the field layer and understorey and consequently their suitability for other species;
- long-term tree-species composition by affecting regeneration.

The effects depend partly on the type of herbivore, but particularly on grazing pressure. Grazing and browsing characteristics of different types of livestock are described in Section 5.4.1. Pigs and wild boar create soil disturbance.

**Fig. 7.13** Grazing, browsing, and open understoreys. Grazing and browsing by domestic livestock and deer can greatly modify the structure and species composition of the field layer and understorey. Moderate to high levels of grazing and browsing remove the understorey, prevent regeneration of trees and shrubs, and encourage grazing-tolerant grasses and bryophytes at the expense of more palatable forbs. In the top photograph heavy sheep grazing of oak woodland is having a beneficial effect in maintaining the open conditions required by the wood's rich bryophyte flora (Borrowdale Woods, Cumbria, England).

Grazing and browsing of the understorey, in the bottom photograph by white-tailed deer, *Odocoileus virginianus*, can, though, also reduce foraging and nesting habitat for many woodland birds (Jug Bay Wetlands Sanctuary, Maryland, USA).

Grazing reduces the density of vegetation in the field layer and encourages low-growing grasses and bryophytes at the expense of taller grasses and forbs (e.g. Cooke and Farrell 2001; Figure 7.13). This reduces the suitability of the field layer for some small mammals. Grazing helps maintain open conditions beneficial for warmth-loving, woodland-edge invertebrates, but also removes the food plants of some species. Grazing helps maintain open conditions along rides and in glades.

High grazing pressure decreases densities of birds that depend on understorey vegetation for nesting and foraging such as, in western North America, orange-crowned warblers, *Vermivora celata*, Rufous hummingbirds, *Selasphorus rufus*, winter wrens, *Troglodytes troglodytes*, fox sparrows, *Passerella iliaca*, and song sparrows, *Melospiza melodia* (Allombert *et al.* 2005), but increases densities of birds that require an open understorey for feeding. For example, grazed oakwoods in western Britain hold higher densities of common redstarts, *Phoenicurus phoenicurus*, European pied flycatchers, *Ficedula hypoleuca*, wood warblers, *Phylloscopus sibilatrix*, and tree pipits, *Anthus trivialis*, than ungrazed woods (see review by Fuller 2001).

Herbivores encourage tree regeneration by providing gaps in which tree seeds can germinate, but inhibit growth of seedlings by browsing. Seedlings can be released by reducing or excluding grazing, but this usually creates a dense, even-aged understorey. Regeneration of tree seedlings is often poor once grazing levels have been reduced, because of the lack of gaps for germination and competition with seedlings from tall grasses and forbs. Medium levels of grazing and variations in grazing levels are most likely to produce patchy and periodic tree regeneration.

Numbers of deer can be reduced by culling, followed by erection of deer fences to prevent re-colonization. Deer fences can, though, kill woodland grouse that fly into them. Collisions can be reduced by marking fences, although they may still result in unacceptable high levels of mortality to small and vulnerable populations of woodland grouse (Baines and Andrew 2003; Figure 7.14).

### 7.4.6 Burning

Burning profoundly influences tree-species composition and structure by killing fire-intolerant tree species, thereby allowing fire-tolerant tree species to attain dominance (Figure 7.15), especially species of fire-tolerant pines, oaks, *Quercus* spp., and *Eucalyptus*.

Fires in forests and woodlands can be divided into:

- surface fires which remove the litter, field layer, shrubs, saplings, and fallen dead wood but do not reach the forest canopy;
- crown fires that burn the canopy and kill large trees, causing a change in dominant tree-species composition.

Crown fires usually develop where high fuel loads result in a more intense fire that ladders up the lower branches of trees to reach the canopy. Prescribed surface fires can be used to:

- maintain the characteristic species composition and structure of fire-prone forests and woodlands and their associated species;
- prevent accumulation of large fuel loads (dead wood and other combustible material on or close to the ground) to reduce the likelihood of larger-scale, catastrophic crown fires (e.g. see Fernandes and Botelho 2003).

## 212 | Forests, woodlands, and scrub

**Fig. 7.14** Deer fencing and woodland grouse. High, wire deer fencing in woodlands can substantially increase mortality of woodland grouse. The fencing is difficult to see and the grouse fly into it at high speed.

Collision rates of grouse can be reduced by marking fences to increase their visibility. One of the most practical, cost-effective, and durable forms of marking involves attaching vertical lengths of wood, known as droppers, to the wire, as shown on the left.

Few other bird species are thought to collide with deer fences in significant numbers (Abernethy, Highland, Scotland).

Using frequent surface fires to reduce fuel loads is known as 'hazard-reduction burning'. Fuel loads can also be reduced by thinning. Suitable precautions should obviously be taken when burning, similar to those described in Section 5.6.

Prescribed surface fires are also used to encourage of flush of re-growth of herbaceous plants to improve grazing. Burning of open habitats can be used to provide suitable conditions for germination and growth of seedlings of tree species, such as Scots pine (Hille and den Ouden 2004; Hancock *et al.* 2005), to aid forest regeneration. The key considerations when using prescribed surface fires are the:

- season of burning;
- frequency of burning;
- area burnt at any one time.

# Managing for conservation | 213

**Fig. 7.15** Fire. This was formerly considered damaging to wildlife and natural fires were suppressed. Since then, research has shown that fire is an important, natural process that removes fire-intolerant plants and maintains characteristic fire-prone forest types. In the top photograph prescribed fire has been used to facilitate reproduction by giant sequoias, *Sequoiadendron giganteum*.

Prescribed burning and cutting and removal of vegetation are also used to reduce fuel loads to decrease the risk, or spread, of larger-scale, catastrophic fires. In the bottom photograph, cutting and burning have been used to remove the understorey in strips either side of a road to create a firebreak/fuel break. The road also provides access for fire-fighting equipment (Yosemite National Park, California, USA).

In practice, prescribed burning is restricted to times of year when it is easiest to control the burn, usually the cool conditions of late winter and spring. This will be at a different time of year to the majority of wildfires, which take place during the hottest and driest periods of the year. Therefore, as in other habitats, prescribed fires burn less intensely than the natural wildfires that the vegetation has previously been subject to, and thereby probably have different effects. Burning should obviously not be carried out during the bird-nesting season. Burning in late winter and spring will probably interfere more with amphibian breeding activity than burning during hot, summer weather when many species shelter below ground. Wetter areas tend to remain un-burnt, providing refuges for species, and result in greater medium-scale variation in vegetation composition (e.g. Lilja *et al.* 2005).

The most common approach to determining the frequency of prescribed burning is to mimic the frequency of natural fires, reconstructed from fire scars on trees. Using fire scars to estimate fire intervals can, though, significantly underestimate the length of the fire rotation (e.g. Baker 2006b).

Frequent surface fires and thinning maintains a more open shrub layer and, by preventing shading, benefits herbaceous plants that require more open conditions and which are tolerant of, or regenerate well following, fire. The short-term reduction in litter, shrubs, and saplings caused by burning usually results in short-term increases in numbers of ground and aerial-foraging birds, but decreases in numbers of ground-nesting birds (e.g. Wilson *et al.* 1995; Artman *et al.* 2001). The open conditions also benefit some small mammals and invertebrates (e.g. Moretti and Barbalat 2004; Converse *et al.* 2006), and provide burnt dead wood for saproxylic invertebrates, which can otherwise be rare in areas that have been subject to fire suppression (Hyvärinen *et al.* 2006). The effects on fire on amphibians are less well understood. The open conditions created by burning and thinning will benefit some species, while others will be disadvantaged by the reduction in fallen dead wood and litter, and possibly by reduction in CWD in streams used for breeding (e.g. Schurbon and Fauth 2003; Bury 2004). The open conditions created by surface fires should also benefit many reptiles (Bury 2004). High frequencies of surface fires in ponderosa pine, *Pinus ponderosa*, forest in the western USA can result in invasion of unwanted, alien/exotic annual, grasses, similar to in dwarf-shrub vegetation in this region (Griffis *et al.* 2001; Dodson and Fiedler 2006; Keeley and McGinnis 2007).

The effects of burning on bird species-composition have been particularly well studied in forests managed to benefit red-cockaded woodpeckers in the southeastern USA (Figure 1.1). Here, burning on a less than 5-year rotation, often accompanied by thinning, has been used to remove broad-leave trees and restore and maintain open pine-dominated forests. This increases the densities of birds typical of open pine-grassland habitats, such as red-cockaded woodpecker, northern bobwhite, *Colinus virginianus*, and blue grosbeak, *Passerina caerulea*, but decreases the densities of birds associated with broad-leaved trees, such as tufted titmouse,

*Baeolophus bicolor* (Wilson *et al.* 1995; Provencher *et al.* 2002). Bachman's sparrows, *Aimophila aestivalis*, benefit from short fire intervals (<3 years; Tucker *et al.* 2004). Prescribed burning is also be used to kill stands of trees in order to encourage desired stages of re-growth, and to stimulate regeneration. In Michigan, USA, burning is used to kill stands of old jack pine, *Pinus banksiana*, to provide young (7–21-year-old) stands suitable for Kirtland's warblers, *Dendroica kirtlandii* (Byelich *et al.* 1985).

Burning can also be combined with other forestry techniques. In managed Norway spruce forests a combination of partial harvesting, creation of dead wood, and burning has been used to restore more natural post-burn characteristics (Lilja *et al.* 2005). In this example the production of large quantities of fallen wood through partial cutting was important in determining the intensity of the fire. It was only where high quantities (60 m$^3$/ha) of fallen wood were created that there was a large enough fuel load to create a crown fire to kill significant proportions of retained trees, and thereby create valuable large-diameter burnt and dead wood.

### 7.4.7 Integrating conservation management with commercial forestry

Forests managed for timber production usually contain only a limited variety of tree species, lack structural diversity, and the trees are harvested before they attain the important features characteristic of older growth. The first consideration when creating and managing commercial forestry for wildlife, though, is avoiding planting trees on valuable open habitat.

High-forest management systems differ in their planting, thinning, and harvesting regimes, resulting in differences in their vertical and horizontal structure, age composition of stands, and tree-species composition. There management systems can be modified to benefit wildlife. Hence, a major way of improving the value of forestry plantation for wildlife is by modifying harvesting, thinning, and planting regimes. The other two main methods of creating these valuable features within commercial forestry are:

- creating and maintaining open rides and their junctions (Section 7.4.1) and planting favoured tree and shrub species along the edges of these rides;
- providing dead wood.

*Avoiding planting trees on valuable open habitat*

Commercially managed plantations are unlikely to ever be of equal or greater value for wildlife than existing semi-natural habitat, so they should obviously not be planted on areas of these. As when creating other areas of habitat, the effects on surrounding habitats should also be considered. For example, plantations may harbour widespread predatory species, such as crows and red foxes, which might detrimentally affect wildlife in surrounding areas. Planting of trees can also disrupt existing drainage systems. Studies suggest that avoiding forestry operation within 10 m of watercourses will be sufficient to protect the watercourse's physical and chemical properties, but that buffer strips of greater than 30 m are probably necessary to maintain suitable habitat conditions within the watercourse and its margins for wildlife (Broadmeadow and Nisbet 2004).

*Modifying planting, harvesting, and thinning regimes*

Systems of high forest management are described below. The most important ways in which these systems differ in the value for wildlife are in:

- their vertical and horizontal structure: variation in horizontal structure is greatest where only individual or small groups of trees are felled at any one time (selection and group-selection systems) and least where very large areas of trees are felled together (clear-cutting);
- the presence of more mature trees: these are only retained in variable retention harvesting.

Other factors that can be modified to benefit wildlife are the:

- densities of trees, which are influenced by planting densities and thinning regimes;
- type and mix of trees, particularly whether broad-leaved or coniferous;
- length of the harvesting rotation, since this will influence the age structure.

Systems in which felling and regeneration take place continually and irregularly throughout the forest are termed continuous-cover forestry (CCF), because they retain a continuous cover of trees over a given area. The main CCF techniques are shelterwood, selection, and group-selection systems. The alternative is clear-felling and strip and wedge felling, in which large blocks of trees are felled and re-planted at the same time. Clear-felling is particularly favoured on windy sites, where the numerous small gaps created by CCF techniques result in greater windthrow. CCF techniques are more suitable for so-called windfirm sites; that is, those not prone to windthrow.

**Clear-felling** and **strip and wedge felling** involve periodically harvesting *all* the trees from either a large (typically more than 0.25 ha) area (clear-felling) or in narrow strips (strip and wedge felling). The next cohort of trees can be planted or the area left to regenerate naturally (Figure 7.16a). Trees are periodically thinned to provide timber and space for retained trees to expand into

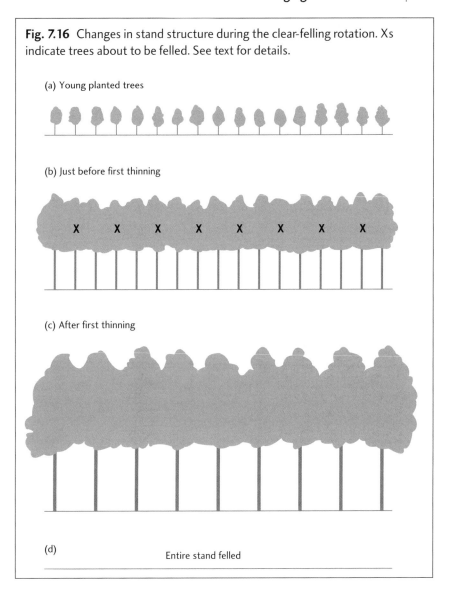

**Fig. 7.16** Changes in stand structure during the clear-felling rotation. Xs indicate trees about to be felled. See text for details.

(a) Young planted trees

(b) Just before first thinning

(c) After first thinning

(d) Entire stand felled

(Figures 7.16b and 7.16c), before the entire stand is eventually felled (Figure 7.16d). These systems create even-aged, single-storied stands of trees.

In the **two-storied high forest system**, light-foliaged trees are thinned to allow natural regeneration or tolerant tree species to be planted beneath them. This creates stands comprising two age-classes of trees. Both age classes are felled at the same time.

**Fig. 7.17** Changes in stand structure during the shelterwood rotation. Xs indicate trees about to be felled. See text for details.

(a) Before thinning

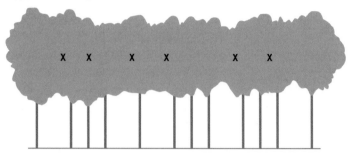

(b) After first thinning

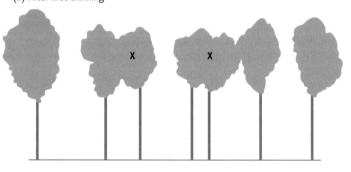

(c) After second thinning with saplings establishing beneath

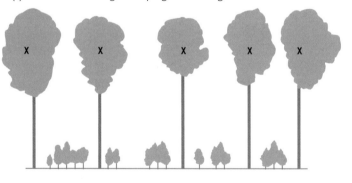

(d) After final felling

In the **shelterwood** system the next crop of trees is established beneath the shelter of an existing, open canopy. The existing canopy is thinned in stages (Figures 7.17a and 7.17b) to provide sufficient light for the establishment and growth of the next crop of trees beneath it (Figure 7.17c). This thinning can be carried out uniformly throughout the stand (uniform system) or in groups around existing patches of saplings (group system). This next crop is usually established through natural regeneration, but sometimes by planting. It is thinned as it grows up and the canopy of retained trees felled. During the establishment of this next crop, the woodland contains both a canopy and a developing understorey. Burning can also be used in combination with the shelterwood management to remove unwanted fire-intolerant tree species, for example to reduce competition with regenerating oaks by tulip poplars, *Liriodendron tulipifera*, and other undesired hardwoods in broad-leaved forests in the south-eastern USA (Lanham et al. 2002).

**Selection and group-selection systems** involve periodically harvesting individual trees (selection) or small groups of trees (group-selection). This creates gaps to allow existing trees with good timber potential to expand in size and natural regeneration of trees to take place (Figure 7.18). Group-selection forest consists of very small patches of even-aged trees. Selection systems create an even more intimate mix of different-aged trees. The difference between group-selection systems and clear-felling is really a matter of scale. Harvesting a very large group of trees, more than about 0.25 ha, would be considered clear-felling.

**Variable-retention forestry/green-tree retention** (GTR) involves retaining a proportion of live trees and snags during harvesting, as well as creating additional high-cut stumps (Figures 7.19a and 7.20a). The next crop is established among these retained trees (Figure 7.19b). The aim of this system is to provide a proportion of more mature trees among the subsequent crop, and increase the quantity of standing and fallen dead wood, both specifically for conservation (Figure 7.19c). Variable-retention forestry usually involves retaining blocks or dispersed trees within relatively large, clear-felled areas.

The complex vertical and horizontal structure produced by the selection and group-selection systems will tend to support a more diverse fauna and flora, especially of birds, than the even-aged, single-storied stands created by large-scale clear-cutting. Selection and group-selection systems more closely mimic the effect of small-scale, patchy tree-falls. Shelterwoods, by maintaining a more continuous canopy cover, will be better at preserving species requiring shady and moist conditions compared to clear-felling (Hannerz and Hanell 1997).

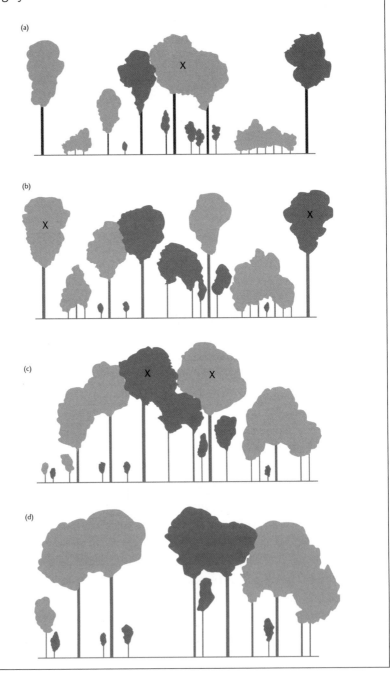

**Fig. 7.18** Changes in stand structure during the selection system rotation. Tree species intolerant of shade are shown in light grey and tolerant species in dark grey. Xs indicate trees about to be felled. See text for details.

## Managing for conservation | 221

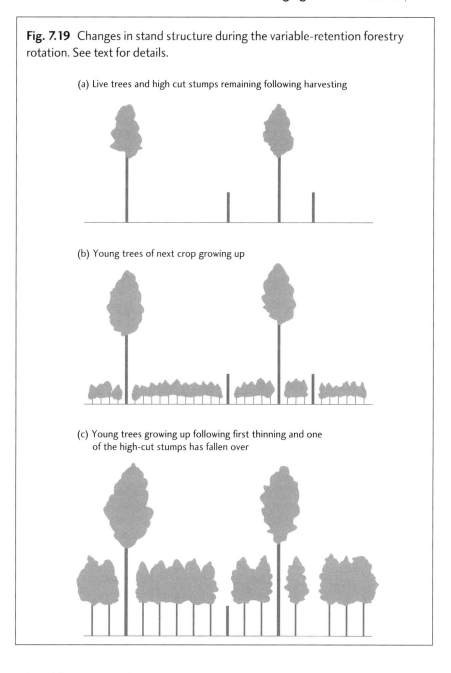

**Fig. 7.19** Changes in stand structure during the variable-retention forestry rotation. See text for details.

(a) Live trees and high cut stumps remaining following harvesting

(b) Young trees of next crop growing up

(c) Young trees growing up following first thinning and one of the high-cut stumps has fallen over

Clear-felling creates large, even-aged, single storied stands, and more closely mimics the earlier stages of re-growth following large-scale disturbance, such as extensive windthrow or large-scale insect herbivory. Clear-felling usually provides less dead wood than other high-forest management systems (Sippola

**Fig. 7.20** Wildlife-sensitive commercial forestry. A number of methods can be used to maximize the value of commercially run forestry operations.

(a) Creation and retention of standing dead trees (snags), and a proportion of unharvested commercially profitable trees, provide habitat for dead-wood invertebrates, feeding areas for woodpeckers, and nest sites for these and other cavity-nesting birds (Yremossen, Västergötland, Sweden).

(b) Increasing tree-species diversity by planting different tree species in small blocks and allowing commercially valuable tree species to regenerate naturally within them. This can increase bird-species diversity, but is unlikely to benefit habitat specialists of high conservation value (Omberg, Västergötland, Sweden).

(c) Sensitive thinning can be used to create an open understorey, thus benefiting the field layer and associated species. Western capercaillie, *Tetrao urogallus*, feed on bilberry, *Vaccinium myrtillus*, in the open understorey of this 50-year-old managed stand of Scots pine (Yremossen, Västergötland, Sweden).

*et al.* 1998). The effects of the other high-forest management systems are intermediate in terms of their effects on structural complexity.

Retention of a proportion of trees during harvesting is now common practice in many areas, but its long-term effects are yet to be fully evaluated. In the first few years of re-growth, areas of forest where patches of trees have been retained support higher densities of breeding birds typical of more mature forest, particularly ground and tree-nesting and forest canopy-gap species, than in forest managed by clear-felling (Annand and Thompson 1997; Merrill *et al.* 1998). A high proportion of retained trees often fall over in the first few years following felling operations, 40% by the end of the second season following felling in a study of Norway spruce forest (Hautala *et al.* 2004).

The diversity of tree species and age composition can also be increased *within* individual stands, and by planting small stands of different trees species (Figure 7.20b). Tree-species composition can obviously be diversified by planting mixtures of tree species, and by retaining naturally regenerated native trees within planted blocks. There is evidence that species richness of birds on conifer plantations is greater if, for a fixed area of broad-leaved trees, these are dispersed throughout the conifers, rather than concentrated in a few large blocks (Bibby *et al.* 1989).

In parts of Western Europe the main bird conservation interest of areas of coniferous plantations on sandy soils is in the bird assemblage associated with felled areas and the very early stages of tree growth (Figure 7.21).

Management to maximize timber production involves minimizing gaps in the canopy to maximize the light intercepted by the trees and converted to timber. This is achieved by planting trees at high density, and then thinning them to allow retained trees to expand into the gaps created. Maximizing canopy cover, though, inhibits development of an understorey, thus reducing the value of the forest for the numerous species dependent on it. The understorey can be encouraged by bringing forward thinning operations to create and maintain a more open canopy, although this will have commercial costs (Figure 7.20c).

*Providing dead wood*

Large-diameter dead wood can be increased in plantations by:

- creating high-cut stumps;
- retaining naturally dying trees;
- lengthening the rotation period to increase the quantity of dead wood produced by aging trees;
- retaining living trees at harvest, which will eventually provide dead wood as the trees age;

224 | Forests, woodlands, and scrub

**Fig. 7.21** Heathland birds in forestry plantations. The conditions following planting of conifers on sandy soils provide suitable conditions for a range of birds typical of Atlantic heathlands. These species require bare, sandy soil (especially wood larks), short, open areas and scattered trees (tree pipits) and woodland edge, rides, scattered trees, and relatively young re-growth (Eurasian nightjars).

Suitable breeding sites for some early successional species, such as wood larks, can only be created by clear-felling moderate-sized blocks of trees (Bowden 1990; Thetford Forest, Norfolk, England).

- minimizing the quantity of fallen dead wood destroyed during preparation for the next crop.

Volumes of smaller diameter dead wood can be increased by retaining a proportion of harvesting residue.

Retaining naturally dying trees and fallen dead wood and creating high-cut stumps provides significant quantities of CWD for a period following harvesting. Retaining naturally dying trees and creating cut stumps are especially cost-effective methods of providing dead wood. Lengthening the felling rotation increases the quantity of CWD towards the later stages of the rotation as trees age, but is a relatively expensive option (Ranius *et al.* 2005). A continuity of CWD throughout the felling rotation can, though, only be provided by also retaining suitable numbers of live, mature trees following harvesting (see variable-retention forestry) to provide a continuity of newly created dead wood (Ranius *et al.* 2003).

High stumps are created by cutting live trees with a harvester, typically to a height of about 4 m. The beetle fauna of high-cut stumps varies in relation to tree species and presence or absence of key decay fungi. Studies have also found total species richness of dead-wood beetles and that of red-listed species to increase with stump diameter and the degree of sun exposure (Lindhe and Lindelow 2004; Jonsell *et al.* 2005). High-cut, sun-exposed stumps tend to support a more limited range of species than natural stumps. This is probably because their uniform method of creation provides a more homogenous dead-wood resource compared to natural stump formation (Jonsell *et al.* 2004). High-cut stumps are also unlikely to support the full range of species associated with other types of dead wood, particularly dead wood in shade and that created by unmanaged, self-thinning deciduous trees and in old-growth (Jonsell 1998; Lindhe *et al.* 2005). The suitability of plantations for rare dead-wood beetles can be increased by leaving a higher proportion of standing live trees following harvesting and, in fire-prone boreal forests at least, by burning retained trees (Hyvärinen *et al.* 2006).

Logs retained during harvesting can support more diverse fungal communities than high-cut stumps, with larger-diameter logs supporting more species-rich assemblages (Lindhe *et al.* 2004) and being more valuable for dead-wood invertebrates (Nitterus *et al.* 2004). A large quantity of the volume of fallen CWD is, though, destroyed during scarification to prepare conditions for the next crop, for example 68% of it in one study (Hautula *et al.* 2004). The quantity of CWD retained through the felling cycle can be increased by:

- employing the least-destructive harvesting methods;
- reducing the use of scarification during preparation for the next crop;
- not removing trees from areas that already contain abundant dead wood.

Whereas all dead wood should be retained in woodlands and forests managed for nature conservation, a more pragmatic approach is necessary at sites where timber production is also an objective. Table 7.1 suggests some benchmarks to landowners on the quantities of dead wood to be retained in semi-natural broad-leaved woodland.

**Table 7.1** *How much dead wood to retain? The following tables give suggested management to provide suitable quantities of dead wood to be retained for conservation in semi-natural broad-leaved woodland managed for timber production (from Butler et al. 2002). Methods for calculating whether moderate, high, or very high levels of dead wood should be retained are given by Butler et al. (2002). These are based on the existing and potential value of woods for their dead-wood fauna and flora and the priority the landowner affords conservation. Butler et al.'s very high category (not shown) involves retaining all dead wood (as should be the case in all woodlands managed for nature conservation) and thinning 50–120-year-old stands to encourage the development of more than 20 veteran trees per hectare.*

| Management | Minimum | Quantity of dead wood | | |
| --- | --- | --- | --- | --- |
| | | Moderate | High | |
| **(a) Stands less than 50 years old** | | | | |
| Retention of dead wood | Minimum of an average of three standing and three fallen stems per hectware | Retain all dead wood where possible | Retain all dead wood | |
| Thinning regime | – | – | Variable thinning to improve stand structure but retain any existing standing dead wood | |
| Cut wood | – | – | Leave some cut wood on site | |
| **(b) Stands of 50–120 years old** | | | | |
| Percentage of fallen wood retained | 100 | 100 | 100 | |
| Retention of standing dead wood | Five per cent of standing stems to be dead or contain significant dead-wood features | Ten per cent of standing stems to be dead or contain significant dead-wood features | Twenty per cent of standing stems to be dead or contain significant dead-wood features | |

# Managing for conservation | 227

| | | | |
|---|---|---|---|
| Thinning regime | Five per cent of native stems per hectare to be retained for perpetuity and canopies freed to allow full crown development | Vary thinning intensity to improve crop while maintaining a sustainable supply of dead wood habitat and veteran trees. Ten per cent of native stems per hectare to be retained for perpetuity and canopies freed to allow full crown development | Only thin where necessary to improve diversity of structure for conservation aims and to maintain sustainable variety of dead-wood habitat and ancient trees. Twenty per cent of native stems per ha to be retained for perpetuity and canopies freed to allow full crown development |
| Percentage of cut wood retained on site | 5 | 10 | 20 |
| **(c) Stands of greater than 120 years old** | | | |
| Percentage of fallen wood retained | 100 | 100 | 100 |
| Retention of standing dead wood | Five per cent of standing stems to be dead or contain significant dead-wood features | Ten per cent of standing stems to be dead or contain significant dead-wood features | Retain all dead wood habitats and 50% of mature trees |
| Thinning regime | Five per cent of native stems per hectare to be retained for perpetuity | Ten per cent of native stems per hectare to be retained for perpetuity and canopies freed to allow full crown development | Identify future veteran trees and allow their crown development |
| Percentage of cut wood retained on site | 5 | 10 | 40 |

# 8

# Freshwater wetlands and water bodies

This chapter discusses the management of open bodies of water and wetlands on seasonally or permanently waterlogged soil. It concentrates on freshwater habitats, but briefly discusses management of brackish habitats, such as coastal grazing marshes, where they form a continuum with freshwater ones. Management of saltmarsh and other saline habitats is discussed in Chapter 9.

There is a wide range of often confusing terms used to describe types of wetland. Fens and bogs are waterlogged habitats, which differ in their main sources of water and associated nutrients. Fens are fed by groundwater and precipitation and described as minerotrophic. They typically contain relatively species-rich vegetation dominated by bulky monocotyledons and often tall, perennial, herbaceous forbs. Bogs receive the majority of their water and nutrients from precipitation. They are described as ombrotrophic. Fens tend to have relatively nutrient-rich and base-rich water, especially where they receive this from calcareous substrates. Bogs have low nutrient levels and acid conditions. There are, though, types of wetland that do not easily fit into these categories. Base-poor fens (known as poor fens) have a *low* pH, while some other fens receive nutrient-poor water. Bogs are always peat-forming, but fens can be either peat-forming or occur on mineral substrates.

There are two broad types of fen. These differ in their water movement. Fens with predominantly vertical movement are known as topogenous, whereas those with predominantly lateral water movement are known as soligenous. Bogs can be convex in shape (raised bogs; sometimes known as raised mires), relatively flat, or sloping.

The term mire is used to describe a range of usually peat-forming wetlands, including fens and bogs. Marsh is a term commonly used to describe waterlogged areas dominated by short grasses (i.e. wet grassland) or tall, bulky, monocotyledons such as common reeds (hereafter referred to as reeds) and bulrushes/cattails. Swamp has two different meanings. It is either used to refer to species-poor vegetation dominated by bulky, emergent monocotyledons on seasonally or permanently *submerged* substrates (i.e. in wetter areas than fens) or wetland vegetation dominated by trees and shrubs. In this chapter we use the first definition and discuss wet scrub and wet woodland separately.

Wet grasslands are those with a high water table and/or which hold surface water. They occur on soils with impeded drainage, often in association with other low-lying wetlands. In areas of high rainfall, wet grasslands can occur at higher altitude and on slopes.

## 8.1 Principles of manipulating water levels

Management of wetlands for conservation commonly involves manipulation of water levels. This requires a basic understanding of hydrology, particularly of whether there will be sufficient water to achieve target levels, and, if not, the volume of additional water that would be needed of estimating these requirements are straightforward, although the detail can be complex. It involves calculating a water balance. This can be carried out for the site as a whole or, where relevant, separately for individual hydrological units; that is, areas that are hydrologically isolated from one another.

A water balance is based on the simple principle that the:

Change in quantity of water stored in a hydrological unit
= the quantity of water *entering* the hydrological unit − the quantity of water *leaving* the hydrological unit

From this it is possible to estimate the change in quantity of water stored within the hydrological unit by estimating the quantities of water entering and leaving it. Because of seasonal variations in target water levels and inputs and outflows of water, water balances are usually calculated separately for different periods of the year. They are often calculated for each month, to determine during which periods of the year there will be an excess, or lack, of water.

The main ways that water can enter a hydrological unit are by:

- **precipitation** falling on it as rain, sleet, or snow;
- **flow from watercourses**; that is, through rivers, streams, and ditches;
- **groundwater flow**.

The main ways it can leave a hydrological unit are by:

- **evapotranspiration** comprising the combined losses of water to the atmosphere through evaporation from the soil surface and transpiration from plants;
- **drainage into watercourses** such as through rivers, streams, and ditches;
- **groundwater flow**, especially seepage to adjacent, drained land with a lower water table.

Inputs of water through overland flow will usually be negligible. Rates of input and output of water will also vary between years. It is therefore usual to calculate water balances taking account of these annual variations and, for example, estimate the quantity of additional water that will be required to maintain target water levels in 75% of years.

In most temperate wetlands there is usually an excess of water in winter when precipitation is higher or similar to that in summer, but evapotranspiration is lower due to the cooler weather and lack of plant growth. In cold climates winter precipitation will be locked up as snow. Wetlands fed by snow-melt will tend to have high water levels in late winter and spring.

Water levels will fall in most wetlands through late spring to autumn as evapotranspiration rates increase and precipitation usually remains similar to, or is lower than, in winter. Where this is the case management will often involve minimizing losses of water to maintain suitably high water levels within the wetland, especially in small sites surrounded by drained land. Water loss can be prevented by installing dams and sluices into watercourses, constructing bunds, reducing seepage losses and providing additional water. When installing dams and sluices, it is important to consider their potential impediment to fish movement (Section 4.2.2). Seepage rates vary greatly depending on the porosity of the soil, which depends on soil type and structure. Seepage rates will be greatest on sands, gravels, and well-structured peat and lowest on poorly structured and compacted soils, especially clays.

Methods of reducing seepage losses include:

- increasing water levels on surrounding land (i.e. creating a buffer) to reduce the difference in water-table height between the wetland and surrounding land and thereby reduce the rate of flow of water from the wetland;
- installing an impermeable membrane or cut-off curtain down to as far as any impermeable soil.

Wetlands can also be designed so that water lost through seepage from one area feeds other areas of wetland habitat that require slightly lower water levels. Additional water for the wetland can be provided by:

- diverting inflows;
- abstracting water from elsewhere, particularly from watercourses;
- storing excess winter rainfall or water abstracted in winter in reservoirs for use in spring and summer.

Methods for avoiding use of poor-quality water when obtaining additional water are considered in the following section.

## 8.2 Water quality

Water quality is a complex subject. For practical site management, the most important aspects are the water's nutrient levels and pH. Pesticide residues may also be an issue at some sites. Salinity is important in saline wetlands (Chapter 9). pH affects the flora and fauna, some species being typical of acidic, neutral, and base-rich conditions.

Nutrient levels are especially important, since they will affect plant growth and consequently the abundance and type of vegetation and its associated fauna. The two nutrients that most commonly limit plant growth in aquatic systems are nitrogen and phosphorus. If one of these is limiting, then increasing its quantity will increase plant growth. The process of increasing nutrient levels is known as eutrophication. Increasing levels of nutrients increases plant growth, favouring more competitive plant species typical of higher nutrient levels. At phosphorus concentrations above about 100 μg/l the water body may continue to support larger, submerged plants, or lose these and become dominated by suspended, microscopic algae (phytoplankton; see Moss *et al.* 1996). This process removes habitat for aquatic invertebrates and food for birds. It is, though, still usually possible to provide botanically poor shallow water habitat that is of high value for wetland birds using relatively nutrient-rich water. At very high trophic states (hypertrophic conditions), the water may become de-oxygenated and cause fish kills.

Eutrophication caused by artificially high levels of nitrogen and phosphorus is widespread in intensively managed lowlands. The main source of high nitrogen levels is run-off of nitrate from fertilizer application. Nitrate is highly water-soluble and leaches readily into watercourses. The main source of high phosphate levels is from treated sewage effluent. Phosphorus binds to colloids, such as clay particles, and is less water-soluble. It often occurs in high concentrations in sediment, which release phosphate into the water above. Because of widespread eutrophication, wetlands with low nutrient levels are rare and generally highly valued in most lowland areas, particularly for their nutrient-poor flora and associated invertebrate fauna.

Nutrient and pesticide inputs into wetlands can be minimized by avoiding poor-quality water. Nutrient levels in water inputs can be tested to determine whether they are within a tolerable range. Nutrient levels vary greatly over time, and with rates of water flow. It is necessary to take a minimum of six samples per year, and preferably more, to obtain a reasonable measure of nutrient levels. In practice, though, there is rarely any choice of water source, and a comparison has to be made between using nutrient-rich water and allowing the wetland to dry out. The only long-term solution for reducing nutrient levels entering wetlands is to reduce inputs of nutrients into their catchments.

Where water is abstracted from rivers for use in a wetland, the timing of abstraction can be adjusted to minimize nutrient inputs. Abstraction is usually only allowed when river flows are high. Nutrient levels in rivers will often be high when a period of dry weather is followed by heavy rainfall that leaches nutrients into it. Nutrient levels will also be high when periods of widespread fertilizer application are followed by heavy rain. It is therefore best to abstract in winter after nutrients have been flushed through. Storage of abstracted water in reservoirs might allow some phosphorus-rich sediment to settle out before the water is used to feed the wetland.

Inputs of nutrient-rich water are particularly damaging to wetlands otherwise fed by nutrient-poor, calcareous, or acidic groundwater. In small, isolated, groundwater-fed wetlands, lowering the substrate to raise the height of the water table relative to its surface (Section 8.8.4) will be preferable to raising water levels using nutrient-rich water.

Nutrient levels in water within catchments can be reduced by:

- stripping phosphate from sewage effluent;
- reducing fertilizer inputs;
- minimizing leaching of sediment and nitrate into watercourses using vegetated filter strips/buffer strips.

The second two measures are incorporated into some agri-environment schemes and conservation programs aimed at reducing so-called diffuse pollution (Section 3.3). Vegetated filter strips/buffer strips consist of vegetated land beside watercourses. They trap sediment washed off cultivated fields and denitrify nitrate run-off before it reaches the watercourse. Vegetated filter strips/buffer strips do not retain phosphorus compound permanently, though. In general, the strips need to be a minimum of 30 m or so wide to significantly reduce sediment and nitrate inputs into watercourses (Hickey and Doran 2004). Vegetated filter strips/buffer strips can themselves be managed to provide good wildlife habitat.

There has been some success in removing nutrients from inputs of water using reedbed treatment systems and sediment traps to reduce phosphate, and by providing sacrificial areas of wetland to remove nutrients before the water enters more sensitive areas.

Attempts to reverse the effects of eutrophication in shallow lakes require first reducing inputs of phosphorus and/or nitrogen. Additional techniques may then be required to switch the phytoplankton-dominated flora into one dominated by vascular plants and stoneworts Charophyta. These include:

- removal of accumulated phosphate-rich sediment that would otherwise continually re-release phosphorus back into the water;

- removal of zooplankton-eating fish (so-called biomanipulation) to increase densities of zooplankton that feed on the phytoplankton;
- reintroduction of submerged, vascular plants to provide refuges from fish predation for zooplankton. This may need to be accompanied by protection of these plants from herbivorous wildfowl.

Restoration of shallow lakes is complex, and its success has been variable (e.g. Gulati and van Donk 2002). Moss *et al.* (1996) provides a good, practical guide.

Growth of algae and cyanobacteria (blue-green algae) can be controlled by adding barley straw (e.g. Everall and Lees 1997; Barrett *et al.* 1999). The straw should be held in loose bundles nets, cages, or bags and, if necessary, attached to floats to prevent it from sinking to more than a metre below the surface. These bundles are commonly known as straw sausages. They should be applied twice a year, in early spring and autumn, at rates of between 10 and 50 g of straw/$m^2$ of surface water (see Centre for Ecology and Hydrology 2004).

## 8.3 Large, deep water bodies

The two principal methods of improving the value of deep water bodies (>1 m) for wildlife are by:

- making their margins shallower by infilling or including shallow margins in their initial design (Figure 8.1) to provide suitable conditions for emergent vegetation and other shallow-water plant species and associated fauna;
- providing islands or rafts for nesting waterbirds.

Fish ponds can be of high value for feeding waterbirds. The main way of increasing their value for birds is by ensuring they also contain suitable nesting habitat (Figure 8.2).

### 8.3.1 Islands and rafts

Islands for nesting waterbirds can created during excavation of water bodies or created in shallow water by deposition of material. The islands can be covered in shingle to provide a suitable substrate for open-ground nesting species, especially terns, and plovers. Coating islands with cockle shells (a by-product of the cockle industry) can be used to encourage nesting terns. In the absence of management, though, islands usually become increasingly vegetated. This will improve their suitability for nesting wildfowl, but make them less suitable for these open-ground nesting species.

Large, deep water bodies | 235

**Fig. 8.1** Shallow margins. The value of otherwise deep water bodies such as gravel pits can be enhanced by including shallow margins and pools around their edges.

The open, silty margins of these gravel pits support a distinctive beetle fauna comprising many species that are rare in this region. These open margins often become colonized by taller vegetation including scrub and trees, and it can be difficult to maintain their open nature in the long term (Dungeness, Kent, England).

One technique for helping maintain open conditions on islands is to design them so they are covered by water in winter and exposed by falling water levels immediately prior to nesting. The flooding helps rot down and disperse vegetation that has grown on them.

Where winter flooding is not practical, vegetation growth can be reduced by covering islands with sheets of geotextile for weed and root control, and then covering this with shingle or other inert material. The geotextile prevents plants from extending their roots deep into the more nutrient-rich substrate below the shingle. However, where large numbers of birds are present, particularly roosting flocks, geotextile sheeting can encourage vegetation establishment by trapping nutrient-rich bird faeces in the surface layer of shingle above it. It can similarly trap silt within the shingle when flooded by high water levels.

Vegetation that becomes established in islands can be removed by cutting, hand-pulling, and burning. However, once sufficient organic matter has accumulated

**Fig. 8.2** Fish ponds. These can provide important habitat for waterbirds, particularly fish-eating species that prey on both commercial and non-commercial fish in them.

The value of fish ponds for breeding waterbirds can be increased by allowing emergent vegetation to grow around their margins to provide nesting habitat. These fish ponds support high densities of breeding great bitterns, *Botaurus stellaris* (Beloe Fish Ponds, Gomel Region, Belarus).

within the shingle or other inert material, even vegetation removal before and after the nesting season may be no longer sufficient to maintain open-enough conditions as vegetation re-grows during the breeding season. In these situations the only option to maintain open conditions is to re-excavate and/or re-coat the islands with shingle or other suitable material. Sometimes, just replacing the surface substrate with material excavated from below can re-create suitable conditions. Islands low enough to be flooded during winter are likely to suffer from erosion, particularly in larger, deeper water bodies, and require periodic re-building.

Even if surrounded by deep water, birds nesting on islands can still be subject to severe predation by mammals that can swim out to islands. Mammalian predators can be prevented from reaching islands by surrounding them with underwater fencing (Figure 8.3).

Islands can be difficult to construct in deep water. Nests on low-lying islands can be vulnerable to flooding on water bodies with fluctuating water levels.

**Fig. 8.3** Underwater fencing. Waterbirds nesting on islands can be protected from mammalian predators by surrounding the islands with fences that have their base in deep water. Approaching mammals find it impossible to jump or climb over the fence from a swimming position (Rye Harbour, Sussex, England).

An alternative method for providing nesting areas for terns is by using anchored nest rafts covered with shingle or other suitable material (see Burgess and Hirons 1992). The rafts need to have raised sides to prevent chicks from falling into the surrounding deep water. Approximately a quarter of Scotland's breeding Arctic loons, *Gavia arctica*, nest on specially constructed, floating rafts. Raft-nesting Arctic loons have far higher productivity than those using natural nest sites, which are often flooded (Hancock 2000).

## 8.4 Large, shallow water bodies

Large, shallow (less than about 1 m), nutrient-rich water bodies are very productive habitats and can be of particular importance for waterbirds. Water levels in large, shallow, nutrient-rich water bodies can be manipulated, sometimes in combination with other forms of management, to:

- increase food supply for waterbirds by periodically drying out water bodies and using moist-soil management to increase supply of seeds for wintering wildfowl;

- provide shallow-water, mud, and open marginal habitat to increase the accessibility of food for waders/shorebirds, herons, egrets, wildfowl, and other waterbirds and provide suitable habitat for marginal plants and invertebrates;
- alter the relative proportions of open water and emergent vegetation.

In addition, nesting islands can also be created and managed for birds as described in Section 8.3.1. Using water levels to manipulate the proportions of emergent vegetation and open water and the effects of other management on emergent vegetation is discussed in Section 8.8.2. Eutrophication is an issue in many shallow, lowland lakes (Section 8.2).

### 8.4.1 Increasing food supply for waterbirds: moist-soil management

Seeds are an important source of food for many wildfowl species in winter. The abundance of suitable seeds can be increased by lowering water levels in spring and summer to provide moist mud for prolific, seed-producing annuals and other valuable wildfowl food plants to germinate and grow on. Some important wildfowl food plants also germinate in very shallow water. Lowering water levels is known as a drawdown. These seeds can be made available to wildfowl by re-flooding in autumn. This technique is known as moist-soil management (Smith and Kadlec 1983; Haukos and Smith 1993; Figure 8.4). Following the initial drawdown the soil needs to be kept moist to maintain suitable conditions for germination and growth of desirable wetland plants and to discourage unwanted dry-ground plants. Areas may subsequently require subsequent irrigation during the summer to keep them suitably moist.

Invertebrate biomass tends to be highest in early successional (i.e. recently flooded) wetlands (e.g. Danell and Sjöberg 1982). This is attributed to high overall productivity fuelled by release of soluble nutrients from freshly inundated soil and decomposition of flooded terrestrial vegetation, and low levels of predation by predatory invertebrates and fish. As the wetland matures, nutrients released from decaying terrestrial plant material decline, numbers of predatory invertebrates and fish increase, and total invertebrate biomass tends to decline. Invertebrate biomass in shallow water bodies can therefore be increased by periodically drying them out and re-flooding them. Drying out will also kill any fish. Re-colonization by fish following re-flooding is usually accompanied by high levels of recruitment of small fish of suitable size for feeding herons, egrets, and other birds.

The most important factors affecting the vegetation that establishes, other than the composition of the seedbank, are the timing and rate of drawdown. The

**Fig. 8.4** Moist-soil management. This is used to maximize seed for wintering wildfowl. It involves lowering water levels during the growing season, in this case in an artificially created impoundment, to provide moist soil conditions for the germination and growth of prolific, seed-producing, annual plants.

Re-flooding in autumn suspends these seeds and makes them available to feeding wildfowl (Blackwater National Wildlife Refuge, Maryland, USA).

timing of drawdown will affect the composition of the vegetation and of desirable seed-producing plants. Drawdowns are usually referred to as being early, mid-, or late season. For example, in the Playa Lakes region of North America, drawdowns in April are recommended to maximize seed production of smartweeds, *Polygonum* spp. (Haukos and Smith 1993). The rate of drawdown will also influence vegetation composition. Rapid drawdowns will produce more uniform soil-moisture conditions across an area of given substrate height and tend to result in more uniform vegetation. Slower drawdowns create more variation in soil-moisture conditions, with different areas drying out at different times of year and under different temperature conditions. Slower rates of drawdown are recommended during the latter part of the summer, when temperatures and evaporation rates are higher and there is a risk that a sudden drawdown will create a large expanse of ground that dries out rapidly and becomes unsuitable for moist-soil plants.

If there is insufficient growth of ruderal vegetation, areas can be quickly disked and planted with additional prolific seed-producing plants, such as sorghum, to bolster seed production prior to re-flooding. Areas managed by moist-soil management can also become dominated by large stands of plants that do not provide valuable wildfowl food, for example cocklebur, *Xanthium* spp., in North America and by woody vegetation such as willows. Most of these problem species are typical of drier conditions. Moist-soil managed areas can also be colonized by unwanted stands of tall moncotyledons, although many of these provide valuable wildfowl food and habitat for other species. Undesired vegetation can be controlled by disking or burning and re-flooding for short periods to drown them. Reduction of unwanted emergent vegetation is also discussed in Section 8.8.2.

To maintain dominance by a desired range of annual plants in the long term, it is common practice to set up a 2–4-year rotation with different timings of drawdowns and re-flooding between years. In practice, moist-soil management is a bit of an art. Even though it is useful to have these rotations, management in any one year needs to be based on an assessment of the conditions and abundance of desirable and undesirable plant species. Introducing summer drawdowns to sites without a recent history of them might not initially result in very prolific growth of ruderal vegetation initially, due to lack of an existing seedbank.

Completely draining water bodies in spring and early summer will conflict with the requirements of most breeding waders/shorebirds, wildfowl, and crakes, that feed in shallow water, and of birds that would otherwise nest on islands free from ground predators. A compromise is to partially lower water levels in spring and early summer to allow germination of ruderal vegetation over a proportion of the area, while still retaining sufficient shallow water for feeding birds and deep-enough water around islands to help protect nesting birds from mammalian predators. This is only feasible if there is sufficient variation in topography. An alternative is to carry out a drawdown in only a proportion of hydrological units. This should take place before birds settle to nest on islands that will become vulnerable to mammalian predators as water levels fall.

A potential problem with moist-soil management is that it may also allow the germination and establishment of perennial emergent plants. While this will be desirable when seeking to increase the area of swamp (Section 8.8.2), it may not be where seeking to maintain areas of open water and mud for waterfowl and invertebrate interest. Grazing by livestock will suppress the growth of unwanted emergents but also that of desirable ruderal vegetation. Grazing by wildfowl, particularly geese, can also reduce or prevent the growth of ruderal vegetation.

Incorporation of plant material into the substrate using these methods can also be used to increase the quantity of coarse detritus for invertebrates to feed on and hence increase their biomass. It can also be used to maintain more open conditions for waders/shorebirds and to increase the ease that waders/shorebirds can probe into what often become a fairly heavily compacted substrate. The benefits of this to waders/shorebirds and other invertebrate-feeding waterbirds will vary between sites. Gray *et al.* (1999) found that soil disturbance during drawdowns *reduces* the biomass of large invertebrate prey for wildfowl the following winter, probably because it reduces the quantity of above-ground detritus for them to feed on.

Moist-soil managed areas should be re-flooded in the autumn/fall, typically September, to suspend the seeds and make them available to returning, wintering wildfowl. Re-flooding playa wetlands in the southern USA in September results in a higher biomass of aquatic invertebrates (predominantly ramshorn snails, Planorbidae) the following winter than does re-flooding in November (Anderson and Smith 2000). In wetlands where the benthic fauna is dominated by non-biting midge larvae, Chironomidae, invertebrate biomass is likely to be higher in winter and early the following spring if they are re-flooded in autumn, while adult midges are still active and ovipositing.

### 8.4.2 Providing shallow-water, mud, and open marginal habitat

Water levels can be manipulated to provide suitable shallow water for waders/shorebirds, dabbling ducks, herons, egrets, and other species to feed in at particular times of year. Water levels are typically lowered, or allowed to draw down naturally, in spring and autumn to provide suitable conditions for migrating waders/shorebirds, but kept high during winter to provide suitable conditions for wintering waterfowl and to flood vegetation on nesting islands (Section 8.3.1). Gradually falling water levels through spring to autumn will also provide a variety of conditions for different marginal plants and invertebrates.

Highest numbers of bird species are typically found in water 10–20 cm deep, with few wader species using water deeper than 40 cm (e.g. Elphick and Oring 1998, 2003). Plovers and some other species feed mainly on bare mud exposed by falling water levels. The range of feeding opportunities available at any one time can be increasing by enhancing topographic variation within the area flooded.

Providing shallow-water habitat by lowering or raising water levels will have slightly different effects on food supply for birds. Lowering water levels has the advantage of concentrating aquatic invertebrate prey, fish, and shrimps, and providing bare mud containing stranded benthic invertebrates on which waders/shorebirds and other birds can feed. Creating suitable water depths by raising water levels to flood new habitat may temporarily raise productivity by increasing the availability of detritus (see above), and provide a short-lived (and probably largely one-off) abundance of displaced terrestrial invertebrates, particularly on grassland.

Where a number of such water bodies are under independent hydrological control, feeding conditions for waterfowl can be optimized by sequentially lowering water levels in different water bodies, to provide a continuity of suitable feeding conditions. It is worth considering lowering water levels at times of year when there is a lack of shallow water available in the surrounding area (Taft *et al.* 2002).

## 8.5 Temporary pools

Temporary pools are water bodies that experience a recurring dry phase at a more or less predictable time of year, or else only fill with water intermittently. Water bodies present only in winter and spring are known as vernal pools. The fauna of temporary pools comprise a selection of more cosmopolitan species also found in permanent water bodies and species largely or completely confined to temporary pools, the latter including a range of crustaceans (e.g. King *et al.* 1996; Williams 1997). These specialists are largely restricted to ancient temporary pools or those close to them. Temporary water bodies also provide critical breeding habitat for amphibians. The regular drying out benefits breeding amphibians by preventing fish and high densities of large, predatory invertebrates becoming established, which would predate their larvae.

The basic principle of managing temporary pools with existing conservation interest is to maintain their historical hydrological regime. In particular, temporary pools should never be drained or deepened to create permanent water bodies. Lengthening the period that pools hold water allows species with longer aquatic stages to complete their life cycles (e.g. King *et al.* 1996), but might allow pools to be colonized by species that out-compete the existing fauna. Reducing the period that ponds hold water might prevent existing species from completing their annual cycle.

Periodic drying out helps retain temporary pools in an early successional state. Some vegetation removal is often necessary to preserve the characteristic, open nature of many temporary pools and prevent them from becoming dominated by tall and rank vegetation. Grazing can also influence the duration of inundation (Figure 8.5).

## 8.6 Permanent ponds and water-filled ditches

Water-filled ditches (dykes or dikes) can support important relict assemblages of invertebrates (e.g. Drake 1998; Watson and Ormerod 2005) and wetland plants, providing they have not been impoverished through past insensitive management and eutrophication. They often support the only remnants of wetland habitat remaining following large-scale drainage. Larger ditch/dike networks

**Fig. 8.5** Vernal pools, climate change, and grazing management. Ancient, vernal pools in California's Central Valley support an endangered fauna comprising branchiopods and the Californian tiger salamander, *Ambystoma californiense*. All are sensitive to changes in the length of inundation.

Experimental manipulations and modelling of Californian vernal pools suggests that cattle grazing as well as changes in precipitation can influence their hydrology (Pyke and Marty 2004). While climate change will influence the quantity and timing of precipitation and evapotranspiration, the period of inundation can also be influenced by whether the pools are grazed. The study found that 3 years after removal of grazing, flooding durations had decreased by an average of 50 days per year. This was probably because there was less soil compaction meaning that water drained from the pools more easily, and also increased loss of water through transpiration from the taller vegetation (San Luis National Wildlife Refuge, California, USA).

comprising open water and areas of swamp and fen vegetation support breeding wildfowl, while their shallow margins can provide suitable feeding conditions for waders/meadow birds nesting on adjacent grassland.

Management of permanent ponds and water-filled ditches are based on similar principles. Networks of water-filled ditches are effectively large, highly branched water bodies containing an enormously high proportion of edge habitat. Some ditches, especially larger ones, have a greater flow and are more similar to canalized rivers.

The basic principle of managing permanent pools and water-filled ditches is to maximize the variation in ditch profiles and successional stages from open water to swamp and fen to maximize the range of conditions for wetland plants and invertebrates. Management may in some cases aim to provide a greater proportion of particular successional stages, vegetation types, and water depths, depending on the specific interest of the site. A range of successional stages can be provided by:

- periodically clearing out sections of vegetation to set back succession;
- if necessary, re-profiling the water body to provide suitable water depths;
- removing vegetation and providing disturbance along the margins to help maintain open conditions and a variety of different microhabitats;
- maintaining desired water depths by manipulating water levels.

Water quality will also be important in influencing the flora and fauna (see, for example, Watson and Ormerod 2004). Water bodies fed by groundwater will usually have higher water quality that those fed by run-off from surrounding agricultural areas. The value of existing ponds and networks of drainage ditches can also be enhanced by excavating new areas of wetland habitat.

Brackish ditches support a more limited range of species than fresh ones, but including a variety of species not found in freshwater, particularly species of beetles. They also support a different aquatic flora, although this tends to be species-poor and lacking many characteristic species. The key to maintaining the interest of brackish ditches is to maintain a range of salinities throughout the entire ditch network. Raising water levels by increasing freshwater inputs has the potential to reduce overall salinity, which will reduce their existing interest, or at least restrict ditches with higher salinity to areas closer to the sea or adjacent estuary.

### 8.6.1 Clearing out vegetation to set back succession

Manual removal of vegetation and sediment is best because it can be carried out sensitively and on a small scale. However, this is impractical in all but the smallest water bodies (but see Section 11.1.3). Larger-scale removal of vegetation from pools and water-filled ditches can be carried out with an excavator, using either a ditch-cleaning bucket, or weed-cutting Bradshaw bucket. Typical ditch-cleaning buckets scoop up both the vegetation and some of the silt beneath it. Weed-cutting buckets only cut and remove the vegetation and their use is considered less damaging to the aquatic fauna and flora. Amphibious weed-cutters can be used in larger channels.

The ideal is to only clear out a small proportion of the vegetation in a pool, or short stretches of any ditch/dike network, at any one time to maximize the chance of plants and animals re-colonizing these cleared areas. As a rule of thumb only clear out about a third of the vegetation and accumulated sediment in pools at any one time. For water-filled ditches, only clear out one side of the ditch at any one time, to enable rapid colonization from the un-cleared side. This is often impractical when managing very narrow ditches using an excavator. For narrow ditches, an alternative is to clear out the whole width of the ditch/dike, but only along very short stretches (Figure 8.6). The potential benefits of these more intricate forms of management have to be set against their higher costs.

**Fig. 8.6** Succession and sensitive ditch/dike management. The flora and fauna of water-filled ditches changes with the length of time since they were last cleaned out. Management of ditch networks usually aims to range of these different stages to help maintain the widest selection of species.

Many submerged aquatic plants, notably in this example sharp-leaved pondweed, *Potamogeton acutifolius*, at one of only a handful of UK sites, are largely restricted to open water in the early stages of succession following ditch clearance (a). Conversely, some species, such as the little ramshorn whirlpool snail, *Anisus vorticulus*, are restricted to mid-to-late-successional ditches with abundant emergent vegetation, but are also intolerant of complete shading. The little ramshorn whirlpool snail is very rare in this region, but occurs in the unexceptional-looking ditch (c). (b) shows a ditch of intermediate successional stage.

Both suites of species require periodic cleaning out of ditches to set back succession and maintain suitable conditions for them. However, the snail is extremely poor at re-colonizing ditches that have been cleaned out along their entirety. The solution successfully trailed at this site is to only clean out 10-m stretches of mid-to-late-successional ditch at a time, to increase the snail's ability to re-colonize stretches as they become suitable for it again (Arun Valley, Sussex, England).

Clearance of ponds and ditches should not be carried out during the bird nesting season. In practice mechanical clearance is usually carried out in autumn after the breeding season and before conditions become too wet for machinery.

The frequency of clearance will depend on the desired proportions of different successional stages. Optimal requirements for aquatic plants and invertebrates can conflict. Much of the conservation interest of aquatic plants in ditches is associated with earlier successional stages, because these tend to support less competitive plants that have declined in lowland areas as a result of eutrophication. Conversely, important assemblages of aquatic invertebrates are often associated with later successional stages dominated by a small number of emergent plants. In practice, though, it is usually possible to cater for both groups by maintaining a range of different successional stages. Maintaining stretches of silted-up, vegetation-choked ditches may, though, conflict with the need to maintain open water for wet fencing (see below) and water transport. One option is to maintain choked conditions in specially excavated stretches of ditch, which do not need to perform these other functions (Section 8.6.5).

The rate at which ditches become choked with emergent vegetation will be faster where nutrient levels are high and ditches shallow and narrow. Succession tends to be slower in brackish water. Ditch-cleaning rotations vary widely, from once every 2 years to as infrequently as once every 15–30 years. In some cases there may be reasons for carrying out ditch/dike clearance on a very short rotation to maintain a high proportion in an early successional stage. However, the benefits of doing this have to be set against the disadvantages of continually removing propagules of vascular plants and stoneworts and the associated risk that they may not re-colonize. Otherwise, since clearing out ditches is always a damaging (and expensive) operation, it is far better to only clear out individual ditches when absolutely necessary to retain the overall desired proportions of different successional stages, rather than simply adhere to a pre-determined rotation.

### 8.6.2 Re-profiling to provide suitable water depths

Re-profiling can be used to provide a range of suitable water depths across the profile of the water body. It is mainly used to introduce shallow, sloping margins on steeper-sided ditches.

Wetland plants and invertebrates vary in their water-depth requirements, but there is typically no marked increase in species richness of invertebrates as water depths increase beyond about 60 cm. Shallow water and muddy margins are favoured by feeding waders/meadow birds that nest on adjacent wet grasslands. It is, though, still useful to provide areas of deeper water to help maintain open water for submerged plants that are not shaded by tall emergent plants and to help maintain at least some water during periods of drought (Figure 8.7). Deeper water is also necessary to maintain the roles of ditches within grassland as wet fences to contain livestock and help transfer water around and off the site. Spoil should ideally be spread away from the margins of ditches and ponds to help maintain an open, shallow profile.

Steeper-sided ditches will also support their own distinctive fauna. Those that are un-grazed or only periodically grazed will provide suitable bankside habitat for water voles, *Arvicola terrestris*. Steeper-sided ditches may also be valuable

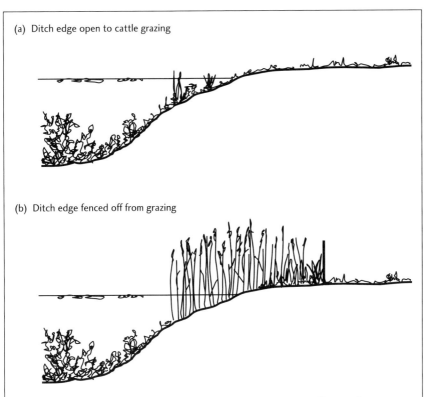

**Fig. 8.7** Profiles of ditch edges and pond margins and the effects of grazing. Creating and maintaining shallow, sloping margins of ditches and ponds provides a variety of habitat conditions in close proximity. The vegetation that develops on these is heavily influenced by grazing or other vegetation removal on their margins.

Grazing shallow margins (a), especially using cattle, creates an open and often diverse mix of marginal and submerged plants and shallow water. Excluding grazing from the margins, for example by fencing (b), usually allows the shallow margins to become dominated by tall, emergent plants, which shade out other marginal and submerged vegetation.

for some invertebrates, such as some ground beetles, that require permanently moist, vegetated conditions. Moisture levels on steep-sided margins remain more constant for a given change in water level than on shallow margins.

### 8.6.3 Management of marginal vegetation

The margins of water-filled ditches can be managed by cutting or grazing, or they can be left unmanaged. If the latter, the margins of ponds and water-filled ditches will usually become dominated by emergent vegetation, rank grasses, and forbs, and in some cases by trees and shrubs. Emergent vegetation will provide suitable habitat for songbirds and invertebrates associated with tall, swampy vegetation and provide nest sites for waterfowl where associated with open water. Scattered, riverside trees and shrubs can be of high value for invertebrates, but will shade ditches and reduce their value for aquatic plants. Scattered trees and bushes on wet grassland will reduce its value for waders/meadow birds and wildfowl (Section 8.12). Rank, grassy vegetation on the margins of ditches will provide suitable conditions for small mammals, including water voles.

As with ditch clearance, the usual approach is to provide a variety of ditch-edge conditions to cater for a range of interests. There will usually be a presumption to maintain open ditch margins in large expanse of grassland important for open-ground breeding waders/meadow birds and wildfowl. A sensible compromise is to maintain grazed, open margins within a large, core area to maximize its value for open-ground birds, while maintaining tall, emergent vegetation along ditches in areas that are otherwise unsuitable for them, such as on their margins, below power lines, and in areas subject to disturbance.

Where ditches run through wet, grazed grasslands their margins can be grazed by livestock. On arable land, cutting will be the only practical method of managing marginal vegetation. Moderate levels of grazing are far better at creating variation in vegetation along ditches than cutting and, because it is less catastrophic, will be less damaging to its fauna. Higher frequencies of cutting, for example once or twice a year, will tend to increase plant species richness compared to less frequent or no cutting (Milsom *et al.* 2004), but will be more damaging to the invertebrate and small-mammal fauna. To minimize these damaging effects, cutting should only be undertaken along short stretches and on only one side of a ditch in any one year. It should also obviously not be carried out during the bird nesting season.

Cattle are far better than either ponies or sheep in creating variation in conditions along the margins of ditches and ponds. Cattle create more small-to-medium-scale variation in vegetation structure, trample more, and will readily enter shallow water and graze and disturb vegetation in it. The overall effect of cattle grazing is to reduce the abundance of tall emergents within their reach, and to replace these with patches of bare ground and lower-growing, grazing-tolerant vegetation (Figure 8.7). Cattle grazing is valuable in providing open shallow-water and marginal habitat for feeding waders/meadow birds and bare and disturbed wet and dry mud for a range of invertebrates, especially beetles and flies.

Sheep are particularly poor at creating variation in conditions on the margins of water bodies. They tend to graze adjacent areas uniformly short (Section 5.4.1) and are reluctant to enter water. However, sheep can be useful where the aim is to graze the surrounding grassland, but retain tall, ungrazed swampy vegetation, especially reeds in ditches. The effects of ponies are intermediate between those of cattle and sheep.

### 8.6.4 Management of water levels

Water levels need to be kept relatively high along suitable profiled ditches to maximize the variety of conditions along them (e.g. Twisk *et al.* 2003). A gradual spring/summer drawdown will provide open mud and damp conditions for feeding waders/meadow birds on wet grasslands and marginal invertebrates and plants.

### 8.6.5 Creating new water bodies within existing networks of ditches

Existing networks of ditches can be enhanced by excavating additional water bodies. This may be particularly valuable in creating ditches that do not need to contain open water to maintain their role as wet fencing or to transport water, and which can therefore be left to become choked with silt and emergent vegetation (Figure 8.8).

## 8.7 Rivers

Most rivers in intensively managed lowland areas have been highly modified to increase their rate of flow and role in land drainage and to maintain suitable conditions for navigation (Figure 8.9). This modification has involved:

- straightening and deepening of their channel;
- isolation of the river from its floodplain to prevent it from flooding farmland and habitation;
- removal of woody debris;
- periodic cutting of weed and dredging.

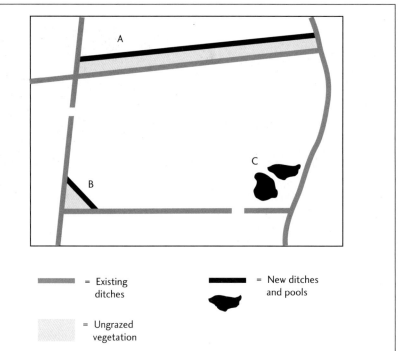

**Fig. 8.8** Enhancing existing ditch/dike networks on wet grassland by excavating new water bodies. Double ditches (A) can be used to isolated strips of marsh from grazing to allow swampy vegetation to develop for nesting waterbirds. Both parallel ditches can be managed, or one left unmanaged, to provide a refuge for invertebrates associated with late-successional, vegetation-choked ditches. Similarly, corners of fields can be cut off (B) to provide ungrazed vegetation and short, vegetation-choked lengths of ditch. Nests in these isolated areas of cover might be easy for predators to find, though.

Shallow, temporary pools (C) can provide habitat for breeding amphibians and invertebrates absent from the permanent water of the ditch/dike network. The excavated material can be used to build or widen crossings over ditches. These isolated pools might have better water quality than surrounding ditches that receive fertilizer runoff.

None of these enhancements should significantly interfere with farming operations, although they will obviously reduce the area available for grazing.

Debris has also been removed with the aim of removing obstacles to fish migration. In practice, fish can migrate past such woody debris during periods of high river flow, and large woody debris is incredibly important in providing suitable habitat for fish by creating heterogeneity within the river channel (e.g. see Middleton 1999).

**Fig. 8.9** Heterogeneity in river channels. Rivers obviously vary enormously in their natural state, but key elements in most are variations in flow and depth that create a variety of conditions for plants and animals. This relatively unmodified stretch of the Rio Almonte in Extremadura, Spain, contains fast-flowing and deep areas in its main channel, shallower cut-off channels and almost stagnant pools, muddy and stony areas, and a variety of herbaceous vegetation and scrub on its margins.

Most management of rivers for conservation and other environmental benefits involves re-introduction of variation in channel morphology, and in some cases re-connection of the river to its floodplain (Figure 2.5). A range of additional methods have been used to improve rivers for fishing interests, especially for salmonid fish. These include:

- restoration of spawning habitat;
- increasing cover for fish;
- removal of artificial barriers to fish migration;
- management of bankside and channel vegetation.

River and floodplain restoration is a large and complex subject outside of the scope of this book.

Otherwise, the best form of management of rivers for conservation is benign neglect, so far as is possible while still maintaining their other uses. Coarse, woody debris in rivers is an especially valuable habitat and should be left intact wherever possible. Smaller woody debris is also very valuable for invertebrates. Frequent weed-cutting is damaging to the aquatic vegetation, while dredging will damage the benthic fauna, especially freshwater mussels, Unionidae. If dredging is necessary to maintain suitable conditions for drainage and navigation, then it should be restricted to the centre of the channel as far as possible. Weed cutting should be minimized and only undertaken in separate and alternating blocks at any one time to maintain refuges for species in uncut areas (e.g. Aldridge 2000; Baattrup-Pedersen *et al.* 2002; Vereecken *et al.* 2006).

Heavy grazing beside rivers can result in excessive erosion and inputs of sediment that may smother gravelly fish spawning habitat. Lack of vegetation beside rivers also allows more run-off of fertilizers and pesticides into the water (Section 8.2). Heavy grazing can also damage valuable riverine scrub, and should therefore be avoided beside rivers, although the benefits of other forms of grazing will depend on the habitats abutting the river.

## 8.8 Swamps and fens

The relative proportions of shallow, open water, swamp, fen, and scrub and the structure of the vegetation will be important in influencing the wetland fauna. Vegetation structure will be especially important in influencing conditions for invertebrates. The overall proportions and distribution of the different successional stages will influence conditions for birds. Densities of breeding wildfowl are highest where there are equal proportions of swamp and open water (Kaminski and Prince 1981; Linz *et al.* 1996; Smith *et al.* 2004a). Densities of breeding waterfowl tend to be highest where there is a high level of interspersion of swamp and open water (e.g. de Szalay and Resh 1997; Kaminski and Prince 1981).

In shallow, moderately to very nutrient-rich wetlands in much of Europe reed most commonly attains dominance in the absence of significant disturbance. These reed-dominated swamps are known as reedbeds and in Europe support a characteristic invertebrate and bird fauna (e.g. Hawke and José 1996; Poulin *et al.* 2002). Much of the experience and knowledge of managing swamps and tall-herb fens in Europe is from managing reed-dominated habitats.

The composition of the avifauna of reed-dominated habitat in Europe is strongly influenced by the following (from Van der Hut 1986; Graveland 1998; Jenkins and Ormerod 2002; Poulin *et al.* 2002; Adamo *et al.* 2004; Brambilla and Rubolini 2004; Gilbert *et al.* 2005a, 2005b):

- extent of open water;
- length and nature of the swamp/open-water interface;

- physical structure and dominant plant species in the swamp and fen;
- duration and timing of flooding;
- extent of scrub.

Management for birds has often focused on restoring and maintaining early successional reed that is covered by water in spring and summer (wet, or water reed) and its interface with open water (e.g. Self 2005), since these contain a characteristic, albeit limited, avifauna. Maintaining a large proportion of wet edge also provide habitat for breeding waterfowl.

Micro-habitats considered to support the richest invertebrate fauna are marginal, transitional, and edge habitats, and area with a significant accumulation of litter, be they in winter-flooded areas with summer dry mud, or other seasonally wet or at most shallowly winter-flooded ground. The presence of scattered scrub adds additional species to the fauna, without the loss of these reedbed invertebrates (Kirby 1992b). Hence, there are potential conflicts of interest when managing particular areas of reedbed to benefit breeding birds and existing invertebrate assemblages. Reed-dominated swamps tend to contain few other plant species. Drier areas of fen in which reed can be an important component can be very botanically rich. There are therefore again potential conflicts between maintaining the botanical interest of reed-dominated habitats and creating wet reedbed and open water to benefit key bird species. In practice, though, the requirements of all these groups can generally be catered for at larger sites, particularly those with variation in topography and associated water regimes.

In North America swamps dominated by cattail/bulrush, *Typha* spp., support a characteristic avifauna, and reed is usually considered an undesirable invasive species, especially in coastal wetlands on its north Atlantic coast (Section 9.4).

The primary considerations when managing swamps and fens will be the:

- annual water regime;
- desired proportions and locations of open water, swamp, and fen, and, where relevant, also of associated woodland, scrub, and grassland;
- desired vegetation structure of these different habitats.

These can be influenced by manipulating water levels and by vegetation removal through grazing, mowing, and burning, as described in the following sections.

### 8.8.1 Manipulating the annual water regime

Water levels in swamps and fens tend to be highest in winter or spring and then fall through late spring until autumn. Often, the rate of spring and summer drawdown is greater than desired, especially in small areas of swamp and fen surrounded by drained land. The rate that water levels fall during this period can be reduced using the methods described in Section 8.1 to:

- maintain high enough water levels to continue to support characteristic vegetation types;
- provide flooded swamp for breeding birds.

Water levels during spring to autumn (i.e. the growing season) will be important in influencing the successional stage of the vegetation and its associated fauna. If water levels are too low then areas will be invaded by drier-ground species, although the rate of succession can be slowed to some extent by vegetation removal (e.g. Fojt and Harding 1995). Maintaining water levels too high for the existing plant assemblage will reverse succession. In particular, introducing surface flooding during the growing season to formerly unflooded species-rich fen will result in its replacement by botanically species-poor swamp.

Maintaining surface flooding in reedbeds in spring and summer will provide wet reed of high value to some breeding bird species and allow fish to penetrate the margins of reedbeds abutting open water and thereby provide suitable feeding conditions for birds such as great bitterns (Gilbert *et al.* 2003). Prolonged flooding of reedbeds in summer probably also increases the invertebrate food supply for some nesting songbirds (Poulin *et al.* 2002), and also affects the availability of nest sites for birds that nest on or close to the ground. In reed-dominated and probably also other types of fen, though, submerging dry reedbed in summer will replace its characteristic and valuable terrestrial and semi-aquatic invertebrate fauna with more ubiquitous, aquatic species of far lower conservation value. Relatively few invertebrate species are characteristic of reed growing in shallow water (Kirby 1992b; Bedford and Powell 2005).

It is unlikely to be necessary to supplement winter water supply in swamps and fens. Hydrological management in reedbeds managed by winter cutting involves *lowering* water levels for a period during winter to allow access. This is quite different to a natural-water hydrological regime.

### 8.8.2 Manipulating the proportions of open water and swamp using periodic drawdowns and year-to-year fluctuations in water levels

Periodic drying out can be used to increase or decrease the proportion of swamp relative to of open water. Emergent plants that form swamp and fen habitat have two methods of spreading: by vegetative growth of existing plants and by germination of seedlings on moist mud or, in the case of some species, in very shallow water. Emergent plants sometimes cease expanding vegetatively, or start dying back, due to herbivory by geese and other wildfowl and muskrats, *Ondatra zibethicus*, or through disease, erosion, or other environmental stresses. Maintaining consistently high water levels over many years has been implicated in regress of reedbeds (Van der Putten 1997). Where emergent vegetation has died back, its extent can be increased by lowering water levels in spring and summer to expose moist mud for seeds of emergent plants to germinate and establish on, in the same manner as described for moist-soil management (Section 8.4.1). Once seedlings have established, water levels can be raised in autumn, taking care not to completely cover and drown them. These plants will then spread through vegetative growth to re-form new expanses of swamp.

If the swamp has expanded more than desired, then the area can be dried out to allow the swamp to be cut, burnt, or grazed to reduce its extent. The area can then be re-flooded to recreate open water. Burning wetland vegetation creates more open ground than mowing, and so tends to result in more growth of ruderal vegetation during subsequent drawdowns (de Szalay and Resh 1997). Kostecke *et al.* (2005) found that the method of reducing the extent of bulrushes/cattails (burning, disking, or grazing) had little or no effect on subsequent invertebrate food supply for birds following re-flooding.

The mosaic of reedbed and open water in the wetlands in the Oostvaardersplassen in The Netherlands is maintained by a combination of fluctuating water levels and grazing by moulting greylag geese, *Anser anser*, which results in periodic expansion and regression of reed. During summers when water levels are high, geese graze back the edges of the reeds from the safety of adjacent water, reducing their extent. In summers when water levels are low, the reeds become surrounded by exposed mud. The moulting (and therefore flightless and vulnerable) geese are unwilling to walk on this mud and so do not graze back the reeds in these years. Furthermore, the damp mud is colonized by reed and willow seedlings and by ruderal plants. When water levels rise again, they flood out the ruderal plants, making their seeds available to wintering wildfowl, create new areas of wet reedbed in the manner previously described, and drown any seedlings of willow that have not grown tall enough to protrude above the water's surface. This results in re-expansion of the reedbed (Ter Heerdt and Drost 1994).

Using periodic drawdowns to manipulate the relative proportions of open water and emergent vegetation will be impractical or unacceptable in many wetlands. It will risk temporary or permanent extinction of less-mobile invertebrates in isolated wetlands, where it is only possible to dry out all, or most, of the habitat. In these situations an alternative is to control succession by removing vegetation and by lowering the surface of the ground relative to that of the water level.

When creating new wetlands, it is worthwhile designing them so they contain a number of separate hydrological units. This will allow periodic drying out of individual units as required, while maintaining suitable conditions for wetland species in others. The range of hydrological conditions within individual units can be further improved by increasing variation in topography within them.

---

**Fig. 8.10** Raising water levels to set back succession. Lake Hornborga in Västergötland, Sweden, is a formerly drained 3500-ha shallow lake that had become almost completely covered with dry reedbed and scrub by the mid-1960s.

An ambitious restoration project began in the early 1990s to restore the lake's importance for waterbirds by increasing the extent of open water by removing reedbed and scrub. The area of reedbed was reduced by 1200–1500 ha by a combination of burning and rotovating the reed using specially designed amphibious machines. Eight hundred hectares of wet scrub and woodland were removed and water levels raised by an average of 0.85 m. This has created shallow water where there was once reed bed and scrub (a) and swampy vegetation where there was wet woodland (b).

However, the raising of water levels has also resulted in a large increase in numbers of greylag geese. Goose grazing is probably now the main mechanism influencing the extent of reedbed, which has now decreased to just 50–100 ha.

Although contrived, creating different hydrological units will in many ways more closely mimic the natural fluctuations in water levels and other forms of disturbance that create diversity in larger, more natural wetlands.

### 8.8.3 Preventing or reversing succession by long-term raising of water levels

Raising water levels will set back succession across an entire hydrological unit, and can be a relatively easy method of setting back succession over a large area (Figure 8.10), providing this does not cause unwanted flooding of adjacent land. The disadvantage is that since most areas of swamp and fen are relatively flat, raising water levels high enough to set back succession in one area is likely to cause detrimental flooding of areas of drier fen within the same hydrological unit. This may be particularly damaging to its existing flora and, in particular, its invertebrate fauna. One option is to hydrologically isolate different areas.

On peat soils, the surface of the wetland may already be higher than areas of surrounding drained, oxidized peat. Where this is the case, the only way to maintain high water levels will be by surrounding the wetland with a buried impermeable membrane to reduce seepage losses of water into surrounding lower land. This is expensive, though.

Raising water levels by supplementing inputs from watercourses or from abstraction has the potential to change the chemistry of the water, particularly by introducing water with higher water levels and diluting the relative contribution of high-quality, base-rich ground water. In areas of spring-fed vegetation which are drying out due to reduction of spring flows, a better option is to carry out small-scale, sensitive lowering of the substrate to a level closer to the water table.

### 8.8.4 Lowering the substrate to set back succession and provide open water

There are a number of terms used to describe different types of excavation used to set back succession and create open water. Sod cutting refers to digging small areas to a shallow depth to provide very shallow water or unflooded ground with a high water table (Figure 8.11). These areas are also known as turf ponds. Bed-lowering refers to excavation or relatively large, slightly deeper areas, principally to provide open water that is relatively quickly re-colonized by swamp vegetation (Figure 8.12). Excavation of ditches is used to provide more permanent areas of open water and to maximize the length of swamp/open-water interface per volume of material excavated. The key decisions when digging turf ponds,

**Fig. 8.11** Sod cutting and turf ponds. Small-scale, removal of the surface layers of peat, known as turf stripping or sod cutting, can be used to set back succession in bogs and fens and expose the buried seedbank.

This area had a thin layer of peat removed from it 20 years ago, and now supports shallow open water and very species-rich fen vegetation. The shallow pools created contain an exceptional dragonfly and damselfly fauna, including the stunning darter dragonfly, *Sympetrum piedmontani*, here on the north-western edge of its European range. Surrounding un-lowered areas are dominated by more species-poor stands of purple moor-grass and patchy scrub (Plateaux, Noord-Brabant, The Netherlands).

sod cutting, and bed-lowering are the depth of excavation and extent of the area excavated.

Many turf ponds created by past small-scale removal of peat for fuel now support highly valued assemblages of plants rare or absent from the adjacent drier un-lowered areas of fen (e.g. Giller and Wheeler 1986). Patchy, small-scale (between less than 1 m$^2$ and several square metres) turf stripping and sod cutting will provide a mosaic of different hydrological conditions and increase small-scale variation vegetation. Providing high-water-level conditions through excavation, rather than raising water levels, has several advantages. If the fen has dried out significantly, then excavation will remove the surface layer of dried out, oxidized surface peat. These would otherwise release high levels of nutrients following raising of water levels and thereby result in the development of

**Fig. 8.12** Large-scale bed-lowering. At this site, large-scale bed-lowering has been used to set back succession on areas formerly prevented from succeeding by small-scale peat extraction.

Small-scale peat cutting formerly maintained a mosaic of strips of land at different heights and supporting different successional stages. The name of this wetland, De Weeribben, is formed from the Dutch words *weer*, meaning turf pond, and *ribben*, meaning the narrow strips of land between the turf ponds on which the extracted land was laid out to dry. This traditional management maintained a mosaic of strips of open water, reedbed, and areas of wet woodland in higher, drier areas (De Weeribben National Park, Overijssel, The Netherlands).

less highly valued, nutrient-rich fen vegetation. Removing the surface peat will also expose any buried seedbank, which might contain propagules of plants not currently present in the vegetation. There is also evidence that re-vegetated turf ponds experience smaller fluctuations in water levels compared to the peat surface than to nearby unexcavated areas of at the same altitude. This might be because the peat that has infilled the excavations is looser, and so contracts and expands as water levels rise and fall. These more stable water levels are thought to benefit the development of species-rich fen vegetation (Giller and Wheeler 1986).

Excavation to a lower level is usually carried out primarily to provide open water and wet swamp, especially reedbed, for birds. The ideal is to lower some areas to a shallow-enough depth that they become colonized by emergent vegetation, while also excavating deeper areas that remain as open water, thus increasing the extent of valuable open water/swamp edge. Leaving reed rhizomes in excavated areas will usually result in the rapid re-establishment of reed, unless the rhizomes are exposed to and damaged by frost during bed-lowering. Grazing, particularly by geese, can restrict or prevent re-growth of emergent vegetation in the open water of lowered areas. Removing all the organic matter and exposing the mineral substrate will provide a relatively infertile substrate. Providing that nutrient levels in the water are low (typically if the site is fed by high-quality ground water), then these lowered areas can provide suitable conditions for colonization by nutrient-poor aquatic vegetation of high conservation value.

When creating open water it is important to consider its suitability for fish. Making areas suitable for fish will benefit fish-eating birds such as herons and egrets, but reduce the suitability of these areas for breeding amphibians and invertebrates vulnerable to fish predation. Suitability for fish will be increased by:

- creating more permanent and deeper water, in which fish can survive during warm weather when shallow water becomes de-oxygenated and during cold weather when it freezes solid;
- connecting newly created areas to other deep water where fish can survive and from which they can re-colonize.

When creating linear ditches, the main considerations are their:

- depth;
- width;
- bank profile.

Blind-ended ditches, particularly those sheltered from the prevailing wind, tend to become more stagnant and therefore may have a different flora and fauna to those with more through-flow of water.

In general, the deeper and wider the ditch/dyke, the longer it takes to vegetate over with emergent vegetation. In the absence of further management (see next section) or grazing by wildfowl or wild mammals, emergent plants will in most cases eventually expand across even deep ditches in the form of floating hover. Wider ditches also have a smaller proportion of their area shaded by fringing emergent vegetation, and are thereby more suitable for growth of submerged plant species. Abundant submerged vegetation provides better habitat for invertebrates and wildfowl and probably also better conditions for fish. The value of ditches in providing open water/wet swamp edge can be enhanced by suitably profiling their margins (Figure 8.13).

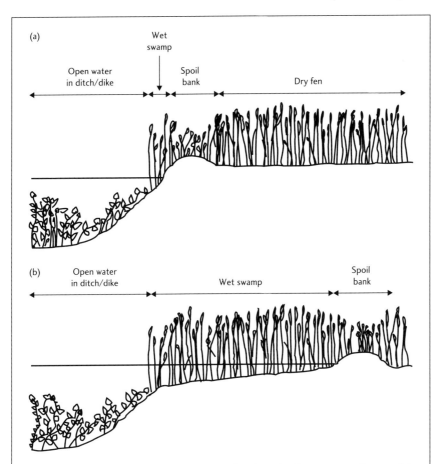

**Fig. 8.13** Ditch/dike profiles through reedbeds and other swamps. Some of the most valuable areas in reedbeds and other swamps, especially for birds, are the margins of wet swamp and open water. This valuable edge habitat can be created and maintained on the margins of ditches, but only if they have a suitable profile.

Clearing out of ditches and depositing the spoil on their margins produces a steep slope with little wet swamp edge (a). The deposited ditch spoil prevents access by fish into the reedy margins, where they can be important prey for birds such as great bitterns.

A larger area of wet swamp can be created by excavating wide, sloping edges to the ditch/dike (b). Fish can enter the margins of the swamp if spoil is placed far back from the ditch/dike edge and if gaps are left in the spoil bank.

Disposal of spoil following large-scale bed-lowering and ditch excavation can be a big practical constraint. Excavated material will reduce in volume as it dries out and decomposes. Leaving banks that protrude above the water in the middle of reedbeds and other swamps provides suitable habitat for water voles, which appear less susceptible to predation by introduced American mink in reedbeds compared to along other water channels (Carter and Bright 2003). Conversely, banks provide higher areas on which scrub can establish, which is undesirable where a principal aim of bed-lowering or ditch creation is to provide open water and open, wet swamp for wetland birds. Raised spoil banks might also provide access for mammalian predators that predate nests. Removing spoil from the site is expensive, though.

### 8.8.5 Long-term maintenance of open water

The basic principles of managing ditches in swamps and fens are largely similar to those in other habitats (Section 8.6), except that:

- because they are surrounded by tall emergent plants, the main objectives of management are usually wet swamp along their margins and open water in their centres;
- regular management is logistically more difficult, since margins are unlikely to be suitable for grazing and mechanical clearance in swamps and fens is more difficult.

There are two approaches: periodically scraping back and removing large quantities of floating swamp vegetation and accumulated silt using an excavator; and cutting encroaching emergent vegetation more frequently using amphibious machinery (see Figure 8.16). The practical difficulties of clearing out vegetation, and the fact that the area of marginal vegetation being cleared usually represents only a small proportion of the area of similar swamp, mean that management is rarely carried out on such a small scale and as sensitively as along ditches on more open habitats. As with excavation of open water, the material removed should be deposited away from the margins of open water (Figure 8.13).

### 8.8.6 Vegetation management: differences between mowing, burning, grazing, and non-intervention

Vegetation removal by mowing, burning, and grazing retards the rate of succession and loss of wet-fen plant species. However, maintaining both suitable hydrology and vegetation management is the only long-term solution for maintaining characteristic types of fen vegetation (e.g. Fojt and Harding 1995).

Compared to non-intervention, removing vegetation by mowing, burning, or grazing:

- in most cases increases plant species richness by preventing dominance by one or a limited number of large swamp or fen plants;
- prevents the accumulation of litter, and thereby reduces the rate of increase in height of the substrate relative to the level of the water;
- prevents or reduces the rate of establishment of scrub, although the effects of grazing on this are quite variable.

Cutting and removal of vegetation and burning also reduces nutrients, thereby favouring less-competitive plant species. Grazing re-distributes nutrients. Differences between grazed, cut, and unmanaged wetland vegetation are shown in Figure 8.14.

The general differences between the effects mowing and removal of vegetation, burning and grazing on the structure, plant species composition, and invertebrate fauna of swamps and fens are similar to those in dry grasslands (Section 5.3). Although all these methods of vegetation removal tend to increase plant species richness by preventing dominance of bulky monocotyledons and forbs, they differ in their specific effects on plant species composition. In particular, grazing encourages lower-growing plants, especially smaller, tillering grasses (e.g. Stammel *et al.* 2003). Continual, heavy grazing converts swamps and fens to open water and wet grassland. Burning tends to remove more litter than mowing, thereby favouring plants that need to frequently re-establish by seed in gaps (Cowie *et al.* 1992; Kost and De Steven 2000). Repeated cutting converts tall fens into short, fen-meadow (Figure 8.14).

Both mowing and burning remove virtually all of the above-ground vegetation in a given area at the same time. This results in relatively uniform vegetation in areas cut or burnt at the same time. The sudden and catastrophic removal of vegetation is damaging in the short term for invertebrates, and possibly has longer-term, detrimental effects on less-mobile species, such as spiders (Decleer 1990; Cattin *et al.* 2003). However, as when managing other habitats by mowing or burning, these negative effects can be minimized by only mowing or burning small areas at any one time (i.e. on rotation). This will also increase larger-scale vegetation diversity by produce a variety of different stages of re-growth. Neither mowing nor burning, though, provide the range of microhabitats for invertebrates produced by grazing, as described below.

Mowing was formerly widespread in parts of Europe to provide reed or sedge for thatching (Figure 8.15), litter, and marsh hay. Many areas of fen have often only survived because they have been managed to provide these products.

**Fig. 8.14** Effects of grazing, cutting, and non-intervention on wetland habitats. The floodplain of the River Prypyat in southern Belarus is considered the least modified in Europe. The vegetation in unmanaged areas is a mosaic of rank swamp, fen, and scrub (a), small areas of open water (b), and wet woodland (c). The main factors contributing to diversity of habitats and wildlife in the floodplain are variations in topography and seasonal and year-to-year variations in water levels. These prevent any one vegetation type from becoming dominant over large areas.

**Fig. 8.14** *continued.*

All forms of vegetation management on the floodplain reduce the growth of trees and shrubs.

Management by mowing for marsh hay (d) and light-to-moderate grazing (e) both convert tall fens into short fen and fen meadow, but differ in their specific effects on the vegetation and its fauna.

Heavier grazing (f) converts fens and swamps into wet grassland and open, shallow water.

Burning has been used more widely in North America (see review by Middleton *et al.* 2006). Harvesting of fen products has become uneconomic in most areas, making it difficult to sustain the frequency of mowing that created and maintained characteristic types of fen vegetation. Where this is the case, the challenge is to determine how *infrequently* fen can be mown to maintain its conservation value, and whether it can instead be maintained or increased through other

**Fig. 8.15** Mowing fen vegetation. Many characteristic and species-rich assemblages of plants have been created by periodic harvesting of fen vegetation.

The flexible leaves of great fen-sedge, (a), are used to cap roofs of houses thatched with common reed, because reed stems are too brittle to bend over the ridge of the roof. Great fen-sedge is usually cut for thatch in summer on a 3–5-year rotation. In mixed fen this management favours great fen-sedge over reed.

(b) shows a calcareous flush containing species-rich sedge and vegetation dominated by marsh lousewort, *Pedicularis palustris*. This low, open community is maintained by annual mowing and removal of cut vegetation. In the absence of this management the smaller plants would be overtopped and shaded out by more bulky ones. Marsh lousewort is partially parasitic on the roots of other plants and so also helps maintain open vegetation (Market Weston and Thelnetham Fens, Suffolk, England).

low-cost means. These include larger-scale mechanical removal and low-cost grazing. Disposal of cut material can be difficult. Burning it for power generation offers opportunities in some areas.

Grazing removes vegetation more selectively than either mowing or burning, with the degree of selectivity varying with livestock type and grazing pressure. Thus, grazing favours unpalatable plants that are avoided by grazers, and those that are well able to tolerate repeated defoliation. Grazing therefore has the potential to create greater variety in vegetation structure. Furthermore, the distribution of livestock, and hence grazing pressure, is also heavily influenced by water levels. Grazing also:

- creates more bare and disturbed ground, especially at high stocking levels, and consequently ruderal vegetation than mowing or burning;
- provides a source of dung for invertebrates.

Wet peat is particularly easily poached by livestock. Concerns have been raised over damage to peat in fens by grazing. Heavy trampling can cause the soil to lose its structure, creating unconsolidated areas devoid of vegetation. Trampling also creates small-scale variations in topography. Creation of shallow, open water and wet mud by grazing, and associated growth of ruderal vegetation, benefits wintering, seed-eating wildfowl and breeding waders/meadow birds that feed in shallow water and muddy margins. Shallow water kept open by grazing and trampling can support abundant amphibians, particularly frogs.

There is little information regarding the effects of grazing on invertebrates. Variation in vegetation composition and structure at medium grazing levels is likely to increase the range of niches available for invertebrates, especially compared to non-intervention, mowing, and burning. Bare mud, shallow, open water, and dung also provide additional niches for invertebrates, which are absent from areas managed by mowing, burning, or non-intervention. At low-to-medium grazing intensities, the effect of grazing should be to provide these additional niches, while still retaining even quite grazing-intolerant invertebrate species within patches of ungrazed or only lightly grazed vegetation (e.g. Ausden *et al.* 2005). Very high grazing levels run the risk of the loss of invertebrates associated with high levels of accumulated litter, removal of nectar sources and food plants, and complete loss of areas of swamp and fen of importance for invertebrates.

*Mowing and burning*

Mowing is only practical when it is dry enough to allow access, when the water is frozen solid, or by using amphibious machinery (Figure 8.16). Burning is only practical in winter when the vegetation is no longer green. Any burning should be carried out using back-fires and taking suitable precautions as outlined in Section 5.6, including the cutting and dowsing with water of suitable firebreaks/fuel breaks.

Mowing to harvest fen products obviously involves removal of cut material. Mowing specifically for conservation also requires removal of cuttings to reduce litter accumulation and to prevent it from smothering re-growth. Cuttings are raked up and often burnt on site or otherwise removed. Piles of litter provide valuable habitat, though, particularly for over-wintering invertebrates, and mimic accumulations of litter washed up by natural floods. It is worth retaining some piles of litter on areas of low botanical interest.

The effects of mowing and burning swamps and fens on plant species composition and structure will depend primarily on their:

- timing, particularly whether during the growing season or in winter;
- frequency.

Increasing the frequency of summer cutting increases plant species richness by further preventing small numbers of more vigorous plant species from outcompeting a larger number of less-competitive species. In general, plant species richness in fens tends to decrease with increasing above-ground vegetation biomass (Wheeler and Giller 1982). A variety of characteristic assemblages of fen plants have been created and maintained by a combination of different mowing and water regimes. Fen and swamp should obviously not be cut during the bird breeding season.

In reed-dominated vegetation both burning and cutting in winter increase plant species richness and result in shorter, thicker and higher densities of live reed stems compared to non-intervention. Burning is more effective at reducing litter and also tends to result in higher flowering densities of reed (Gryseels 1989a; Cowie *et al.* 1992). This might be important for species that feed on their flowers and seeds. Experiments have found little or no difference in the invertebrate fauna between cut and carefully winter burnt wet reedbeds (Ditlhogo *et al.* 1992). It is important not to burn in dry conditions in summer or autumn, though, when the fire will be hotter and burn deeper into the litter. This will probably be more damaging to invertebrates. Despite the results of this research, many site managers still consider burning more damaging to invertebrates than cutting. Areas of cattail-/bulrush-dominated swamp can be reduced by combinations of burning followed by disking or heavy grazing (e.g. Kostecke *et al.* 2004).

The effects of timing and frequency of cutting are fundamental to the management of reedbeds in Europe. Cutting to harvest dried reed stems for thatch is carried out in winter, after their lower leaves have dropped. This management perpetuates dominance by reed. Cutting reed in summer reduces the dominance of reed and creates a more species-rich mixture of it and other tall monocotyledons and forbs (e.g. Gryseels 1989b). Rotational summer mowing also favours great fen-sedge over reed (Figure 8.15), whereas annual or biennial summer mowing favours reed sweet-grass, *Glyceria maxima*, over reed. Cutting reed underwater in summer using a reciprocating mower prevents it from transferring oxygen down to its rhizomes. This can kill it and is a useful method of creating shallow, open water.

Cutting reed in winter usually requires lowering of water levels to allow access. This can compromise attempts to achieve suitable water levels for breeding birds in early spring.

Although cutting or burning in winter may be necessary to prevent, or reduce the rate of, succession in many in reedbeds, winter cutting also leaves areas open and unsuitable for nesting birds the following spring. In southern Europe, where re-growth of reed is rapid, winter reed cutting only reduces densities of early nesting, resident passerines (Poulin and Lefebvre 2002). In Northern Europe, where re-growth is slower, winter reed cutting also reduces densities of later-arriving migrant warblers (Graveland 1999). Winter cutting also eliminates moth larvae which overwinter in reed stems, and which are important prey for some reedbed songbirds. Studies have produced conflicting results regarding the effects of winter reed cutting on total biomass of invertebrate prey of reedbed songbirds (Poulin and Lefebvre 2002; Schmidt *et al.* 2005a).

The frequency of winter reed cutting used to arrest succession varies between sites, but is typically once every 5–15 years. Commercial reed cutting for thatching takes place on a 1–2-year rotation, known as single or double wale, to provide high densities of strong, straight reed stems. Annual cutting of large areas of reedbed is detrimental to some nesting birds for the reasons just described, and is also damaging to its invertebrate fauna. Therefore, a compromise between the needs of commercial cutting and conservation is to cut only a proportion of the reedbed for thatching in any one year, carry this out on a 2-year rotation, and cut other areas on a longer rotation or not at all. Where there are limited resources for reed cutting, cutting should be concentrated on areas where it will provide the greatest benefits, such as in maintaining early-successional, wet reed and maximizing edge, while leaving larger areas of drier areas unmanaged or only infrequently cut. There is a variety of machinery that can be used to cut and remove reed and other vegetation at sites where commercial cutting is no longer economic, or desirable on account of its short rotation (Figure 8.16).

## *Grazing*

Grazing of swamps and fens is only practical using cattle, water buffalo, and ponies. Sheep and goats are unsuitable for grazing very wet habitats, sheep being susceptible to foot-rot.

An initial consideration when considering swamps and tall-herb fens is the composition of habitats within the grazing unit. All livestock, even those considered particularly suited for use in wetlands, require access to dry ground to lie up on and woodland or scrub for shelter. Hence grazing units in wetlands need to also contain sufficient areas of these other habitats.

**Fig. 8.16** Mechanised cutting and removal of common reed. Where small-scale mowing of fens for thatch and litter is no longer economic, larger-scale machines can be used to cut and remove vegetation to retard succession.

(a) The 'Truxor' is an amphibious machine that can be fitted with a variety of tools. These include a reedcutter unit for cutting common reed and other vegetation underwater, and a reedrake for collecting, transporting, and piling up cut vegetation.

(b) The Softrack is a tracked, low-ground-pressure vehicle for use on soft and shallowly flooded ground. It cuts reed and other vegetation with a flail and blows the chopped material into a loading bin on its back.

The effects of grazing wetlands can be difficult to predict, because of often large spatial variations in grazing pressure. The main factors to consider when setting up a grazing regime are the desired/likely:

- proportions of open water, swamp, fen, and grassland vegetation created and maintained through grazing;
- dominant plant species and structure of the swamp and fen areas resulting from grazing and their associated fauna, particularly birdlife.

The main factors that will influence the overall effects of grazing are the:

- type of livestock;
- overall grazing pressure;
- timing of grazing, whether summer-only or year-round;

- spatial and year-round variation in water levels and how they influence the distribution of livestock.

Grazing pressure within a given area can be difficult to predict, particularly in mosaics of swamp, fen, open water, and grassland. It will vary in relation to the:

- relative preference for the particular type of vegetation relative to that elsewhere in the grazing unit;
- physical access to the particular area by livestock;
- experience of familiarity with the site of livestock;
- territorial behaviour of different groups of livestock.

The often patchy nature of grazing caused by differences in water levels during and between years can create abundant variation in vegetation composition and structure (Figure 8.17).

Livestock often, but not always, prefer grazing drier, grassy areas to entering water and grazing swamp and fen vegetation, although their grazing preferences vary according to the range and palatability of different types of vegetation in their grazing unit (e.g. Duncan and D'herbes 1982; Vulink et al. 2000; Menard et al. 2002). Where livestock prefer drier grassland, they tend to concentrate any grazing of swamp and fen in areas close to this dry ground. They often also concentrate grazing nearer areas of shade (in summer) and shelter. Livestock usually avoid entering stands of tall, dense fen vegetation. They can be encouraged to graze swamp and fen vegetation by cutting the vegetation first to provide areas of short, succulent re-growth and by minimizing the quantity of alternative forage in drier areas.

The distribution of livestock and hence grazing intensities can be influenced by manipulating water levels to alter grazing regimes. Crossing points can be constructed over deep ditches that would otherwise prevent access to particular areas. Livestock are often initially wary of using these.

Livestock tend to increase the area over which they forage as they become more familiar with the site. Their grazing influences the vegetation, which in turn affects their grazing patterns. Hence the distribution of livestock and vegetation composition and structure often change substantially over a number of years. Ponies in extensive and naturalistic grazing systems often form strong, social groups, which also influence grazing patterns.

Grazing is often used to create and maintain mosaics of wet grassland and reedbed. There is, though, a danger of these mosaics not providing suitable habitat for either wet-grassland- or reedbed-specialist birds. The presence of small blocks and strips of reed can make areas unsuitable for open-grassland birds that

# 272 | Freshwater wetlands and water bodies

**Fig. 8.17** Extensive grazing and variable water levels. Many large wetlands are managed by a combination of variable water levels and extensive grazing which, in combination, can produce a dynamic mosaic of open water, wet grassland, and swamp (Parque Nacional de Doñana, Huelva, Spain).

prefer an unobstructed view, whereas small areas of reedbed will not be large enough to support reedbed-specialist birds either.

*Type of livestock*

Basic differences in the effects of cattle and pony grazing on vegetation composition and structure are similar to on dry grasslands (Section 5.4.1). Cattle are less selective and at moderate grazing intensities produce more tussocky vegetation. At moderate grazing intensities ponies typically create mosaics of closely cropped lawns interspersed with largely avoided areas of tall swamp, fen, and other vegetation (Figure 8.18). Water buffalo are similar to cattle in their feeding habitats, but also create wallows. They are also better at swimming off across deep water and escaping.

Cattle and ponies have broadly similar food preferences in wetlands, although cattle eat a higher proportion of broad-leaved plants and are likely to survive less well than ponies in grassy habitats when food is limiting (e.g. Duncan and D'herbes 1982; Menard *et al*. 2002). Grazing and trampling by cattle and ponies in spring and summer reduces the extent of reed relative to most other tall monocotyledons such as sea club-rush/alkali bulrush, cattails/bulrushes, and reed sweet-grass (e.g. Duncan and D'herbes 1982; Kostecke *et al*. 2004; Ausden *et al*. 2005). In drier areas moderate to heavy grazing in summer by cattle or ponies results in the replacement of reed by grassland (e.g. Vulink *et al*. 2000). Year-round moderate and high levels of grazing and trampling by ponies also reduces the extent of sea club-rush/alkali bulrush relative to more open vegetation (e.g. Bassett 1980).

Positive attributes of breeds of cattle and ponies for use in wetlands are small size, which helps prevent them becoming stuck in mud, and tolerance of insect bites and internal parasites. A preference for entering water is desirable, unless the aim of grazing is to keep the vegetation on dry ground short and open, while retaining ungrazed, tall emergent vegetation in water. Livestock do not need to be particularly hardy if only used for summer grazing, since the quality of forage in complexes of grasslands and swamps/fens is usually high. More hardy breeds are required for year-round grazing due to the lower quality of forage and generally wetter and colder conditions in winter. Highland cattle (Figure 4.7) are good for year-round grazing of fens.

*Grazing pressure*

The overall grazing pressure heavily influences the structure and composition of the vegetation. When livestock first start grazing a patch of fen, they usually concentrate on grazing its margins, creating a border of trampled and grazed and

**Fig. 8.18** Effects of pony grazing on swamp and fen vegetation. Grazing by Konik ponies has created a wide variety of vegetation types and structures in this one small (15 ha) grazing unit at Minsmere, Suffolk, England. These include the following.

(a) Tussocks of common reed, sea club-rush/alkali bulrush, and grasses, created by patchy opening up of areas formerly dominated by swamp and fen vegetation.

(b) Heavily grazed, short, dry grassland alongside almost completely ungrazed, tall emergent vegetation in the shallow ditch to the right. The water in the ditch was only 2–10 cm deep when this photograph was taken (and not much deeper during most of the rest of the year), but the ponies still avoided grazing areas of it they could not reach from dry land.

(c) Complete eradication of areas of sea club-rush/alkali bulrush by grazing and trampling. The tall sea club-rush/alkali bulrush remains dominant within the grazing exclosure.

often quite tussocky vegetation. Further grazing continues to reduce the abundance of the bulky, palatable emergents and tall forbs. Continued heavy grazing eventually converts the area to short grassland with or without taller, unpalatable plants. Often, the short grassland created by heavy grazing of fen vegetation retains short, heavily grazed tall emergent vegetation, which can re-grow and re-assume dominance if grazing pressure is reduced.

Swamps and fens are highly productive and can support relatively high grazing intensities in summer. However, since the aim of grazing is usually to only open up a proportion of the swamp and fen vegetation, while leaving much of it ungrazed or only lightly grazed, overall grazing intensities are usually lower than that which the habitat could potentially support. Grazing intensities of between 15 and 70 livestock unit days/ha per year are typically used for year-round grazing.

Moderate-to-high levels of livestock grazing prevent the establishment of swamp and fen vegetation. If the intention is to create mosaic of swamp, fen, and grassland that are subsequently maintained through grazing, it is important to first allow the areas of swamp and fen to establish *before* grazing is subsequently introduced to maintain this mosaic.

### Timing of grazing

Plants vary in their relative palatability during the year and this influences the effects of grazing. Some tall emergents, such as reed and sea club-rush/alkali bulrush, are most heavily grazed and controlled by grazing in spring and early summer when they are more palatable. Cattle eat a higher proportion of reed in summer than ponies. Ponies excavate and eat the rhizomes of reed more in winter than cattle (e.g. Vulink *et al.* 2000).

In winter livestock feed more on evergreen plants, particularly grasses on higher ground, evergreen sedges, and nutritious rhizomes. Ponies and cattle also eat bark during winter when other food is in short supply and this can kill scrub (e.g. Vulink *et al.* 2000). The reduction in suitable forage in winter can cause livestock to wander more widely in search of food and hence have quite a different spatial effect on the vegetation. High water levels in winter are also likely to reduce the area of habitat available to them and there will be an increased need for accessible dry ground. Poaching is also likely to be greater in winter when soils are wetter.

### Encouraging livestock to graze swamp and fen vegetation

Livestock can be encouraged to graze swamp and fen vegetation by reducing the quantity of alternative forage in otherwise preferred areas, particularly on drier grassland which is often favoured by livestock over wet swamp. This can be achieved by only allowing access to a small area of dry grassland within the grazing unit and by reducing the quantity of forage on it. This can be done by, for example, using sheep to graze the grass in dry areas short, to force cattle within the grazing unit to seek food in wetter areas. Animal welfare considerations obviously need to be taken into consideration when deciding the trade-off between providing too little food on the grassland to encourage animals to graze tall emergent vegetation, and maintaining them in suitable condition, particularly in winter.

Livestock can be encouraged to enter and graze swamp and fen vegetation by cutting it first to provide succulent re-growth. Cutting paths through areas of tall fen and even vehicle tracks through fen vegetation encourage livestock into new areas. Mineral licks can also be used to entice livestock into new parts of the fen.

*Modifying grazing patterns by manipulating water levels*

The general avoidance of submerged areas by livestock can be used to manipulate grazing patterns by controlling water levels. They can be raised to discourage access or lowered to encourage it. Livestock are often more reluctant to enter water on peat, because it is soft and unstable, compared to firmer clays.

Large, seasonal drawdowns and between-year variations in water levels are a feature of many extensive, near-natural wetlands, such as those in Mediterranean climates which have relatively high rainfall and extensive flooding in winter, and then dry out during the hot summer. In dry years livestock enter and graze swamp and fen vegetation, thereby creating structural variation in it and increasing the length of its edge. Flooding of these areas during subsequent wetter years provides a diverse mixture of scattered swamp, interspersed with open water, and provides plenty of edge.

A particular combination of summer drawdown and summer grazing is used around the margins of shallow, reed-fringed margins, to create a feature known as a Blue-Border. This is considered the most productive zone of shallow, reed-fringed lakes for birds in parts of Northern Europe (Figure 8.19). It is a particularly simple and cheap form of management for providing a mixture of valuable wetland habitats in close proximity: deep, open water; ungrazed reedbed; shallow, open water; and wet and dry grassland. The fact that the method only relies on grazing animals entering shallow water in summer means that commercial livestock (usually cattle) can be used, rather than more specialist, hardy breeds. This method relies on a substantial drawdown, and is thus useful in wetlands where it is difficult to maintain high water levels throughout the spring and summer. Blue-Border management can also be designed into newly created wetlands. However, to maximize its benefits, the site needs to be designed to have a suitable topography in relation to the extent of drawdown (Figure 8.20). A disadvantage of Blue-Border management is that it does not provide dry reedbed, which is valuable for invertebrates.

**Shallow lake viewed from above**

(a) No surrounding grazing and no Blue-Border

(b) Livestock allowed access to the lake margins have grazed back the reed to create a ring of open water known as a Blue-Border

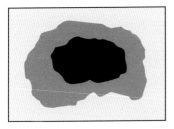
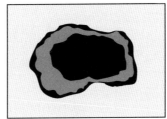

 = Grassland    = Reedbed     =Open water

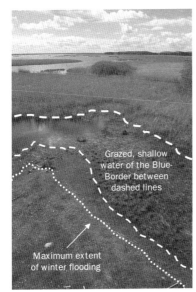

**Fig. 8.19** Blue-Border management. Allowing livestock to graze the margins of shallow, common reed-fringed lakes creates a ring of open water known as the Blue-Border (a and b). The photograph on the left shows how livestock have entered the shallow water and grazed back the reed to form, in this case, a relatively narrow Blue-Border. A more typical, wider Blue-Border is shown on the right. The Blue-Border contains grazing tolerant plants, such as low-growing grasses and sedges, and often an abundance of aquatic and marginal plants that would otherwise be shaded out by the tall reed (Lake Tåkern, Östergötland, Sweden).

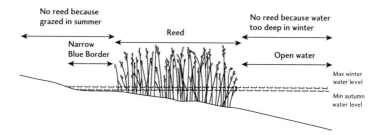
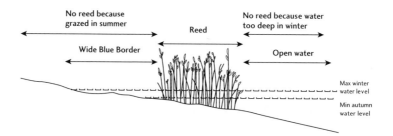
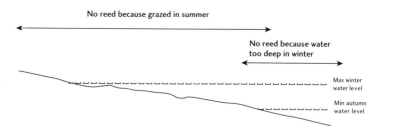

**Fig. 8.20** The width of the Blue-Border, and hence its value for birds in particular, is determined by the gradient of the shore and extent of the drawdown. For a given gradient, a small drawdown results in a narrow border, while a large drawdown allows livestock to eat all the reed.

## 8.9 Bogs

Vegetation growth and succession are slow in the acidic and wet conditions of pristine bogs. The only possible acceptable management is very light grazing, which can help to provide variation in vegetation structure and microhabitats. However, any grazing needs to be carried out sensitively. The slow-growing vegetation and surface layers of peat are fragile and easily damaged.

In practice there are few, if any, pristine bogs remaining in most areas. Many have been degraded through combinations of drainage, burning, and heavy grazing (overgrazing). Drying of the upper layers of peat results in oxidation and release of nutrients, making areas unsuitable for existing bog plants and allowing colonization by more competitive plants, including scrub. The artificial drainage may also cause erosion, while steep-sided drainage ditches can be dangerous for livestock and chicks that get trapped in them. Restorative management of degraded bogs involves:

- restoring a near-natural hydrology by blocking artificial drainage;
- removing any colonizing scrub.

The only long-term solution to maintaining the long-term value of bogs is to restore near-natural hydrology on a large-enough scale to restore active *Sphagnum* moss growth and peat formation. This will often require both blocking of artificial drainage on the bog itself, as well as raising water levels on surrounding land to maintain sufficiently high water levels on its periphery. Blocking of artificial drainage can be a large task where numerous, shallow drains (often known as grips) have been installed. There are a number of different methods used for blocking drainage, the optimal design at a particular site depending on the size of the drain and the availability of materials. Smaller drains can be blocked using solid dams constructed from:

- scoops of saturated peat and associated vegetation taken from the sides or bottom of existing drains and their margins (dried out and de-natured peat will not retain water);
- interlocking, plastic sheet-piling driven into the peat.

Smaller drains can also be partially blocked using bales of heather cut from adjacent areas. These help slow the flow of water, trap any sediment and increase vegetation growth within the drain. Shallow drains should become filled in with *Sphagnum* moss quite quickly following blocking. Larger drains will require more substantive dams, including those made of stronger, interlocking metal sheet piling driven deeply into the ground (Figure 8.21) or constructed from wood and stone. Care should obviously be taken to minimize damage to the peat when using heavy machinery. Dams should be installed so that their tops are slightly higher than the peat surface, to maintain the water level as close to

**Fig. 8.21** Blocking artificial drainage on bogs. The near-natural hydrology of bogs can be restored by blocking artificial drains using dams. Where these drains traverse sloping ground, series of stepped dams need to be installed at frequent intervals along their length to maintain water levels close to the peat surface.

A common method for blocking larger drains involves driving interlocking, metal sheet piling into the peat, as shown (a; Mondhuie, Highland, Scotland).

(b) Shows the characteristic shallow dome of a raised bog. This site is being restored by blocking artificial drains and removing scrub. The paler line from left to right across the centre of the photograph shows the more grassy vegetation that has established along the drier edges of one of the smaller drains (Ford Moss, Northumberland, England).

the peat surface as possible in winter, and within about 20 cm of it throughout the year. It is worth considering incorporating overflows or soakaways to reduce water pressure on dams during periods of high flow and the risk of them failing.

## 8.10 Wet scrub

Wet scrub can be a valuable component of fens. It also occurs on the margins of watercourses, and includes valuable riparian scrub in otherwise arid areas. In fens, the presence of scrub will increase the total number of breeding bird species, mainly through addition of generalist scrub species (e.g. Hanowski *et al.* 1999), but reduce it for specialist wetland birds. Scrub and trees, though, can provide important nest sites for colonial waterbirds such as herons, egrets, and ibises. Scattered scrub, with sheltered areas of fen among it, will be richer for invertebrates than large areas of scrub-free fen. Patches of scrub will also provide shade for grazing animals. Scrub will eventually succeed to wet woodland, which can itself be of high conservation value.

Scrub does not require any management to maintain it. As in drier habitats, there is the potential to diversify the age and structure of existing scrub by cutting small patches close to ground level (coppicing) and creating glades within it, to increase structural diversity and the length of edge. Mixtures of structurally diverse, coppiced, wet scrub and areas of herbaceous vegetation can contain extremely high densities of breeding songbirds (Wilson 1978).

The most common form of wet-scrub management is its removal to prevent it from out-competing more botanically rich vegetation (Figure 8.22), and to maintain open habitat for wetland birds. Removal of scrub and wet woodland can be viewed as long-term rotational management: allowing swamps and fens to succeed to scrub before setting back succession. Any decision to remove wet scrub needs to compare the benefits of doing so against the potential value of the wet woodland to which it will eventually succeed. Established scrub can be removed by:

- cutting and removal (usually requiring treatment of stumps with herbicide to prevent re-growth or grinding of the stump to at or below ground level);
- burning;
- clearing in winter when the ground is sufficiently frozen using a modified blade on a bulldozer (shearing).

Cutting and removal of scrub will usually require follow-up spraying of re-growth the first year. Grinding of stumps will be essential where the restored fen will be subsequently managed by mowing. Cutting scrub can be undertaken at any time

**Fig. 8.22** Large-scale, scrub removal. Fen vegetation can be restored by removing colonizing scrub. Scattered scrub and wet woodland, though, can be important habitats in their own right, and a balance needs to be struck between the desired proportions and distribution of these.

The photograph shows fen vegetation in the third growing season following removal of dense willow scrub and alder carr, similar to that shown in the background of the photograph. There has been virtually no re-growth of scrub at this site. This is probably because annual winter flooding has prevented willow rooting from cut branches and establishing from seed (Upton and Woodbastwick Marshes, Norfolk, England).

of year outside of the bird breeding season, although wet conditions will usually make it impractical in winter.

The key to successful restoration is being able to maintain high-enough water levels for maintenance of the re-instated swamp and fen vegetation following scrub removal. Minimizing ground damage, particularly on peat, is probably also important. Damage can be reduced by designing the pattern of machinery movement to reduce the number of passes and amount of turning, particularly in wetter areas. Turning causes considerably more damage than movement in a straight line. Where there are numerous blocks of scattered scrub it is best to move machinery back and forth along straight lines, rather than to and from specific areas of scrub, to reduce the amount of turning. Also ensure that important access routes are not unnecessarily damaged, because this will greatly reduce working efficiency. Areas churned up by vehicle movement will vegetate less rapidly, but will nevertheless provide additional heterogeneity within the restored habitat. Care should be taken to remove brash and to minimize creation of small hollows, since these can be hazardous to future mowing and to grazing animals. It is probably best to wait at least 2 years until above- and below-ground vegetation have established before any grazing is introduced.

The surface of the substrate can also be lowered to raise water levels relative to the ground surface and expose any buried seedbank, although this will be expensive and it may be difficult to dispose of the material removed.

## 8.11 Wet woodland

Wet woodlands support distinctive and valuable assemblages of birds, amphibians, and invertebrates in particular, but are rarely managed specifically for them.

Coppicing (Section 7.4.2) has been used in some wet woodlands. Small-scale, patchy coppicing of wet woodlands can help maintain a variety of different stages of tree re-growth and consequently variation in the openness of conditions within the understorey. Its benefits to wildlife are probably limited. Abandoned coppice will eventually become more structurally diverse as trees die and fall over.

The other type of management of wet woodlands is their use as greentree reservoirs. These are impounded areas of oak-dominated, forested wetlands managed for timber in the Mississippi and associated valleys in the USA, which are artificially flooded in winter to provide suitable habitat for ducks, such as mallard, *Anas platyrhynchos*, and wood ducks, *Aix sponsa* (e.g. Reinecke *et al.* 1989). The artificial flooding makes acorns and invertebrates accessible to these ducks. The pools left following artificial flooding also provide breeding areas for amphibians. This artificial winter flooding influences the structure of the woodland by maintaining an open understorey, with only seedlings and saplings able to survive prolonged periods of winter flooding being able to grow.

The only management required to maintain or enhance the conservation value of most wet woodlands is to maintain or restore their natural hydrology (Figure 8.23).

## 8.12 Wet grasslands

The general principles of vegetation removal on wet grasslands are similar to those of dry grasslands except that, unsurprisingly, there are no fire-prone types of very

**Fig. 8.23** Natural hydrology and tree regeneration in wet woodlands. The only management necessary to maintain the conservation value of forested wetlands, like this bald cypress, *Taxodium distichum*, swamp, is maintenance or restoration of natural flooding regimes. Fluctuations in water levels are particularly important in influencing tree regeneration. Like many wetland trees, bald cypresses require flood water to disperse their seed, and periods of low water levels so that seeds can germinate on mud exposed as the flood waters recede. However, if water levels then rise too highly they kill the young seedlings, which cannot survive submergence for more than a few days (Lake Fausse Pointe State Park, Louisiana, USA).

wet grassland. Hence, management by burning is rarely an option, although it is occasionally used in restoration management of damp grasslands. The key difference is that much of the conservation interest of wet grassland is associated with moist ground and permanent and seasonal water bodies, including associated water-filled drainage ditches. Manipulation of hydrology is therefore key to managing wet grasslands for conservation. Two particularly ingenious forms of artificial flooding have been developed on wet grassland, which both inadvertently provided good habitats for wetland wildlife: water meadows and washlands (Figures 8.24 and 8.25).

Agriculturally unimproved wet meadows and pasture can support species-rich vegetation of high conservation value, which is dependent on continuation of a similar management regime and lack of use of inorganic fertilizers, as in drier

# Wet grasslands | 285

**Fig. 8.24** Water meadows. These are meadows which have calcareous spring or river water flowed over them in the early spring via an elaborate network of channels, in this case involving a raised watercourse (a). The water deposits nutrients and warms the soil, encouraging early spring grass growth.

These water meadows support an impressive display of the rare meadow saffron, *Colchicum autumnale* (b). Very few actively managed water meadows survive (Plateaux, Noord-Brabant, The Netherlands).

**Fig. 8.25** Washlands. These are artificial floodplains constructed to aid land drainage. They consist of a flat, embanked area lying between a river and artificial relief channel, or between two artificial relief channels, into which water is diverted and stored during periods of high flow.

The Ouse Washes in Norfolk and Cambridgeshire, England, was constructed in the seventeenth century to help drain the English Fens. Its winter floodwaters (a) provide habitat for large numbers of wildfowl. As floodwaters recede they expose grassland (b) used by breeding waders/meadow birds and wildfowl and which supports a valuable relic wetland flora and rich invertebrate assemblage.

Carefully designed, new washlands can both alleviate flooding and provide valuable wildlife habitat, although their benefits for wildlife will depend on their hydrological regime. In particular, flooding during the breeding season will disrupt nesting (e.g. Ratcliffe et al. 2005).

grasslands. Agriculturally unimproved grasslands are rare in many lowland areas. Where they so occur a primary aim of management will be to maintain the existing botanical interest by maintaining similar cutting and grazing regimes and water-level regimes to those that created and formerly maintained them.

Many wet grasslands have been agriculturally improved. This greatly impoverishes their flora and fauna, as it does on dry grasslands (Section 5.9). Agricultural improvement has involved drainage, addition of inorganic fertilizer, addition of lime to raise the pH of acidic grasslands, and in some cases reseeding with agriculturally productive grass species. Application of inorganic fertilizers will be especially damaging on wet grasslands where there is even greater potential for nitrates to leach into and raise nutrients levels in associated ditches and other water bodies. Although installing drainage in damp grasslands is often successful at increasing grass growth in spring, it can have an opposite effect later in the season, by reducing the quantity of water available for plant growth during drier periods of the year, even on 'badly drained' soils in the relatively wet south-west of England (Tyson *et al.* 1992). From an agricultural perspective, though, grass growth in spring is often at a premium.

Any botanical interest of agriculturally improved grasslands is invariably restricted to their edges, especially to water-filled ditches and their margins, and to wetland and ruderal vegetation in areas of temporary flooding. Scattered scrub and trees will add to the wildlife interest of wet grasslands, particularly old willows that provide a continuity of decaying wood. However, their presence may be detrimental to some ground-nesting birds, particularly breeding waders/meadow birds. Trees and tall scrub will provides nest sites and look-out posts for crows and raptors which predate eggs and chicks (e.g. Green *et al.* 1990b). Enclosed fields will tend to be avoided by breeding and wintering waders/shorebirds and other waterbirds (Milsom *et al.* 1998, 2000).

Larger areas of wet grasslands are often of particular conservation value for their breeding and wintering waterfowl. The main breeding bird interest comprises breeding waders/meadow birds on the grassland itself and breeding waterfowl and swamp/fen songbirds along water-filled ditches and other areas of shallow water and swamp and fen vegetation. Common aims of wet grassland management are therefore to:

- provide suitable conditions for wintering wildfowl and other waterbirds;
- in central, Western, and Northern Europe provide suitable conditions for breeding waders/meadow birds (e.g. Ausden and Hirons 2002) and other waterbirds.

Suitable conditions for these birds can be created on agriculturally improved grassland of little or no existing botanical value. On agriculturally unimproved grassland *raising* water levels (as opposed to maintaining existing high water levels) has the potential to damage the existing botanical interest, especially if it involves introducing surface flooding during the growing season. Creating surface flooding during the spring and summer can also damage the invertebrate interest associated with areas of wet but unflooded peat. However, on most drained and agriculturally improved grasslands with little or no existing invertebrate or botanical interest, raising water levels and introducing patchy, surface flooding to benefit wetland birds should also provide additional habitat for wetland invertebrates and vegetation. Extensive flooding of grasslands will, not surprisingly, eradicate terrestrial small mammals (e.g. Jacob 2003).

The main areas of wet grassland themselves tend to support few invertebrates of high conservation value. Most of the scarcer species are found in associated fen, swamp, open water, and marginal habitats, including saltmarsh vegetation on coastal grazing marshes, and habitats along associated water-filled drainage ditches and their margins (e.g. Drake 1998). Hence another method of maximizing the conservation value of wet grasslands is by sympathetically managing associated water-filled ditches for their plant, invertebrate, breeding waterbird, and other interest as described in Section 8.6.

Wet grasslands often lack suitable habitat for species of amphibians that require temporary pools for breeding. Ditch networks usually contain permanent water, while temporary winter flooding often dries out too early in spring for them to complete their larval life cycle. Additional pools for breeding amphibians can be created as shown in Figure 8.8.

### 8.12.1 Maintaining existing botanically rich swards

General principles of grazing and cutting to maintain the existing botanical interest of wet grasslands are the same as those described for dry grasslands in Section 5.3. Sheep, though, are unsuitable for grazing very wet conditions as in swamps and fens. As with drier grasslands, managing areas as agriculturally unimproved hay meadows will maintain their high plant species richness, but be damaging for invertebrates. The timing of cutting needs to be set back to minimize loss of nests and chicks of ground-nesting birds. Other methods of minimizing loss of nests and chicks during mowing are described in Sections 5.5.4 and 8.12.3.

Grazing usually take place from spring to autumn (i.e. summer grazing), with winter grazing a less practical option, especially on very wet grasslands. Grazing regimes also need to take account of potential trampling of the nests of waders/meadow birds and other ground-nesting birds (Section 8.12.3).

Species-rich wet meadows are typically flooded for periods during the winter. Flooding by rivers deposits nutrient-rich sediment that can help maintain suitable nutrient levels for maintenance of species-rich grassland. Water-level management during the growing season should seek to maintain the hydrological regime, which created and maintained the characteristic vegetation type. Raising water levels to introduce surface flooding for significant periods during the growing season will be especially damaging to botanically rich swards, and result in their replacement with more species-poor aquatic and swamp vegetation and inundation grassland.

## 8.12.2 Providing suitable conditions for wintering wildfowl and other waterbirds

Wintering wildfowl and other waterbirds can be attracted to wet grasslands by providing extensive areas of shallow flooding (between a few centimetres and about 30 cm deep) and maintaining high water levels to increase the accessibility of larger soil invertebrates for them to feed on. Large, undisturbed areas of shallow water can provide important daytime roosts for dabbling ducks, which feed at night on areas often otherwise disturbed during daytime. High water levels will force earthworms closer to the soil surface where they are more accessible to birds and make the soil softer and easier for long-billed waders/shorebirds to probe. Most birds that feed on large soil invertebrates prefer relatively short (less than about 10 cm), open swards, presumably because it makes their prey easier to detect (e.g. Milsom *et al.* 1998). Wintering geese also prefer shorter, unflooded swards for feeding, typically between about 5 and 20 cm (Vickery and Gill 1999). Grasslands need to be appropriately grazed or cut in late summer and autumn to achieve these sward conditions in winter.

Flooding grassland in winter and spring encourages a range of plants that produce seed that ducks feed on during winter. These include perennial plants such as many sedges and rushes and prolific, seed-producing annuals that require bare ground for germination. Production of seeds from annual plants can be maximized using moist-soil management (Section 8.4.1). Re-flooding these areas in autumn and winter suspends these seeds and makes them available to dabbling ducks. Regular inundation also encourages some grass species favoured by herbivorous wildfowl, notably creeping bent, *Agrostis stolonifera*.

The prey available to birds as floodwaters recede depends on the origin of the floodwater. If the grassland is flooded by river water or overflow from another permanent water body, then fish and aquatic invertebrates will exploit the shallow floods but also become trapped in small pools as the floods recede. This provides an abundant, but temporary concentration of prey for fish-eating and invertebrate-eating waterbirds. If the flooding is created by rainwater lying on fields, with little or no connection to rivers or substantive water bodies, then it will lack fish and aquatic invertebrates. In both situations, though, receding floods will still, though, expose often bare or sparsely vegetated ground containing earthworms and other soil invertebrates that have survived the flooding.

### 8.12.3 Providing suitable conditions for breeding waders/meadow birds

Suitable conditions for breeding waders/meadow birds and other waterbirds can be created by:

- maintaining high water levels and shallow flooding in spring and early summer to provide suitable feeding conditions;
- maintaining a suitable sward height and structure for nesting and feeding waders/ meadow birds by sward management during the previous summer/autumn and, where grass continues to grow throughout the year, also during the winter or early spring;
- minimizing loss of eggs and chicks during grazing and mowing.

Removal of trees and shrubs may be necessary to providing the open conditions preferred by breeding waders/meadow birds and to remove perches for avian egg and chick predators. This removes potentially valuable invertebrate habitat, though. An option is to pollard trees instead. This maintains their invertebrate interest and helps provide a continuity of decaying wood in their trunks, while preventing the trees from becoming tall enough to support nests of these species and reduces their suitability as look-out posts.

Lime has been added to already agriculturally improved meadows in The Netherlands to raise soil pH with the aim of increasing earthworm biomass for breeding waders/meadow birds.

*Maintaining high water levels and shallow flooding*

There are two main ways of providing suitable hydrological conditions for feeding waders/meadow birds.

- The first is maintaining a high field water table to keep the upper soil wet and therefore soft enough for common snipe and black-tailed godwits, *Limosa limosa*, to probe for large soil invertebrates, mainly earthworms and leatherjackets, Tipulidae, in the soil.
- The second is providing shallow pools and their margins for waders/meadow birds, especially northern lapwings and common redshank, *Tringa totanus*, to feed on aquatic invertebrate and those in the damp, sparsely and unvegetated mud on their margins.

Maintaining large areas of soft, moist soil in which waders/meadow birds can continue to probe throughout their breeding season is only possible on permeable soils. Permeability varies with soil type and structure and is highest on well-structured peat. Water tables within fields on permeable soils are intimately linked to those on surrounding, water-filled ditches. On peat it is possible to maintain a high field water table throughout the breeding season by maintaining high water levels in surrounding ditches (Figure 8.26b), particularly if these ditches are closely spaced. This keeps the upper soil wet and therefore moist enough for common snipe, *Gallinago gallinago,* and black-tailed godwits to probe for food (Green 1988; Green *et al.* 1990a). On peat soils maintaining a water table within 20–30 cm of the soil surface is recommended for maintaining suitable conditions for breeding snipe (RSPB, EN & ITE 1997). Even on these soils, though, it is important to provide at least some shallow pools as well. The level of control over field water tables can be increased using shallow drains, often known as foot-drains or grips (Figure 8.27).

On less permeable soils, such as compacted soils, especially clays, it is not possible to maintain soft, moist upper soil over large areas of the field by maintaining high water levels in surrounding ditches (e.g. Armstrong and Rose 1999; Figures 8.26c and 8.26d). These soils are generally unsuitable for breeding common snipe and black-tailed godwits. Northern lapwings and common redshank are usually the commonest species and these feed on a variety of invertebrates taken from the sward, soil, and shallow water and its margins (Ausden *et al.* 2003). On these soils shallow, surface flooding is more important in providing suitable conditions for feeding waders/meadow birds, particularly towards the end of their breeding season (Milsom *et al.* 2000, 2002; Ausden *et al.* 2003; Smart *et al.* 2006). Flooding benefits these species by reducing sward height and providing shallow water and soft, muddy margins for them to feed on invertebrates. However, these breeding waders/meadow birds also feed on larger soil invertebrates during the early part of the breeding season, and winter flooding greatly decreases the abundance of these (Ausden *et al.* 2001). The ideal is therefore to contain a mosaic of short, unflooded grassland, winter flooded grassland and a succession of sequentially drying out shallow pools and open ditch edges present until the end of their breeding season.

Retention of shallow floods during the breeding season also provides feeding areas for breeding wildfowl, although more densely vegetated ditches are usually more valuable for brood rearing.

Overall, it is best to maintain a suitable variety of hydrological conditions to help ensure there are always suitable feeding conditions present, and this will be easiest on areas with a varied topography. This also maximizes the range of conditions available for wetland invertebrates and plants (Figure 8.28). However, many grasslands are relatively flat. Raising or lowering water levels a small amount on these either floods or dries out large areas at the same time.

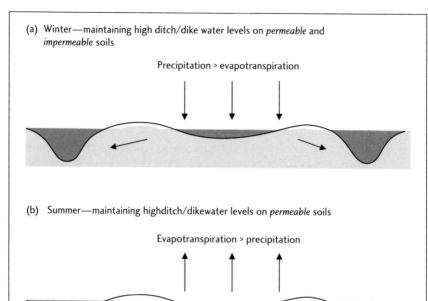

**Fig. 8.26** Field water tables on permeable and impermeable soils in summer and winter. In winter, inputs of water on to the field through precipitation are greater than losses of water from it through evapotranspiration. The soil becomes saturated and there is a net movement of water out of the field and away into surrounding ditches (a). Surface ponding may occur, particularly on impermeable soils, such as poorly structured clays, because water only moves slowly away through the soil into surrounding ditches.

As the weather warms and plants start growing, then loss of water from the field through evapotranspiration begins to exceed inputs of water on to it through precipitation. On permeable soils such as well-structured peats, the water table within the field can be kept high in late spring and summer by maintaining high water levels in surrounding ditches. This allows water to flow back through the soil into the field and partially replace water lost through evapotranspiration (b). On impermeable soils there is less flow of water back through the soil from surrounding ditches to replace water lost through evapotranspiration in late spring and summer, and field water levels rapidly fall (c). The only way to maintain wet conditions on the surface of impermeable soils under these conditions is by flowing water over the field surface to create surface flooding (d).

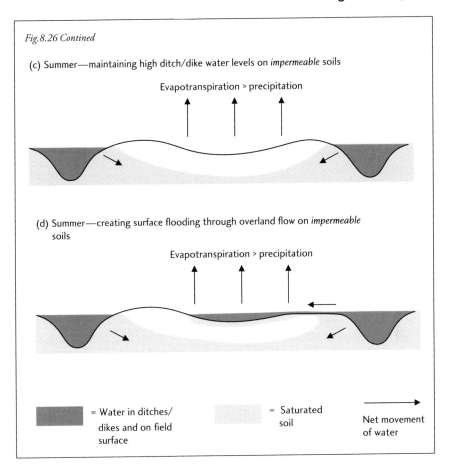

Fig. 8.26 Continued

(c) Summer—maintaining high ditch/dike water levels on *impermeable* soils

(d) Summer—creating surface flooding through overland flow on *impermeable* soils

It may also be impractical to raise water levels sufficiently due to lack of water or because raising water levels will flood adjoining farmland. Where this is the case, the only way of providing shallow-water and marginal habitat is by excavation.

The maximum length of shallow water and muddy edge for quantity of material excavated is created by digging shallow, linear drains. These can also be used to feed water to low areas to maintain shallow flooding and damp soils. However, breeding waders/meadow birds seem to prefer feeding in shallow pools on grassland in preference to along linear drains. Nesting northern lapwings often nest close to the edges of isolated foot-drains, where they are sometimes subject to higher levels of predation than when nesting further away. This is probably because mammalian predators such as red foxes follow these linear features when hunting. This might be less of a problem where there are high densities of foot-drains and more variation in their patterns. Creating foot-drains of different

**Fig. 8.27** Foot-drains and grips. These are narrow, shallow drains, which were originally excavated at many sites to drain water off fields into surrounding ditches. Maintaining high water levels in surrounding ditches can be used to feed water back along these foot drains into the interior of fields. This helps maintain a high water table and the interior of the field moist and soft enough for breeding black-tailed godwits and common snipe to probe for earthworms and leatherjackets. Lowering water levels in surrounding ditches can be used to drain water out of the field to drain it for agricultural operations after the end of the breeding season.

This field is covered in closely spaced foot-drains. One can be seen stretching from the middle foreground to middle back of this picture, while the lines of cut grass mark the high ridges between individual foot-drains. This site supports high densities of breeding black-tailed godwits (De Pine, Freisland, The Netherlands).

depths, including varying depths along their length, will again maximize the likelihood of their being some suitable areas of shallow water and freshly exposed mud being present whatever the height of overall water levels. The disadvantage of increasing densities and patterning of foot-drains is that it can make agricultural operations more difficult.

*Providing suitable sward conditions*

Moderate grazing by cattle is considered the best type of vegetation removal for creating suitable tussocky vegetation and patches of poached, bare ground

**Fig. 8.28** Providing shallow water for breeding waders/meadow birds. Optimum conditions for breeding waders/meadow birds on clay and other mineral soils can be provided by creating a mosaic of shallow water, bare mud, and unflooded grassland. This is easiest to achieve by raising water levels on areas with existing variation in topography, such as unlevelled grazing marsh, which still retains its former saltmarsh channels (a). This coastal grazing marsh supports some of the highest combined densities of breeding northern lapwings and common redshank in the UK. The profile of flooded channels often flattens out over time and their margins can become steep and cliffed. Occasional re-profiling is sometimes needed to maintain their suitability (Elmley Marshes, Kent, England).

(b) has a mosaic of shallow water and high-water-table grassland created by maintaining high water levels in its closely spaced ditches. These high densities of ditches were originally excavated for drainage (Veijlerne North Jutland, Denmark).

for breeding waders/meadow birds. This variation in structure should also support a more diverse invertebrate fauna than more uniform grassland. Different species of wader/meadow bird prefer different-lengthed swards. Northern lapwings prefer swards less than about 5 cm high for nesting and foraging, while common snipe, common redshank, and Eurasian curlews require taller swards (up to 25 cm) for concealing their nests in. All these species avoid unmanaged or only lightly grazed or irregularly cut vegetation with dense litter, with the possible exception of common snipe.

Grazing levels during summer and autumn will be important in determining sward conditions for breeding waders/meadow birds the following spring, where there is little or no vegetation growth in winter. Additional winter grazing by livestock may be required where vegetation continues to grow through the winter and is not suppressed by winter flooding. Summer cattle grazing can be used to produce the desired tussocky structure combined with winter sheep grazing used to keep areas short. Winter cattle grazing can be used to provide light poaching to increase birds' access to earthworms and leatherjackets.

Grassland that will be flooded in winter should be grazed relatively short. Flooding to a depth that covers the grass will suppress its growth and provide bare mud and sparsely vegetated ground for feeding waders/meadow birds and other birds as water levels recede in spring. If grass protrudes above the floodwater it will continue to grow and, in the case of creeping bent, often the dominant grass in nutrient-rich, winter-flooded grasslands, produce a dense, floating mat. Flooding rank grass will leave behind a thick, dense mat of grass and thick layer of anoxic, decaying plant litter, containing few larger soil invertebrates and will be of limited value for feeding birds. Grazing by wintering, herbivorous wildfowl can be important in maintaining suitably short, open areas for feeding waders/meadow birds the following spring.

A common issue when managing wet grasslands for breeding waders/meadow birds is extensive, dense growth of rushes, especially on grazed, agriculturally improved grassland and former arable land. The presence of rushes will provide cover and produce seed for wintering wildfowl, but dense stands will make areas unsuitable for breeding waders/meadow birds. On coastal grazing marshes sea club-rush/alkali bulrush can form dense stands on the margins of shallow water. Again, small stands add diversity, provide valuable habitat for invertebrates, seed for wintering wildfowl, and cover for nesting wildfowl, but extensive areas make grasslands unsuitable for feeding waders/meadow birds.

Rushes and sea club-rush/alkali bulrush can be controlled to some extent by grazing in spring when their fresh growth is more palatable, but grazing is usually avoided or only carried out at low densities in spring to minimize trampling of birds' nests (see next section). Other options for reducing the cover of rushes include heavy grazing using hardy livestock in winter when there is little alternative forage, repeated cutting ideally followed by grazing of re-growth and cutting followed by application of systemic/translocated herbicide to the re-growth. Sea club-rush/alkali bulrush can be controlled by cutting underwater, herbicide application, and heavy grazing, and will be further disadvantaged by low water levels in summer. On agriculturally improved swards where excessive growth of these species is an issue, occasional dry years should be taken advantage of and used to manage the vegetation as hard as possible to reduce their abundance.

## Minimizing trampling of birds' nests by livestock

The proportion of birds' nests trampled vary according to the type of livestock and grazing pressure (Figure 8.29). There are three main approaches to reduce nest loss:

- excluding grazing from areas with nesting birds during the breeding season, but where necessary providing suitably open conditions for birds by heavily grazing adjacent fields without nesting birds;
- reducing grazing levels during the nesting season and accepting a low level of nest trampling;
- continuing grazing while protecting individual nests.

The benefits of reducing or excluding grazing during the nesting season have to be balanced against the potential disadvantages of allowing the sward to become too tall and dense for birds that require shorter, more open conditions. Grazing can be excluded from entire fields supporting high densities of nesting birds, or just from key areas of them using electric fencing. Nidifugous birds (those whose young leave the nest very soon after hatching) that require short swards can then take their chicks to feed in these more heavily grazed areas. Swards that have grown tall through exclusion of livestock during the nesting season can be difficult to return to their desired condition. Re-introducing livestock to tall swards tramples down much of the vegetation. This can form a matted layer of litter and live grass that smothers germination gaps and prevents access by birds to soil invertebrates. It can be difficult to remove this litter layer, or thatch, by grazing, although using combinations of cattle and sheep can be effective (Section 5.4.2). Power harrowing is a useful method for lifting up and dissipating thatch.

An alternative is to graze at low stocking densities during the nesting season. Some nests will be trampled, but hopefully the decrease in nest survival will be offset by increased chick survival due to the improved sward conditions provided by the grazing.

Nests of species that are easy to locate can be protected from trampling using nest protectors or nest exclosures. Nest protectors consist of raised metal grilles placed over the nest (e.g. Guldemond *et al.* 1993). There is concern that predators such as crows and red foxes might learn to associate the presence of the relatively conspicuous nest protectors with nests. Where this is a potential problem, nests can be protected from both trampling and *larger* predators using nest exclosures. These have a roof and sides constructed from plastic-coated steel bars. Incubating birds can enter through the gaps in their side but these are too narrow

**Fig. 8.29** Nest trampling rates. Grazing animals trample birds' nests. The rate of nest trampling depends on the type and density of livestock, and also varies between bird species.

The graph shows the estimated survival of nests of different wader species/meadow birds in fields with homogenous vegetation grazed throughout the entire egg-laying and incubation period at a grazing pressure of one livestock unit per hectare (see Fig. 5.8 and Table 5.1 for an explanation of grazing units). It was calculated using daily nest survival rates from Beintema and Müskens (1987). Adult dairy cattle trample a far lower proportion of nests per quantity of vegetation removed than sheep, mainly because it requires about 12 sheep (i.e. 12 times as many feet) to remove the same quantity of vegetation as one cow. Yearling cattle trample a far higher proportion of nests than adult dairy cattle, because they run about more. Northern lapwings and Eurasian oystercatchers, *Haematopus ostralegus*, suffer lower nest trampling rates than common redshank, probably largely because northern lapwings and Eurasian oystercatchers are able to successfully defend their nests against approaching livestock.

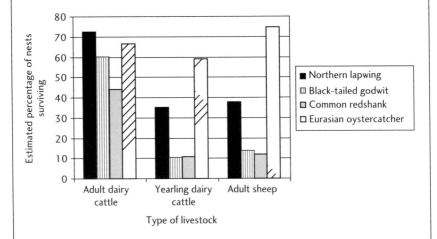

Trampling rates increase with stocking levels. Trampling rates probably also vary to some extent according to the heterogeneity of the grassland. This might occur by birds preferentially nesting in vegetation subject to higher or lower grazing pressure than the rest of the field. Even if nests are only partially destroyed by trampling, predators often take the remainder of eggs. Otherwise, birds will often continue incubating the remainder of partially trampled clutches.

to allow access by larger predators such as gulls, crows, European badgers and red foxes. Exclosures have been found to increase nest survival of killdeers, *Charadrius vociferus*, northern lapwings, and common redshank, but to also *increase* predation rates on incubating common redshank. These sit tight on their nest and only fly off once a predator if very close. Nest exclosures probably impede their escape (Johnson and Oring 2002; Isaksson *et al.* 2007).

## Minimizing losses of birds' nests and chicks during cutting

The most widely used method for minimizing loss of birds' nests and chicks is delaying cutting until birds have finished breeding (e.g. Kruk *et al.* 1996). Alternative methods include:

- marking nests and mowing around them;
- using flags to deter birds with chicks from entering fields that are about to be mown; the flags are made out of bamboo canes with blue or white plastic bags attached to their tops (Kruk *et al.* 1997).

The pattern of mowing can also be altered to reduce chick loss (Figure 5.12).

Marking nests and mowing around them can only be used for species whose nests are easy to locate (e.g. Guldemond *et al.* 1993). Nests remaining in isolated patches of un-mown grassland might be particularly conspicuous and subject to high levels of predation. In The Netherlands strips of un-mown grassland are left in otherwise cut agriculturally managed fields to provide corridors for black-tailed godwit and other waders/meadow birds to safely move their broods to suitable un-mown grassland in nearby nature reserves.

# 9
# Coastal habitats

This chapter discusses a range of intertidal habitats, mudflats, saltmarsh, and sandy and shingle beaches, together with habitats beyond the usual upper tidal limit, which are affected by coastal processes, saline water, or exposure from salt-laden winds. These include sand dunes, cliffs, saline lagoons, brackish (between freshwater and seawater) marshes, and grasslands.

The successional stages of intertidal habitats are largely determined by changes in coastal geomorphology (coastal processes), and in most cases less so by grazing or other forms of vegetation removal. Most principles of managing coastal habitats above the usual upper tidal limit are similar to those of their inland equivalents.

A fundamental consideration when managing coastlines is the extent to which wider coastal processes are allowed to operate, or are constrained to protect existing valuable habitats. In a fully natural dynamic soft coastline (i.e. one comprised of soft, easily erodable cliffs and sedimentary deposits) the mixture of habitats such as sand banks, saltmarsh, coastal lagoons, and brackish marshes will be in a state of flux. These different habitats will be formed, modified, and destroyed by changes in geomorphology, some on a relatively short timescale. Along many coastlines, though, these coastal processes cannot operate in a fully natural way, because of the presence of hard sea defences (dikes and seawalls), which prevent or constrain landward migration of coastal habitats. Removal of sediment by dredging and disruption of transportation and deposition of sediment by structures will also influence natural coastal processes. This can mean that valuable habitat can be lost in one area, but similar habitat not re-formed elsewhere. Relative sea-level rise and predicted changes in storm activity will accelerate rates of coastal change.

## 9.1 Sea cliffs

Vegetation succession on sea cliffs is prevented by exposure and salt deposition. Principles of managing grassland and heathland on the tops of cliffs are similar to

**Fig. 9.1** Soft cliffs. These are composed of soft deposits such as clays and shales, which erode and create bare ground through landslips and slumping. This produces an ever-changing variety of successional stages, from bare ground, cracks and crevices, to ruderal vegetation, grassland, and scrub.
The small-scale mixture of habitats, sunny slopes, often together with small wetlands formed by water seeping through on to the cliff slopes, can support many restricted-range invertebrates. Soft cliffs are particularly rich in solitary bees and wasps, several groups of beetles, and crane flies. Such cliffs should be left unmanaged. This photograph shows bare and disturbed ground and a small area of common-reed-dominated seepage wetland on cliffs along the south Dorset coast near Worth Matravers, England.

those when managing these habitats elsewhere, although the reduced rate of vegetation growth can mean that less management is required than to maintain similar conditions. Soft cliffs can be of high value for invertebrates (Figure 9.1).

## 9.2 Intertidal mudflats, saltmarsh, and other tidal marsh vegetation

Intertidal mudflats do not require active management to maintain their conservation value, other than the control of unwanted alien/exotic plants. In Europe the hybrid common/English cordgrass, *Spartina anglica*, has been considered a problem where it covers important areas for feeding waders/shorebirds, and has been controlled using herbicide or mechanical disturbance (Frid *et al.* 1999). Along the north Atlantic coast of the USA, disturbance and restriction of tidal influence have encouraged expansion of vigorous strains of common reed at the expense of cordgrass, *Spartina* spp., meadow vegetation, which supports a more important avifauna (Benoit and Askins 1999). Herbicide can be used to control this common reed and restore cordgrass-dominated vegetation and its associated invertebrate fauna (e.g. Gratton and Denno 2005).

Saltmarshes consist of vegetated flats, interspersed with creeks and in some cases pools. Largely unvegetated salt pans also occur within saltmarsh where evaporation of saline water in warmer regions increases the salinity of the substrate. Flats support a limited array of specialist grasses, forbs, and small shrubs that can tolerate periodic inundation by saline water. In many lowland areas saltmarsh is the vegetation type least modified by human activity. Saltmarsh can grade into tidal brackish and freshwater marshes, tidal woodland, or terrestrial habitats, although these transitions are often prevented by the presence of hard sea defences. Some of the plants characteristic of high levels of saltmarsh have now spread inland along the edges of salted roads.

Saltmarsh is of particular value for feeding wildfowl and its limited, but largely specialist, assemblage of breeding birds. Waders/shorebirds also use its creeks and pools. Saltmarsh with a high tidal amplitude, such as on the Atlantic and North Sea coasts of Europe, does not support resident mammals. None can withstand the high frequency of tidal inundation. However, sheltered areas of saltmarsh with a low tidal amplitude can support high densities of small mammals, and consequently provide important feeding areas for raptors. Introduced nutria/coypus, *Mycocastor coypus*, are important agents of erosion of saltmarsh in parts of the USA. Saltmarsh is unsuitable for most amphibians and reptiles. The invertebrate fauna can be moderately diverse, comprising mainly marine species in the mud and more terrestrial groups within the vegetation (Mason *et al.* 1991). Saltmarshes also provide valuable habitat for some fish species, including probably nursery areas for juveniles of some species (Desmond *et al.* 2000; Able *et al.* 2001).

The composition of saltmarsh vegetation is largely determined by the frequency of tidal inundation, and consequently varies with elevation. Saltmarshes can usefully be divided into:

- **low marsh** between the mean high water level of neap tides and mean high water;
- **middle marsh** between mean high water and the mean high water level of spring tides;
- **high marsh** above mean high water level of spring tides.

In general, the number of plant species increases with elevation. High marsh and transitions with terrestrial habitats are rare along many coastlines, these higher areas having been easier to claim for agriculture. Specialist plants and invertebrates that would be found at these higher elevations are often restricted to areas of brackish seepages in and around seawalls and on margins of saline and brackish water bodies. Saltmarshes in temperate Europe typically contain a far higher diversity of vascular plant species than those in temperate North America. The latter are mainly dominated by a small number of cordgrass species.

Although saltmarsh does not require intervention to maintain it, management can be used to influence its structure and plant species composition, which can in turn influence its suitability for birds and invertebrates. Management may also be used to reduce erosion of saltmarshes and increase accretion of sediment on their seaward side. Probably the majority of the higher areas of saltmarshes in Europe have been grazed by livestock or mown for hay during some period in history. Some areas have been fertilized and heavily grazed to produce grass turfs. In North America high marsh has been mown for the production of salt hay. Burning is also widely used on the eastern seaboard of the USA to provide feeding areas for wildfowl.

Some areas of saltmarsh, particularly in the USA, have been impounded to provide suitable shallow water habitats for wildfowl. This is referred to as structural marsh management. In warmer regions saltmarsh has also been embanked to create salinas/salt evaporation ponds/solar ponds for salt production. Impoundment of saltmarshes and brackish marshes generally increases their value for waterfowl, but decreases it for saltmarsh-specialist bird species (see review by Mitchell *et al.*, 2007). Impoundment will destroy the intrinsic value of the saltmarsh. Management of existing impoundments and salinas/salt evaporation ponds/solar ponds is discussed in Section 9.5.

The structure of saltmarshes on the eastern seaboard of North America has also been modified with the aim of reducing numbers of salt marsh mosquitoes, *Aedes sollicitans*. Early attempts involved ditching, whereby narrow, parallel ditches were cut to drain ephemeral pools in high marsh used by mosquito larvae. In practice the drainage ditches often became blocked, rendering this method ineffective. This technique was replaced by open-marsh water management (OMWM), in which ephemeral pools are deepened, or connected to other more permanent pools, to allow fish to live in them and eat the mosquito larvae.

There are only two types of vegetation management that can be used on existing areas of saltmarsh to maintain or enhance their conservation value: grazing and burning. Unlike in grasslands, mowing is rarely if ever used for *conservation* purposes. There is little experience of introducing burning regimes to formerly grazed saltmarsh or vice versa.

## 9.2.1 Grazing

The main effects of introducing grazing to unmanaged saltmarsh are to:

- influence plant species composition;
- influence vegetation structure;
- reduce litter depth;
- remove nectar sources and reduce or prevent seed production;
- reduce sedimentation rates;
- maintain areas of open water that would otherwise become covered by dense vegetation.

The most important considerations are the type of livestock and grazing pressure. There are usually few options concerning the seasonality of grazing. In most cases it is only practical in summer. The start of summer grazing can, though, be delayed, or livestock numbers maintained at a low level in spring, to minimize trampling of birds' nests. Livestock obviously need to have access to high ground at high tide and during storms.

Wild geese can also exert heavy grazing pressure. Where this is accompanied by rooting out of tubers and rhizomes, as by snow geese, *Chen caerulescens*, in North America and greylag geese in Europe, this can create bare areas known as eat-outs (Giroux and Bédard 1987; Esselink *et al.* 1997).

The effects of introducing grazing on vegetation composition and structure varies according to its elevation. In north-west Europe ungrazed middle to high marsh is usually dominated by sea couch, *Elytrigia atherica*, with less-saline areas often dominated by common reed. Grazing and trampling reduces the abundance of sea couch and also sea-purslane, *Atriplex portulacoides*, allowing them to be replaced by lower-growing, grazing-tolerant plant species, usually common saltmarsh-grass, *Puccinellia maritima*, and red fescue, *Festuca rubra* (Bos *et al.* 2002). Grazing also reduces the abundance of common reed, as in freshwater wetlands (see review by Bakker 1998). These general effects are similar to those on terrestrial grasslands dominated by large, competitive grasses and other plants in the absence of grazing. Grazing in middle and high marsh benefits plants able to exploit the open and muddy condition caused by grazing and trampling. These include glasswort/pickleweed, *Salicornia* spp., which in the absence of grazing are largely confined to low marsh.

Grazing the lowest levels of saltmarsh tends to have little, if any, effect on plant species richness (Bouchard *et al.* 2003). This is because plant species in low marsh comprise more stress-tolerant species, with no one vigorous species competitively excluding the others. Hence vegetation removal does not prevent dominance by more competitive plants.

Excluding, or reducing, heavy grazing on mid to high marsh does not necessarily have the opposite effect to introducing it. Instead, it usually just allows the existing sward of grazing-tolerant grasses to grow taller, with other grazing-intolerant plants often being slow to re-colonize. Hence there is usually a presumption to continue grazing on existing saltmarsh.

Grazing also reduces the quantity of sediment trapped within vegetation. For example, Andresen *et al.* (1990) found that annual accretion rates in ungrazed common saltmarsh-grass dominated marsh in northern Germany were 2.3 cm/year compared to 1.7 cm/year in heavily grazed areas at similar elevation. This accretion further encourages plant species typical of higher marsh.

### Type of livestock

Three types of domestic animal are used to graze saltmarsh: cattle, sheep, and ponies. The general effects of these on vegetation composition and structure are similar to those in dry grasslands (see Section 5.4.1). Cattle are usually preferred for conservation grazing because they create more medium-scale variation in vegetation structure.

### Grazing pressure

Grazing pressure will have a profound effect on vegetation composition and structure, similar to that described for other grass-dominated habitats. Grazing at moderate intensities will maximize structural and plant-species diversity in middle and high marsh by causing moderate levels of selective vegetation removal (Bouchard *et al.* 2003; Figure 9.2). Grazing at very high intensities can significantly reduce or eradicate some plant species, especially tall, palatable forbs. Heavy grazing typically results in a species-poor sward of low-growing, grazing-tolerant grasses, bare ground, and annual plants. Heavily sheep-grazed saltmarsh can be particularly short and uniform in vegetation structure and plant species composition.

Optimal stocking levels for saltmarsh are, though, often difficult to define. This is because most grazing units are relatively large and often contain a variety of different vegetation types related to differences in elevation. Grazing pressure can also vary widely within an individual grazing unit. Grazing pressure tends to decrease with distance from seawalls and other areas of high ground (e.g. Esselink *et al.* 2000), while deep creeks can prevent livestock from accessing many areas. As a guide, though, medium stocking levels for summer grazing

**Fig. 9.2** Saltmarsh grazing. Grazing levels can have a profound effect on the vegetation composition and structure of saltmarsh. These photographs show the effects of three grazing intensities on adjacent areas of saltmarsh of similar elevation (Frampton Marsh, The Wash, Lincolnshire, England).

(a) Ungrazed: dominated by dense stands of sea couch with abundant litter. Most small pools and creeks are covered by matted vegetation.

(b) Moderately grazed: a mixture of tall, patchily grazed sea couch and shorter, common saltmarsh-grass-dominated areas. Most small pools and creeks are relatively open.

(c) Heavily grazed: almost entirely dominated by a short sward of common saltmarsh-grass. Pools and creeks are very open.

(April to October) in middle to high marsh in Western Europe are typically between 60 and 120 livestock units per hectare per year (see Figure 5.8 and Table 5.1 for an explanation of how to calculate grazing pressure).

Livestock increase the area of saltmarsh over which they forage as they become more familiar with a site. They can be encouraged to roam over larger areas by including a proportion of animals already familiar with the site from a previous year, to lead other livestock around. Bridges can be built across creeks to allow livestock access to otherwise ungrazed areas. However, variation in grazing levels through avoidance of some areas can also be beneficial. Patches of ungrazed saltmarsh within a larger grazed area will add variety.

As well as creating general botanical and structural diversity, other common specific aims of grazing saltmarsh in north-west Europe are as follows.

- To provide a short sward in middle and high marsh dominated by palatable common saltmarsh grass and red fescue at the expense of unpalatable sea couch to benefit wintering geese and Eurasian wigeon, *Anas penelope*, and maintain open pools for waterbirds. This can be achieved by moderate to high levels of grazing;
- To provide a mixture of short and long vegetation in middle and high marsh to provide suitable nesting habitat for common redshank, the main breeding species of high conservation value on these saltmarshes (Norris *et al.* 1997). This can also be achieved by moderate levels of grazing, although the optimal level will depend on the existing variation in vegetation structure.

There has been little specific investigation into the effects of grazing on saltmarsh invertebrates. However, the principles are probably similar to when managing other types of grassland and herbaceous vegetation. Moderate grazing pressure that provide a diversity of vegetation composition and structure will probably provide suitable habitat for the widest range of above-ground invertebrate species, although it will reduce the abundance of litter-dwelling species. Very heavy grazing pressure that creates a short structurally uniform, botanically poor sward will impoverish the above-ground fauna by removing nectar sources and palatable forbs. Heavy grazing will, though, benefit invertebrate species associated with shallow pools with open margins.

Overall, then, the best option at most sites will be to graze at moderate levels, thus creating:

- a range of structures and plant species composition;
- short swards of palatable grasses for grazing wildfowl;
- open pools for waterfowl;
- where relevant, mixtures of dense and open vegetation for breeding common redshank.

Grazing pressure needs to be low enough that it does not eliminate grazing-sensitive plants, remove most nectar sources and litter used by invertebrates, and remove seedheads used by invertebrates and wintering seed-eating songbirds. Providing grazing intensities are not extremely high, though, this can be relatively easy to achieve due to patchy use of the marsh by grazing animals, particularly where deep creeks prevent access to some areas. It can also be valuable to retain areas of tall, ungrazed saltmarsh for their intrinsic value and invertebrate interest. One option is to graze different units at different intensities, to ensure a variety of conditions at a larger scale.

## 9.2.2 Burning

Prescribed fire is mainly used along the eastern seaboard of the USA to provide suitable conditions for wintering wildfowl. It benefits geese by stimulating succulent re-growth of their favoured foodplants and increasing their access to nutritious rhizomes, and seed-eating wildfowl by promoting the growth of preferred waterfowl foodplants, stimulating seed production, maintaining open water, and providing other open loafing and feeding areas (Chabreck *et al.* 1989). This burning also provides open feeding areas for icterids. Burning has also been used to control invasive common reed. As in some other habitats, prescribed burning can be used to reduce fuel loads and thereby the risk of large-scale, catastrophic fires started close to roads and other areas of public access.

Burning is typically carried out in autumn or early winter to provide suitable conditions for arriving waterfowl. Any benefits to waterfowl have to be set against short-term, detrimental effects on songbirds requiring cover. These include breeding marsh wrens, *Cistothorus palustris*, and sedge wrens, *Cistothorus platensis*, and wintering Nelson's sharp-tailed sparrows, *Ammodramas nelsoni*, and seaside sparrows, *Ammodramas maritimus*. Use of these areas by songbirds, though, typically returns to pre-burn levels by a year or so after burning (Van't Hul *et al.* 1997; Gabrey *et al.* 1999, 2001). Hence *annual* burning of large areas will be damaging for many songbirds, but burning smaller areas on a rotation of 2 or more years is probably compatible with both wintering waterfowl and saltmarsh songbirds. There is some evidence that in saltmarshes in the southern USA, which commonly experience natural lightning-induced fires, infrequent burning is important in preventing the vegetation from becoming too dense for some nesting sparrows. It is not known whether this is the case in more northerly marshes (Mitchell *et al.* 2007).

Other effects of the timing and intensity of burning are likely to influence its effects, but there has been little other critical evaluation of this (Mitchell *et al.* 2007). There is also little known about the effects of burning on small mammals, reptiles, and invertebrates, although the general principles are likely to be similar to when burning other types of grass-dominated habitats (see Chapter 5).

### 9.2.3 Reducing erosion and increasing sediment supply

Loss of saltmarsh through erosion can be prevented or reduced by protecting the saltmarsh from wave action. Eroded saltmarsh can be built up by supplying additional sediment from dredging operations elsewhere.

Wave action can be reduced by positioning wavebreaks offshore to protect exposed saltmarsh, and within intertidal creation areas to reduce the force of incoming or internally generated waves. Wavebreaks can be solid structures extending above water level, which provide impermeable barriers to wave action, or sub-surface structures that reduce water depth and cause waves to break before they reach the shore. Offshore wavebreaks are usually aligned parallel to the shore. Their orientation and spacing needs to be designed carefully, ideally based on

**Fig. 9.3** Restoration of saltmarsh by sediment deposition. The area on the right was open water a year before the photograph was taken, the original saltmarsh having been lost due to erosion initiated by introduced nutria/coypu. The saltmarsh has been restored by pumping sediment into areas cordoning off using wooden stakes with haybales tied to them and planting this with smooth cordgrass, *Spartina alterniflora* (Blackwater National Wildlife Refuge, Maryland, USA).

modelling, to ensure they do not cause increased unwanted erosion elsewhere along the coast.

There are two methods of providing additional sediment. Fine sediment can be deposited offshore, to allow coastal processes to then re-distribute it on saltmarsh and mudflats towards the shore, in a process known as trickle charging. Alternatively, more cohesive sediment can be pumped into contained areas. These are usually left to vegetate naturally, but can be planted (Figure 9.3). Sediment re-charge is a complex procedure, requiring in-depth knowledge of water movements and sediment supply.

## 9.3 Management of intertidal habitat-creation sites

While wholesale re-creation of intertidal habitat is outside of the scope of this book, the following sections will briefly consider the range of hydrological regimes that can be employed at intertidal recreation sites, and to restore saltmarsh in existing impoundments. It will also emphasize the potential to enhance re-created saltmarsh by re-introducing topographic variation.

### 9.3.1 Tidal regime

There are two types of tidal regime that can be used in intertidal habitat-creation areas:

- *full* tidal exchange through *managed realignment*;
- *regulated tidal exchange* (RTE).

Managed re-alignment involves the creation of intertidal habitat by *breaching* existing coastal defences, to allow the development of fully intertidal habitat on previously claimed land behind them (Figure 9.4). RTE involves *maintaining* existing sea defences, but allowing a *regulated* flow of tidal water through them (Figure 9.5).

Flow of water into and out of the RTE areas can be controlled using self-regulating tidegates and more expensive computer-controlled electronically operated tidegates that can regulate water levels to within a few centimetres. A tidegate consists of a door hinged to the top of the end of a culvert or other opening, which can only open seaward. With self-regulating tidegates the force of the incoming tide pushes the door closed and prevents ingress of seawater. Floats can be attached to the door to change the tidal regime within the RTE area by altering the time within the tidal sequence that the door closes. Computer-controlled tidegates are used where it is necessary to control water levels very precisely in areas subject to complex variations in water levels caused by large inflows of water from rivers interacting tides.

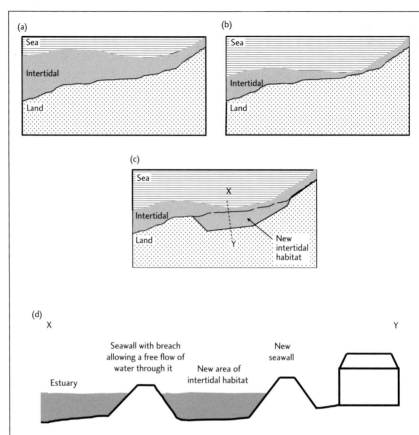

**Fig. 9.4** Coastal squeeze and managed re-alignment. Coastal squeeze occurs where existing intertidal habitat is lost at its seaward side through sea-level rise and/or falling land levels, but replacement habitat is prevented from being formed to its landward side due to the presence of hard coastal defences (a, b). New intertidal habitat can be created to offset these losses. This can be done by:

- constructing a new line of defence further inland and allowing tidal water to pass through the former sea defences to allow intertidal habitat to develop in the area in between (c & d); or by
- breaching the existing defences and allowing intertidal habitat to develop between these former defences and high ground further inland.

This process is known as managed re-alignment. The intertidal habitat created, in this case created between the former seawall and a newly constructed one further inland, experiences a similar tidal regime to that of surrounding intertidal areas of similar elevation.

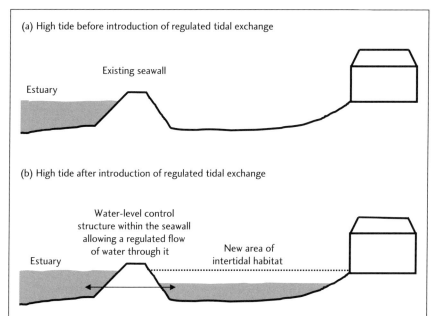

**Fig. 9.5** Regulated tidal exchange (RTE). In RTE the existing sea defences are retained but modified to allow a regulated flow of water through them. RTE can be used to produce a different tidal regime within the created or restored intertidal area to that in surrounding fully tidal areas of similar elevation. The main benefit of RTE is in reducing the maximum level of tidal water within the habitat-creation area, in situations where the higher water levels allowed by full tidal exchange would necessitate construction of a costly inner seawall to protect infrastructure. The dotted line indicates the maximum water level that would occur in the new area of intertidal habitat if full tidal exchange had been allowed through managed re-alignment.

The vegetation and other environmental conditions that develop under RTE in areas below the elevation at which tides are prevented from entering will be similar to that which would develop at similar elevations under full tidal exchange. Above this height there will be a more rapid transition to more terrestrial vegetation than would occur under full tidal exchange, because of the sudden curtailment of tidal influence.

A further variant of RTE involves designing it to provide flood storage (Figure 9.6). Alternative tidal regimes using RTE are described in Section 9.4.

### 9.3.2 Re-introducing topographic variation

Variations in topography on saltmarsh provided by creeks, pools, and salt pans are important in providing variation in drainage and consequently in vegetation. The

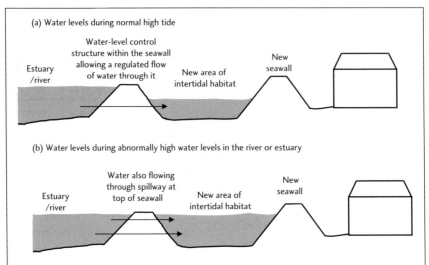

**Fig. 9.6** Use of RTE for flood storage. This design incorporates a spillway in the top of the seawall. This allows water to flow rapidly into the new area of intertidal habitat during storm surges and other periods of high water levels.

creeks and pools themselves are important in transferring water, sediment, and nutrients, and as habitat for birds, fish, and crustaceans. This topographic variation is usually removed following claiming of saltmarsh for arable land, although at least relic patterns are often retained following claiming of land for grazing marsh.

Following managed re-alignment or introduction of RTE to formerly drained, levelled land, creeks only tend to form within soft sediment accreting in lower-lying areas on its seaward side, particularly in areas with a steeper gradient in elevation. Higher areas will accrete less sediment, and the flow of water into and out of these higher areas usually has insufficient energy to cut creeks into the usually hard, consolidated, formerly drained soil. Development of more natural saltmarsh morphology and function can be accelerated by excavating a rudimentary creek network in these higher areas prior to breaching (Wallace *et al.* 2005). These will elongate and aid the development of associated secondary and tertiary creeks by channelling water towards them. Excavation of creeks and pools is logistically near-impossible once these areas have become intertidal.

It is best to position re-instated creeks in as close a location to where they occurred prior to land-claim, while ensuring that they are linked to the breaches in the former seawall and water inflows/outflows in RTE areas. It may be possible to identify the location of at least the major former creeks from depressions on the ground, or identify their locations from aerial photographs if the area has been claimed relatively recently. It is also be important to in-fill any existing

## Restoration of impounded saltmarsh | 315

**Fig. 9.7** Increasing topographic variation in newly created intertidal habitat. At Goosemoor in Devon, England, artificial creeks have been excavated in formerly flat, grassy field prior to introduction of regulated tidal exchange (RTE) and the spoil used to create higher islands for birds to roost on. This has created a diverse range of conditions from permanent saline water, to intertidal mud, saltmarsh, and non-tidal grassland. This photograph was taken 2 years after the introduction of RTE.

linear drains within the area prior to breaching, to prevent water from being channelled along these instead. The physical characteristics of these creeks should be designed to mimic those that would occur naturally at the site. Suitably constructed creeks can rapidly develop similar assemblages of fish to those found in natural saltmarsh (Williams and Zedler 1999) and probably increase use of the saltmarsh by shrimps (Minello *et al.* 1994).

Higher areas can also be created using spoil from these excavations, to provide areas exposed at high tide for roosting birds, and allow valuable transitional habitat to develop between high marsh and only irregularly flooded and unflooded ground (Figure 9.7).

## 9.4 Restoration of impounded saltmarsh

RTE can be used to re-introduce exchange of tidal water to restore brackish conditions and saltmarsh vegetation to impounded saltmarsh, which has become either too fresh or too saline. Many areas of saltmarsh have had ingress of tidal regime restricted or prevented by the construction of roads, causeways, and bridges, or through impoundment to provide standing water for wildfowl and in the north-east USA for the commercial production of salt hay. These embankments and dikes usually contain control structures that allow freshwater to flow out of the marsh into estuarine areas at low tide, but prevent tidal water from flowing back into them at high tide. The consequent reduction in salinity has converted these saltmarshes into freshwater to brackish marshes. In the north-eastern USA these freshwater to brackish marshes have often become dominated by common reed, narrowleaf cattail/lesser bulrush, *Typha angustifolia*, and broadleaf cattail/bulrush, *Typha latifolia*, at the expense of cordgrass-dominated vegetation (Roman *et al.* 1984; Sinicrope *et al.* 1990; Warren *et al.* 2002). These water-level-control structures also impede fish movement (Section 4.2.2). Conversely, areas of impounded saltmarsh that lack an appreciable input of fresh water can, due to evaporation, become more saline than seawater.

Re-introduction of tidal exchange through control structures in existing seawalls/dikes can be used to raise salinity to impounded saltmarsh, while still maintaining the existing embankment/dike to provide protection from high sea levels. Re-introduction of tidal exchange has been successfully used to reduce the abundance of common reed and cattail and restore the intertidal flora and fauna (Sinicrope *et al.* 1990; Brawley *et al.* 1998; Swamy *et al.* 2002; Warren *et al.* 2002; Buchsbaum *et al.* 2006). Impoundment and drainage reduces the elevation of marshes through oxidation of the substrate, and also prevents further accretion of tidal sediment. This reduces the elevation of formerly impounded saltmarsh relative to that of adjacent, unimpounded tidal marsh (Bryant and Chabreck 1998). Following restoration these marshes therefore retain tidal water for longer periods than marshes that have never been impounded and drained. This will increase use by waterbirds but reduce the area of high marsh (Slavin and Shisler 1983), at least initially until the elevation increased through accretion. Any excavation of drainage ditches and infilling of saltmarsh creeks following impoundment will also influence their hydrology following re-introduction of tidal exchange.

Water-level control structures can be used to vary the hydrological regime within the impounded area at different times of year. At Little Creek Wildlife Area in Delaware, USA (Figure 9.8), water levels are held constantly high between October and the end of January to cover most of the marsh and provide permanent water for wildfowl. RTE is then used during the rest of the year, but with water levels kept lower than on adjacent Delaware Bay such that only between 25 and 50% of the marsh is covered with water. The exposed mud attracts large

**Fig. 9.8** Restoring impounded saltmarsh by reducing salinity. At Little Creek Wildlife Area, Delaware, USA, regulated tidal exchange has been used to restore saltmarsh to an area that had originally been impounded to control salt marsh mosquitoes. With no direct input of freshwater, this impoundment had become hypersaline due to evaporation, resulting in an almost complete loss of saltmarsh vegetation. RTE was the used to restore tidal flow and reduce salinity. The area now supports smooth cordgrass in the lower marsh (middle and back of photograph) and upper marsh is dominated by saltmeadow cordgrass, *Spartina patens* (foreground). Details of the water regime are given in the text.

numbers of feeding waders/shorebirds, especially when the surrounding mudflats are covered at high tide. Water levels are sometimes held high for periods of 3–4 weeks between April and October to reduce the area of smooth cordgrass and help maintain areas of open mud.

There is also the potential at RTE sites to maintain intertidal conditions for most of the year, thereby helping provide a high biomass of benthic invertebrates, but to then maintain water levels constantly high during the bird breeding season to provide a temporary, productive saline lagoon with islands for birds to nest on. Although seemingly unnatural, such a regime would probably mimic a naturally dynamic coastline along which storms periodically create temporary, saline water bodies. A further variation would be to periodically maintain artificially extended periods of low tide within the impoundment, thereby providing extra feeding areas for waders/shorebirds when it is high tide on the adjacent estuary. Prolonged periods of immersion will, though, reduce the biomass of invertebrate prey in the mud, any such reductions probably taking place more quickly during hot, dry weather.

## 9.5 Saline water bodies

These comprise a range of salty coastal pools, lagoons, and impoundments with no, or only very limited, tidal variation in water levels. They also include salty silt lagoons created for the deposition of dredgings and salinas/salt evaporation ponds/solar ponds used for commercial salt production (Figure 9.9). Saline water bodies can also occur away from coastal areas in areas with high evaporation. Brackish is used to describe water with a salinity (salt content) between that of fresh and normal marine water. Hypersaline is refers to water with a salinity greater than that of normal, marine seawater. Salinity is usually measured in parts per thousand (ppt or ‰), and is approximately equal to $0.64\times$ conductivity in

**Fig. 9.9** Salinas/salt evaporation ponds/solar ponds. These are artificial, shallow, saline lagoons used for salt production in hot climates. Seawater is passed through a series of lagoons, during which process it becomes increasingly saline due to evaporation. Eventually the salt crystalizes out on the final set of lagoons and is collected. Salinas support a similar assemblage of specialized birds and invertebrates to those found in natural, high-salinity lagoons (Margherita di Savoia, Puglia, Italy).

milliSiemens (mS) (Jones *et al.* 2006). Marine seawater typically has a salinity of 35 ppt, but is lower in some areas, for example the Baltic Sea.

Saline water bodies can support large numbers of wintering and passage waterfowl. Islands in coastal, saline lagoons can support nesting waders/shorebirds and important colonies of gulls and terns. Deeper water supports mainly diving ducks and grebes. Species characteristic of shallow saline water bodies include species of flamingos, stilts, and avocets.

The invertebrate fauna of saline water bodies comprises a relatively small number of species, but includes a high proportion that are restricted to saline as opposed to freshwater habitats. The bare and sparsely vegetated margins of saline water bodies support a very rich invertebrate fauna, comprising a diverse range of ground beetles, rove beetles, and flies in particular, whose importance has often been overlooked (Figure 9.10). The flora of saline lagoons comprises a

**Fig. 9.10** Unassuming saline pools and their margins. Re-building of the shingle bank from its landward side has been used to create and maintain these small, shallow saline lagoons. These are percolation lagoons, which maintain a fairly constant, high salinity by percolation of seawater through the shingle bank. Despite their unassuming appearance, these small lagoons contain a number of rare and specialized invertebrates, including the starlet sea anemone, *Nematostella vectensis*, in the water and a large number of rare and specialized beetle and fly species on their bare and sparsely vegetated margins. These particular pools have since been lost following breaching of the shingle bank; see Fig. 4.10 (Walberswick, Suffolk, England).

very restricted range of vascular plants, algae, and stoneworts, but also includes a number of species restricted this habitat. Widgeon grass/beaked tasselweed, *Ruppia maritima*, and stoneworts are important food for herbivorous and seed-eating wildfowl.

In terms of management, there is a major distinction between:

- water bodies with a regular exchange of seawater, but which still retain a significant proportion of their water at low tide; these consequently maintain a relatively constant and high salinity, and are primarily of conservation value for their specialized lagoon invertebrate flora and fauna;
- shallow-water bodies of either low salinity or, due to changes in inputs of freshwater and seawater and evaporation, widely fluctuating salinity; these are primarily of conservation value for waterbirds and, as discussed below, rarely support important assemblages of specialized saline lagoon invertebrate and plants. However, important and diverse assemblages of invertebrates can occur on their margins. Shallow lagoons specifically created for waterbirds are often referred to as scrapes.

Salinity has a profound effect on the abundance and composition of the invertebrate fauna (Bamber *et al.* 1992; Joyce *et al.* 2005), and also influences vegetation composition. The abundance of invertebrates will in turn affect the suitability of lagoons for waders/shorebirds and the abundance of fish will influence their suitability for herons, egrets, and other fish-eating birds. The quantity and type of submerged and emergent vegetation will influence conditions for wildfowl.

The key principle of managing saline water bodies to maximize their conservation value is to maintain salinities within the tolerance ranges of the desired suite of species, to maintain characteristic suites of species, and to maximize the biomass of invertebrate prey for birds.

As in freshwater bodies, water levels can also be manipulated to provide:

- shallow water and mud for feeding waders/shorebirds, herons, egrets, wildfowl, and other waterbirds;
- moist mud for ruderal and emergent plants to germinate on in spring in low-to-medium-salinity lagoons, whose seeds can subsequently be made available for feeding ducks by raising water levels the following winter (moist-soil management);
- bare and sparsely vegetated margins for a range of insect species.

Water levels will also influence salinity levels, especially in shallow lagoons, where evaporation in hot conditions can cause rapid increases in salinity. Excessively high nutrient levels are assumed to be a problem in saline water

**Fig. 9.11** Reducing nutrient levels in saline lagoons. Nutrients levels in this lagoon are high due to large inputs of agricultural run-off in the freshwater that drains into it. Two projects are being used to address this. The first involved increasing exchange of water through a shingle bank that separates the lagoon from the sea, to help flush out nutrients that have accumulated in the lagoon. The second involved creation of a 54-ha shallow freshwater lake and 14 ha of grassland on former agricultural land within the catchment. This aims to reduce the area of arable from which nutrient-rich run-off can occur and to retain nutrients within the run-off before they reach the saline lagoon (Tryggelev Nor, Fyn County, Denmark).

bodies, as in freshwater ones, but there is little quantified information on this. Methods aimed at reducing nutrient levels include reducing nutrient inputs within the catchment, increasing exchange of seawater, and constructing nutrient traps (Figure 9.11). There is also little information on the success of these measures, though.

In addition, islands can be created and managed to provide breeding sites and safe roosts for waterbirds and marginal vegetation controlled to maintain open conditions.

### 9.5.1 Manipulating salinities

Salinities can be manipulated by controlling inflows of fresh and saline water. There are several potential sources of saline water: channels, pipes, and culverts connecting the lagoon to the sea, overtopping of banks, and percolation of seawater water through sand and shingle. In water bodies that lack continual inputs of seawater, salinities tend to be highest in summer due to increased evaporation and, at sites that receive inflows of freshwater, reduced freshwater inputs. Water bodies that contain areas of deeper water are better buffered against large changes in salinity caused by evaporation. Increasing the rate of exchange of seawater not only helps maintain more constant salinities, but probably also increases the likelihood of colonization by lagoonal specialist invertebrates via the sea.

Although the salinity tolerances of individual lagoon invertebrate species differ from one another, and are in many cases poorly understood, it is possible to broadly divide the fauna of temperate lagoons into those characteristic of:

- **low salinities** (less than about 8 ppt), in which the most abundant invertebrate prey are usually non-biting midge larvae in the mud, and water boatmen Corixidae and opossum shrimps *Neomysis* spp. in the water column;
- **medium-to-high salinities** (between about 8 and 40–70 ppt); in these the most abundant species in the mud are polychaete worms, non-biting midge larvae, molluscs, and amphipods, with opossum and other shrimps in the water column;
- **hypersaline conditions** in salinas and natural high-salinity lagoons above about 70 ppt. These are dominated by brine shrimps, *Artemia* spp., and brine flies, Ephydridae. Brine shrimps can occur at extremely high densities and withstand salinities up to approximately 320 ppt (Britton and Johnson 1987).

Management should therefore aim to maintain salinities within the ranges required by these different assemblages. In the UK maximum biomass of a non-biting midge larvae/water boatman and ragworm, *Hediste diversicolor,*/mud shrimp, *Corophium volutator*, fauna occur at, respectively, about 6 and 24 ppt (Robertson 1993). Lagoons supporting the highest-value specialist lagoonal faunas typically have a salinity of 20–35 ppt, and a high rate of exchange of their water with the sea: 35–50% of it on each tide (Bamber *et al.* 1992, 1993). In lagoons supporting a hypersaline invertebrate fauna, the highest densities and diversity of waterbirds have been found to occur at salinities between 100 and 200 ppt (Warnock *et al.* 2002). Fish, and consequently fish-eating birds, are usually more restricted to lower-salinity lagoons, although some species found in salinas can reproduce at salinities of 75 ppt (Lonzarich and Smith 1997).

The ability of populations of lagoonal invertebrates to recover following periods of unsuitable salinity, and hence the importance of maintaining suitable salinities, will vary between species. Non-biting midge larvae quickly re-colonize via winged adults, and have several generations per year. They can thus rapidly increase to high densities when conditions become suitable for them during spring to autumn. Biomass of polychaetes and bivalves may be slow to recover following a period of adverse salinity, because both take several years to reach maximum size.

Salinity levels will also influence the flora, although are rarely specifically controlled to benefit particular species or suites of plants. The important wildfowl foods dwarf spikerush, *Eleocharis parvula*, and widgeon grass/beaked tasselweed are typically found in water with a salinity of 5–20 ppt. Higher plants are absent from very-high-salinity lagoons, for example those more than 64 ppt in southern France (Britton and Johnson 1987).

## 9.5.2 Manipulating water levels

Principles of manipulating water levels to provide feeding habitat for waterbirds are similar to those in shallow freshwater bodies, as follows.

- The highest numbers of bird species occur in water 10–20 cm deep (Section 8.4.2), but with many waders/shorebirds requiring water shallower than this and some, such as plovers, preferring bare mud. Therefore, the range of bird species present can be maximized by providing a variety of water depths.
- Receding/falling levels will provide suitable invertebrate-rich shallow water and damp mud containing stranded invertebrates suitable for feeding waders/shorebirds (Taylor 2004). Raising water levels, although creating apparently suitable shallow water habitat, will not produce suitable feeding conditions until aquatic invertebrates have had time to colonize it.

The typical water regime used to maximize the value of shallow lagoons for waterbirds and probably also marginal invertebrates is described below. In practice, this mimics or accentuates the regime that would occur naturally in lagoons without significant tidal exchange.

Water levels are typically held relatively high in winter to provide open water for wintering wildfowl and to kill off vegetation on islands. They are then lowered in spring to:

- expose un-vegetated areas on islands for nesting birds;
- provide suitable conditions for germination of plants on damp mud (see below) and shallow water and bare mud for migrating waders/shorebirds.

Gradual lowering of water levels during summer, ideally at varying rates in different parts of a saline lagoon complex, will help ensure a continuity of shallow water and freshly exposed, moist mud for feeding waders/shorebirds. It will also provide a variety of different soil moisture conditions and vegetation types on the lagoon margins, thus increasing the variety of niches available for marginal invertebrates. Water levels should obviously not be lowered to the extent that they allow access of ground predators to nesting islands.

Further lowering of water levels in autumn can be used to provide additional feeding habitat for waders/shorebirds during migration. Water levels are then raised in autumn to winter levels. This also suspends seeds and makes them available to feeding ducks.

Principles of moist-soil management in shallow, low-to-medium-salinity lagoons are also similar to those in shallow freshwater wetlands (Section 8.4.1), although the details will differ according to the range of plants that management seeks to encourage or discourage (Figure 9.12). Management used to reduce

**Fig. 9.12** Management of brackish impoundments for wildfowl. Drawdowns can be used in brackish impoundments to provide suitable conditions for germination and growth of preferred wildfowl foods, similar to as in moist-soil management in freshwater wetlands.

In this impoundment partial spring and summer drawdowns are used to produce extensive, low-growing carpets of prolifically seed-producing dwarf spikerush and widgeon grass/beaked tasselweed, and to expose mud and shallow water for feeding waders/shorebirds. The tall vegetation is saltmarsh fleabane, *Pluchea odorata*, which germinates in areas that dry out quickly once exposed, and is not considered a valuable wildfowl food (Edwin B. Forsythe National Wildlife Refuge, New Jersey, USA).

the cover of moist-soil-managed species that are poor food for wildfowl include disking to reduce saltmarsh fleabane, and drying out in late winter, burning, and re-flooding to control big cordgrass, *Spartina cynosuroides*. As with large-scale moist-soil management in freshwater habitats, the disadvantage is that it removes large areas of shallow-water habitat for breeding waterbirds and, if the drawdown is undertaken over a very extensive area at the same time, does not provide a *continuity* of freshly exposed marginal habitat for invertebrates.

Whereas periodic drying out, either specifically as part of moist-soil management or for other reasons, can be used to increase invertebrate food supply for birds in freshwater lagoons, there is little known of its effects on invertebrate biomass in saline water bodies. Drying out kills fish, and high densities of fish can reduce densities of benthic invertebrates. Set against this, small fish are themselves important prey for birds such as herons and egrets. Experiments have found varying effects of addition of dead plant matter, as would occur through moist-soil management, on invertebrate biomass. Effects varied between invertebrate taxa and depending on the initial organic content of the substrate (Robertson 1993). As mentioned in Section 9.5.1, biomass of polychaetes and bivalves in higher-salinity lagoons may be slow to increase once they have been killed during a sustained period of drying out, whereas that of non-biting midge larvae in less-saline lagoons should recover more quickly. Disking/rotovation can be used to break up compacted mud and possibly improve feeding conditions for probing waders/shorebirds. Burning patches of emergent vegetation prior to re-flooding to increase interspersion of brackish swamp and open water has been found to increase the biomass of non-biting midge larvae, whereas increasing interspersion by burning had no effect (de Szalay and Resh 1997).

## 9.5.3 Controlling marginal vegetation

In some shallow, less-saline lagoons management of emergent plants might be required to maintain open, shallow water and bare mud for feeding waterbirds. This may include control of common reed, sea club-rush/alkali bulrush and mare's-tail, *Hippuris vulgaris*. The methods used will depend on the species involved. They can include cutting stems underwater, periodically drying the lagoon out with or without cutting or burning, and grazing the lagoon's margins. Vegetation growth tends to be slower at higher salinities, and there is unlikely to be any need to control vegetation in very-high-salinity lagoons.

## 9.5.4 Creating and managing islands

The value of saline water bodies for nesting and roosting waterbirds can be increased by providing suitable bare or sparsely-vegetated islands, as described for freshwater bodies (Section 8.3.1). Because of the coastal location of most saline lagoons, management of islands is likely to be aimed at benefiting a different range of breeding birds, especially colonies of gulls and terns. These generally require more open and un-vegetated conditions than breeding wildfowl.

## 9.6 Beaches and shingle

Beaches do not need any management to maintain their conservation interest. The only habitat-related issue on beaches, other than human disturbance to nesting birds, is the importance of retaining tidal refuse. This provides important habitat for invertebrates (Figure 9.13).

Shingle also does not require management to maintain its conservation interest, although some areas have traditionally been lightly grazed. Undisturbed, vegetated shingle is a particularly rare and fragile habitat, supporting characteristic

**Fig. 9.13** Tidal refuse. This can support important assemblages of invertebrates, especially of flies and beetles. The richest faunas are associated with a continuity of refuse on sites backed by dunes or other semi-natural habitat. Sea-soaked carrion and driftwood have their own distinctive faunas. On popular tourist beaches refuse is often cleared away using beach-cleaning machines. Driftwood is usually removed for firewood and rarely left to accumulate in any quantity (Mersehead, Solway Firth, Dumfries and Galloway, Scotland).

assemblages of higher plants, lichens, and associated invertebrates. The main issue on vegetated shingle is preventing damage to the fragile vegetation, especially lichens, and to the patterns of ridges and troughs of different-sized stones, by human activities. These patterns of sorted stones and associated patterns of vegetation types cannot be recreated.

## 9.7 Dunes

Dune systems contain a range of vegetation types of different successional stages. These include:

- fore dunes consisting of bare and specialized dune-forming plants that trap wind-blown sand and initiate the process of dune formation;
- vegetated yellow (low organic content) or grey (higher organic content) dunes supporting grassland, heathland, or other dwarf shrub and containing varying quantities of bare and sparsely vegetated sand;
- blow-outs (Figure 9.14);

**Fig. 9.14** Blow-outs. These occur in dunes where areas of loose sand, usually initially created by excessive human trampling and rabbit burrowing, are then rapidly enlarged by wind action.

Blow-outs are often regarded as damaging and in need of stabilization. They are, though, important in setting back succession and necessary for the formation of secondary dune slacks (Oriñón, Cantabria, Spain).

- dune slacks, where dunes have been eroded by the wind down to the level of damp sand, usually following a blow-out;
- scrub and woodland.

Dune grasslands can support a rich flora and fauna similar to that of other, open, sandy grasslands. As with these, the richest floras typically occur on base-rich substrates which, in the case of dunes, are found on wind-blown shell sand. Stable, especially calcareous dunes, are often especially rich in lichens. Dune heaths occur on acidic sands and support a similarly limited range of plant species, but rich invertebrate fauna, to those found in other types of heathland. The range of successional stages present in most dune systems supports a rich diversity of insects, and a range of warmth-loving reptiles. In north-west Europe the richest assemblages of dung beetles are usually found on grazed dunes. Dunes slacks can support characteristic and very species-rich wetland vegetation, particularly during their earlier successional stages, and provide valuable habitat for invertebrates. Temporary pools in dune slacks provide breeding sites for amphibians.

Dunes do not require management to maintain them. However, the proportions of the different successional stages present within a dune system can be manipulated by vegetation removal. Management can also be used to influence the structure and plant species composition of dune grasslands, heathlands, other types dwarf-shrub communities, and dune slacks. The types of management used are grazing on the dune system as a whole, mowing, and sod cutting in humid dune slacks and, where necessary, cutting and removal of scrub and control of alien/exotic plants. Small-scale rotovation and turf-removal can be used to provide small patches of bare and disturbed sand where this is lacking, and management of access points and fencing used to control excessive trampling and erosion by people.

### 9.7.1 Grazing

The principles of grazing dune grasslands, heathland, and other types of dwarf-shrub communities on dunes to provide suitable conditions for open-ground species (e.g. WallisDeVries and Raemakers 2001) are similar to those when managing these habitats elsewhere on light, sandy soils (Chapters 5 and 6). Cattle, ponies, sheep, and goats can all be used. Donkeys are particularly well-suited to the usually warm and dry conditions of sand dunes. European rabbits are important grazers and agents of soil disturbance on many dune systems in Europe.

Individual grazing units often contain a wide variation in habitat conditions caused by small-scale differences in slope, aspect, organic content of soils, and hydrology across the dune system. Hence it is usually possible to provide a variety of conditions by enhancing this variation through low-intensity, extensive grazing.

More tightly controlled systems are rarely needed. Stocking levels depend on the relative proportions of different habitats present, but are typically between 0.05 and 0.30 livestock units per hectare.

### 9.7.2 Bare ground and erosion

Much former management of dune systems has involved preventing and stabilizing erosion, in parts of Europe and North America often through planting of marram, *Ammophila arenaria*. Bare and disturbed ground is an important feature of dune systems, as it is in other types of grassland on light soils and on heathlands. There is unlikely to be any reason for planting to stabilize dunes for conservation, unless there is a risk of very large-scale loss of dune habitat.

There are, though, situations where erosion caused by human trampling might be excessive. Where this is the case, areas can be fenced off and signs erected to explain why this is necessary. Most people only walk a short distance from car parks and the distribution and extent of recreational pressure can be heavily influenced by the location of these and other access points (Figure 9.15).

**Fig. 9.15** Controlling public access on dunes. Where there is excessive disturbance to dune systems by people, it can be reduced by managing access. Here, the dunes on the left are fenced off from the busiest areas and car parks at the nearby holiday resort of Matalascañas. People can still visit and enjoy the dunes, but only by making a special effort to enter them via access points situated further away (Parque Natural des Dunas, Huelva, Spain).

Conversely, excessive stabilization and lack of grazing and trampling by wild or domestic animals or people can result in a lack of bare ground, especially in the inland areas of large dune systems. Even where dune systems contain a high proportion of bare sand, this is often concentrated in relatively large blocks, such as those created by blow-outs. These large blocks can be of limited value to warmth-loving invertebrates and reptiles that prefer a smaller-scale mixture of bare and sparsely vegetated ground and denser cover. More intimate mosaics of bare and sparsely vegetated ground can be created by sensitively rotovating small areas, or, if large quantities of vegetation and organic matter are present, by removing turves/sods. Care should obviously be taken to avoid initiating unwanted blow-outs.

### 9.7.3 Management of dune slacks

The principles of managing the wetland vegetation in humid dune slacks by mowing and sod cutting (Grootjans *et al.* 2001) are similar to those in other short, fen vegetation (Sections 8.8.4 and 8.8.6). In some cases pools in dune slacks have been created or existing ones deepened, particularly where water levels within the dunes are falling. However, any benefits have to be set against the damage to the near-natural topography they will cause. Deepening seasonal pools so they hold water permanently will also damage their existing invertebrate, amphibian, and vegetation interest (Section 8.5).

# 10
# Arable land

Arable (also known as rowcrop fields) is land that is cultivated regularly for production of food and, increasingly, bioenergy. The value of arable land to wildlife is heavily influenced by the presence of permanent grassland and other uncropped habitats along its boundaries, such as hedgerows, scattered trees, water-filled drainage ditches, grass strips, and stone walls. Management of these boundary features is therefore also discussed. Management of short-rotation coppice used for production of bioenergy and other systems containing trees is described in Section 7.4.3.

There are two approaches to managing arable and other farmland to benefit wildlife:

- farm in a wildlife-friendly manner to maintain or increase the value of farmland for its associated wildlife, but which may result in lower yields and thereby *increase* pressure to convert other natural/semi-natural habitats for food production;
- intensify farming to maximize yields, with a consequent reduction in the value of the farmland for its associated wildlife, but in doing so *reduce* pressure to convert other natural/semi-natural habitats for food production; this is called land-sparing.

The relative merits of these approaches will depend on the demand for the agricultural products produced and how populations of species on farmed land change with respect to agricultural yield. The ideal solution would be to develop farming practices that produce both relatively high yields and high-quality farmland for wildlife. Evidence from a range of taxa in developing countries suggests that intensifying farming to maximize yields may be the best long-term option for maintaining the persistence of the widest range of species, where the majority of biodiversity is associated with uncropped habitats (Green *et al.* 2005). High-intensity farming can, though, also have negative indirect effects on other habitats. These can be caused by pollution from pesticide and fertilizer run-off and abstraction of water for irrigation that reduces the quantity available for semi-natural wetlands (Bradbury and Kirby 2006; Jones *et al.* 2006).

## 332 | Arable land

Farmland is considered of relatively high existing or potential conservation value in large parts of Europe, where it contains a higher proportion of declining bird species than any other habitat (Tucker and Heath 1994; Birdlife International 2004). Unintensively managed arable land can also support assemblages of annual plants (so-called arable weeds) that are rare or absent from both more intensively managed arable land and surrounding semi-natural habitat.

Several long-established, low-intensity agricultural systems in Europe are considered of particularly high conservation value. These include the machair of north and west of Scotland and western Ireland, the pseudosteppes of the Mediterranean region (Figure 10.1), and the wooded dehesas (Spanish)/montados (Portugese) of the western Mediterranean (Figure 1.3 and Section 7.4.4). Machair is the Gaelic term for mosaics of low-input rotationally cropped arable, wet and dry grassland, and associated habitats on calcareous wind-blown sand.

**Fig. 10.1** Pseudosteppes. These comprise flat or gently undulating semi-arid areas containing mixtures of extensively grown, non-irrigated cereals, pulses, grazed fallows (shown above), dry grassland, and dwarf-shrub vegetation. They support the majority of the populations of many bird species of high conservation priority whose more semi-natural steppe habitat has almost entirely disappeared. These include little bustards, *Tetrax tetrax*, great bustards, *Otis tarda*, black-bellied sandgrouse, *Pterocles orientalis*, and pin-tailed sandgrouse, *Pterocles alchata*, and several species of lark. Pseudosteppes are also important for many wintering farmland birds from further north (Suárez *et al*. 1997; between Trujillo and Monroy, Extremadura, Spain).

The Scottish machair is of particular importance for its high densities of breeding waders/shorebirds and breeding corncrake population.

The general mobility of birds makes them better suited to exploiting the rapidly changing habitat conditions on arable land through the cropping rotation. The short lifespan of arable crops does not allow small mammals to build up to high densities. Instead, any small-mammal interest of farmland is mainly confined to uncropped field boundaries and land left fallow.

Most of the specific techniques for benefiting biodiversity on arable land have been developed in Europe, especially the UK. Information on these consequently dominates this chapter. Since the majority of arable land is farmed privately for profit, management prescriptions to benefit wildlife have been designed so that they can be incorporated into conventional farming systems and paid for using agri-environment schemes. Methods have focused on maintaining traditional, biodiversity-rich farming systems and, in more conventionally managed farmland, on specific techniques for reducing risks from, for example, pesticide use, and benefiting wildlife through adoption of specific management prescriptions (Ovenden *et al.* 1998; Directorate General for Agriculture and Rural Development 2005; Llusia and Oñate 2005). Methods to benefit breeding birds have focused on reversing declines of species associated with agricultural intensification in Western Europe (Chamberlain *et al.* 2000; Donald *et al.* 2001b, 2006), and minimizing predicted future declines in Eastern Europe (e.g. Sanderson *et al.* 2007). The prescriptions used on conventional farmland often involve removing specific, small areas of the crop from intensive management. This has the benefit of improving the value of the arable land for wildlife, while minimizing the reduction in agricultural production.

The main conservation priority in open habitats in North America is birds associated with grasslands. The emphasis of environmental programmes there has been on *removing* land from crop production, either in large blocks or in strips along field boundaries, for soil conservation, water protection, and restoration of wildlife habitat. In the USA this has been achieved principally through the Conservation Reserve Program (Section 3.3).

An additional aim of managing farmland land for wildlife has been to provide a harvestable surplus of gamebirds for hunting, especially grey partridges, *Perdix perdix*, in Western Europe and northern bobwhites in North America.

The main techniques for benefiting wildlife on farmland involve:

- maintaining farming systems that provide high land-use diversity and growing specific crops that benefit farmland wildlife, especially birds;
- minimizing pesticide and fertilizer use on field margins to benefit breeding birds, arable weeds, and invertebrates;

- providing cultivated but unsown areas within fields for birds that prefer to nest and feed on bare and sparsely vegetated ground;
- minimizing destruction of birds' nests during mechanical operations;
- providing unharvested crops for birds to eat;
- where relevant, manipulating flooding regimes to maximize benefits to waterbirds;
- creating and sympathetically managing uncropped habitats along field borders, mainly hedgerows and other scrub and grass strips, to benefit a range of wildlife that may or may not also utilize the adjacent arable;
- leaving land fallow and reverting it to grassland;
- introducing arable systems back into grass-dominated areas to increase land-use diversity and the range of feeding and nesting opportunities for birds.

In North America wholesale reversion of rowcrop fields to grassland, primarily through the Conservation Reserve Program in the USA, has benefited many grassland birds, although some enjoy only low breeding success in Conservation Reserve Program fields (e.g. Patterson and Best 1996; Best *et al.* 1997, 1998; Ryan *et al.* 1998; McCoy *et al.* 1999). Principles of managing such grasslands for birds and other wildlife are discussed in Chapter 5.

## 10.1 Farming systems and crop types

Most of the information on the use of different vegetation types in arable farmland by wildlife is from studies of birds in Europe. The value of different crops for birds is thought to be mainly due to differences in:

- vegetation structure for nesting;
- abundance of food (invertebrates, seeds, other plant material, and in some cases vertebrates) in the growing and harvested crop and vegetation structure for accessing it;
- timing of farming operations that can cause nest loss, and especially the timing of harvesting.

There are a number of different farming systems and vegetation types that have been shown to, or are predicted to, benefit birds and other farmland wildlife. These are described below.

### 10.1.1 Low-input, mixed, and organic farming

Many farmland bird species utilize a variety of arable crops, fallow, pasture, and uncropped habitats during the year and therefore require a heterogeneity of field types (Evans 1996; Wilson *et al.* 1996, 1997, 2001; Atkinson *et al.* 2002; Benton *et al.* 2003; Laiolo 2005). European hares, one of the only mammals considered of high conservation priority on farmland, have similar requirements (Smith *et al.* 2005b).

Heterogeneity of field types is best provided by farming systems that include both livestock production and arable rotations. This is known as mixed farming (Figure 10.2). Machair and pseudosteppes are both low-input, mixed-farming systems that provide a variety of vegetation types.

**Fig. 10.2** Low-input, mixed farming. Mixtures of different crop types and grassland, as found in mixed farming such as this in northern Poland, typically supports a high diversity of farmland wildlife. The village just behind these fields regularly holds between 36 and 45 pairs of white storks, *Ciconia ciconia*.

Patterns of farming and wildlife are intimately linked with economic conditions and past history. This and other areas of north and north-east Poland retained their small farms, while in most of the rest of Eastern Europe they were amalgamated to create large, collective farms. Farming in Poland is now likely to intensify following its accession to the European Union, and result in a loss of its farmland-associated wildlife. Agri-environment schemes could help retain some of this wildlife interest.

Ironically, in the Russian Federation a kilometre or so north of here, a decrease in the intensity of farmland is resulting in land being abandoned and succeeding to forest. This will also reduce its value for wildlife associated with low-intensity, farmed habitats although increase it for a range of other species (Żywkowo, Warminsko-Mazurkskie, Poland).

The scale of the optimal mosaic of field types will differ between species and also vary between the breeding and non-breeding seasons. Small songbirds are only able to travel shorter distances from their nest to forage, and so require a suitable variety of suitable foraging habitats within very close proximity (e.g. Field and Anderson 2004). Nidifugous species (those whose young leave the nest very soon after hatching) also require suitable chick-feeding areas relatively close to nest sites. For example, northern lapwings prefer nesting on spring-tilled land that is adjacent to pasture on which their chicks can feed (Wilson *et al.* 2001). Chick survival of northern lapwings nesting on cereal fields is greater where they have direct access to pasture, compared to where broods have to cross areas of unsuitable habitat to reach this favoured chick-rearing habitat (Galbraith 1988). Breeding raptors are able to forage over larger areas. During the non-breeding season most, but not all, bird species are able to range more widely and exploit temporarily suitable food sources.

Many organic farms practise mixed farming to maintain soil fertility and minimize pest outbreaks without the need for inorganic fertilizers and pesticides. A number of studies have demonstrated that organic farming generally benefits farmland birds, insects, and plants compared to conventional farming (Bengtsson *et al.* 2005; Hole *et al.* 2005). Most organic farming, though, involves a wide range of measures that benefit wildlife. These include mixed farming, reduced herbicide and fertilizer use, and sympathetic management of uncropped habitats. This makes it difficult to differentiate which of these specific measures are the main causes of the beneficial effects of organic farming on wildlife. It is worth noting that some organic farming practices are actually damaging to wildlife, especially frequent mechanical hoeing of crops.

So far, there are few examples of the introduction of specific crop types having demonstrable benefits to wildlife, although the increase in area of oilseed rape in Western Europe has inadvertently providing suitable habitat for breeding common whitethroats, *Sylvia communis*, Eurasian linnets, *Carduelis cannabina*, and reed buntings.

### 10.1.2 Spring-sown crops and winter stubbles

In Western Europe spring-sown crops benefit both birds nesting in the crop and by providing weedy, winter stubbles (Figure 10.3). Spring-sown cereals tend to have shorter, more open vegetation in spring and early summer than autumn-sown cereals. In Western Europe birds that nest in arable crops, such as northern lapwings and Eurasian sky larks, prefer more open conditions within the crop for nesting. Autumn sown cereals tend to be too tall and dense for nesting northern lapwings in spring, and are too tall for Eurasian sky larks' later nesting attempts

**Fig. 10.3** Weedy, winter stubbles. Weedy stubbles, particularly those with a high proportion of bare earth, provide an important source of seeds of annual plants and spilt grain for finches, buntings, sparrows, and larks in winter. Families of plants thought to be especially valuable for these are grasses, Gramineae, knotgrasses and persicarias, Polygonaceae, the pink family, Caryophyllaceae, goosefoots, Chenopodiaceae, composites, Asteraceae, brassicas, Brassicaceae, peas, and vetches Fabaceae (Wilson et al. 1999; Apaj, Pest County, Hungary).

that are thought necessary for them to maintain a stable population (Hudson *et al.* 1994; Wilson *et al.* 1997; Donald *et al.* 2001a). In Northern Europe, though, where there is little vegetation in spring due to slower vegetation growth and the majority of crops being spring-sown, breeding Eurasian sky larks are associated with areas containing over-wintered vegetation. It therefore appears to be the absence, rather than excess of vegetation, that limits their numbers there (Piha *et al.* 2003). In general, providing a range of different developmental stages of crops throughout the season will be best for most species, especially those that rely on multiple nesting attempts per season to maintain stable populations.

Stubbles only remain over the winter if the subsequent crop is sown in spring. With autumn-sown crops, the stubble is ploughed in immediately following harvesting in late summer and the next crop planted that autumn. A wide variety of weedy stubbles are suitable, including those from cereals, peas, oilseed rape, and linseed (Donald and Evans 1994; Evans and Smith 1994; Wilson *et al.* 1996; Buckingham *et al.* 1999; Moorcroft *et al.* 2002; Hancock and Wilson 2003). Oat and barley stubbles contain more weed seeds and support higher densities of birds in winter than wheat stubbles (Delgado and Moreira 2002; Moorcroft *et al.* 2002). Sugar beet stubbles provide important feeding areas for pink-footed geese, *Anser brachyrhynchus*, that feed on the waste sugar beet tops left after harvesting (Gill 1996; Gill *et al.* 1996).

Introduction of winter stubbles, often in combination with other measures, has proved successful in increasing numbers of seed-eating songbirds. Numbers of cirl buntings, *Emberiza cirlus*, rose in England following introduction of management agreements providing weedy winter stubbles along with other measures including grass margins around arable fields (Peach *et al.* 2001). Encouragement of a range of changes in field use (increases in the extent of winter stubbles and under-sown spring cereals), together with conservation headlands (Section 10.2), wild-bird cover (Section 10.5), and grass margins (Section 10.7.2) in a pilot scheme in England, has also shown benefits. These have resulted in increases in productivity of grey partridges and of abundance of wintering seed-eating songbirds and several other groups at a farm-scale in one of the two study areas (Bradbury *et al.* 2004). One-kilometre squares containing winter cereal stubbles have been found to have more positive population trends of breeding seed-eating songbirds, than 1-km squares without stubbles (Gillings *et al.* 2005).

### 10.1.3 Fallow, set-aside, and grassland creation

Fallow is land that is temporarily rested from cropping as part of an arable rotation. Set-aside is arable land temporarily taken out of agricultural production in return for grants, with the aim of reducing agricultural over-production.

The value of fallow and set-aside land for wildlife will vary primarily in relation to:

- the length of time since cropping;
- whether it is sown or left to regenerate naturally;
- any subsequent management.

When cropping ceases on arable land, the vegetation that establishes will depend on the composition of the seedbank and the conditions for its germination and subsequent growth. It will initially consist mainly of annual forbs and grasses, a smaller proportion of perennial species, and often some volunteer crops (i.e. crops growing from seed already present in the soil). The first winter of set-aside is similar to a weedy stubble (see Section 10.1.2), and may also include rare arable weeds. Weedy set-aside (and also weedy fallow) is especially valuable for breeding and wintering songbirds (e.g. Millenbah *et al.* 1996; Buckingham *et al.* 1999).

In subsequent years, in the absence of disturbance, the vegetation tends to become increasingly dominated by perennial vegetation, mainly grasses (e.g. McCoy *et al.* 2001a), with a general increase in species richness of species characteristic of uncropped habitats. It effectively becomes an early-successional grassland, in most cases rich in nutrients from previous fertilizer use. It will tend to be dominated by competitive perennials, mainly widespread grass species, and therefore be of low conservation value for its flora.

If the aim is to provide early-successional, weedy areas for breeding and wintering birds, then a short rotation using natural regeneration will be most appropriate. Most breeding farmland birds also occur at higher densities during the first year of set-aside, probably because it is more patchy, species-rich, and structurally complex than older swards (Henderson *et al.* 2000). If the aim is to provide habitat for species more typical of permanent grassland, then allowing older grassland to develop through non-rotational set-aside will be best (e.g. Bracken and Bolger 2006). However, as mentioned below, restrictions on management to comply with objectives for set-aside may mean it is impossible to manage these areas suitably for key bird and other species.

Set-aside and fallow takes 1–2 years to be colonized by reasonable densities of small mammals, with some species, such as Eurasian harvest mice, *Micromys minutus*, being characteristic of the early-successional, ruderal phase (Churchfield *et al.* 1997; Tattershall *et al.* 2000). Therefore, if areas are managed on an annual rotation, they will not have time to become of value for small mammals, and the predators that feed on them, before being converted back to arable. The composition of the small-mammal fauna changes with age. Eurasian harvest mice are typical of early successional stages dominated by annual and perennial forbs (Churchfield *et al.* 1997). Field voles, an often abundant small mammal in rough grassland in parts of Europe and an important prey species for raptors, only begin to colonize set-aside after 2 years or more when significant quantities of grass and litter have developed (Tattershall *et al.* 2000).

Sowing fallow land with a perennial grass mix will reduce the abundance of annual plants (Critchley and Fowbert 2000). It also alters the species composition of the perennial vegetation than subsequently becomes established. The extent to which it does so depends on how well the sown species are suited to the specific conditions, and on competition from species establishing from the seedbank. Seeding is often used to minimize the establishment and setting seed of plants that might cause agricultural problems when the land is returned to agricultural production. Bird species differ in their preference for areas sown with warm- or cool-season grasses (e.g. Delisle and Savidge 1997; McCoy *et al.* 2001b; Henningsen and Best 2005).

Any subsequent management of fallow and set-aside has to take account of the need to control pernicious agricultural weeds that might become a problem once it is returned to arable production. In the case of set-aside, there are also restrictions on its management to prevent the land from contributing to agricultural production. Mowing is allowed, but since the cuttings cannot be used, they are usually left on the field and this reduces the value of the developing grassland for plants and animals (Section 5.5.2). In some low-intensity rotational farming systems, such as in pseudosteppes, fallow land is grazed; that is, there is effectively a rotation of arable and pasture.

### 10.1.4 Undersowing cereal crops

Undersowing cereal crops with grass leys benefits birds that feed on invertebrates during the breeding season, notably grey partridges (Potts 1986). Sawfly larvae are important prey for chicks of a range of bird species. The lack of cultivation between undersowing and the following spring probably results in higher survival of sawfly pupae overwintering in the soil from July onwards (Barker *et al.* 1997). However, undersowing cereal crops probably reduces their value as stubble for seed-eating songbirds in winter by reducing the amount of bare soil and abundance of arable weeds (Moorcroft *et al.* 2002).

### 10.1.5 Non-inversion tillage

Non-inversion tillage is used to describe a range of methods of establishing a crop without inverting and deeply burying the upper soil to bury the surface residue that occurs when using using a conventional mouldboard plough. Instead, it uses various types of cultivation to disturb the soil surface to create a seedbed. Non-inversion tillage is also known as and conservation tillage (particularly in North America), minimum tillage (min till), reduced tillage, no-till, and ECOtillage, and also includes direct-drilling. Non-inversion tillage has mainly been developed in North America primarily to maintain soil structure, reduce erosion, and protect watercourses. A number of studies suggest that it might be beneficial to some bird species.

The reduced physical disturbance caused by non-inversion tillage tends to result in higher earthworm biomass compared to conventional mouldboard ploughing. The effects on weed seed abundance are more variable and depend on the method of non-inversion tillage and subsequent weed control. Most non-inversion tillage methods dos not bury seeds lying on the soil's surface as conventional mouldboard ploughing does, so may increase their accessibility to seed-eating birds. Set against this, the reduced soil disturbance of non-inversion tillage means that there is bare and disturbed ground for seed-bearing annual plants, including rare arable weeds, to germinate in (Albrecht and Mattheis 1998). In non-inversion tillage there is also a greater reliance on herbicides to control weeds. If very effective weed control is undertaken, then there will be no benefits to seed-eating birds.

Non-inversion tillage leaves more straw and other plant debris at the soil surface, and this might increase the habitat available to some invertebrate groups. The effects of non-inversion tillage compared to mouldboard ploughing on abundance of beetles and spider are unclear (Cunningham *et al.* 2004).

Several studies have shown higher densities of ground-nesting songbirds on non-inversion tillage land compared to on conventionally ploughed land. This is probably due to its greater vegetation cover, particularly that of litter, early in the breeding season when conventionally managed fields are more devoid of vegetation. However, productivity of birds nesting in minimum tillage fields is often low (as in conventionally ploughed fields). This is mainly due to high predation rates but also to some extent to the albeit less frequent mechanical operations. Hence there is concern that although non-inversion tillage may confer benefits for some breeding bird species, it may also act as an ecological trap for others that would otherwise nest more productively in prairie grasslands (Lokemoen and Beiser 1997; Martin and Forsyth 2003). There is evidence that fields cultivated using non-inversion tillage support higher densities of seed-eating songbirds and gamebirds in winter (Cunningham *et al.* 2005).

## 10.2 Minimizing pesticide and fertilizer use on field margins

Many of the declines of farmland birds (and of arable weeds) in Europe have been associated with the increased use of herbicides, insecticides, and fertilizers (e.g. Chamberlain *et al.* 2000).

Herbicides are obviously designed to be damaging to arable weeds. The use of both herbicides and insecticides decreases the availability of weed seeds and arthropod prey for birds, particularly for chicks of some species (Potts 1986; Freemark and Boutin 1995; Campbell *et al.* 1997; Newton 1998). Increased use of fertilizer (primarily nitrogen), together with the introduction of higher-yielding crop varieties, has increased crop growth and density. Less competitive arable weed species are unable to compete for light with the more vigorous crop (Kleijn and van der Voort 1997). Increases in crop growth and density, particularly combined with a switch from spring to autumn sowing, have reduced the availability of sparse, open crops favoured for nesting by several bird species as described in Section 10.1.2. There are two general approaches to reducing, or maintaining reduced, pesticide and fertilizer use:

- reducing their use across entire fields/farms (through organic farming or continuation of low-input farming systems; Section 10.1.1);
- only reducing their use on the margins of otherwise conventionally managed fields.

This second option aims to maximize wildlife and other environmental benefits, while minimizing the overall reduction in crop yield. It involves leaving an unfertilized strip around the margins of the field that is not sprayed with broad-spectrum herbicide or insecticides. Fungicides are usually allowed, which are not considered damaging to vascular plants, invertebrates, and birds. Conservation headlands can also be sprayed with approved selective herbicides before the spring to control specific, pernicious agricultural weeds. These unsprayed strips are often called conservation headlands, and were pioneered to increase productivity of grey partridges for shooting in the UK (Rands 1985; Potts 1986; Boatman *et al.* 1989). Breeding songbirds and gamebirds benefit from the lack of spraying, because it does not reduce the abundance of key invertebrate prey of their young.

Avoiding fertilizer use is likely to further benefit less-competitive arable weed species by reducing competition from the crop. Alternatively, competition with arable weeds can be completely removed by just tilling the ground annually without sowing any crops (Figure 10.4). This will provide suitable conditions for germination of arable weeds and prevent the establishment of perennial vegetation. Arable weeds vary in being mainly spring- or autumn-germinating. If a crop is also sown, then spring cultivation is preferable for the majority of arable weeds that germinate in both spring and autumn, because the sparser growth of spring-sown crops provides less competition.

Locating the unsprayed strip on the margins of the field has a number of benefits. It minimizes the loss in yield, because field margins typically have lower crop productivity than the rest of the field. This is due in part to soil compaction caused by more frequent passage and turning of farm machinery and less-efficient herbicide and fertilizer application. Field margins also tend to support higher densities of arable weeds than the rest of the field, probably because of this lower crop productivity. It is also logistically easy to leave this strip unsprayed and unharvested. An additional benefit is that a small number of often rare, short-lived perennial, and non-competitive annuals are also characteristic of irregularly cultivated land between the regularly cultivated crop edge and the uncultivated perennial vegetation of the field boundary (Wilson and Aebischer 1995).

Locating conservation headlands adjacent to hedgerows and scrub also benefits a number of farmland birds (Hinsley and Bellamy 2000). Many species that feed their young on invertebrates from within the crop nest and feed in hedgerows, in grassland, and among other tall vegetation along field boundaries. Many farmland birds that feed on weed seeds will only forage close to the shelter of hedgerows and scrub, probably to reduce predation risk (Henderson *et al.* 2004).

**Fig. 10.4** Arable weeds. Cultivated land can support a diverse array of ruderal plants commonly known as arable weeds. Many have become rare since the advent of modern herbicides. Arable weeds and associated invertebrates can be conserved within conventional farming systems by maintaining an unsprayed, cultivated strip (headland) on the field margin.

This 8 m-wide unsprayed strip is on the margin of an otherwise conventionally managed cereal field. It is managed specifically for its annual flora, which on the sandy soils of this area—known as Breckland—contains several plant species rare in, or absent from, the rest of the UK (Cherry Hill, Suffolk, England).

There is little information on the specific benefits of conservation headlands to invertebrates in their own right, other than for some butterflies. Weedy conservation headlands benefit adult butterflies of mobile species that can feed on the nectar of arable weeds. It also benefits the relatively few butterfly species whose caterpillars can complete their development on annual arable weeds before harvest, such as some species of whites, Pieridae, that have more than one generation per year (Feber and Smith 1995).

The width of conservation headlands is typically between 6 and 24 m. Six-metre strips have traditionally been used, since these can be created by turning off the outer 6-m length of a conventional spray boom unit. In most cases a mown, herbicide-sprayed sterile strip is maintained between the conservation headland and field boundary to prevent ingress of any pernicious agricultural weeds from the field boundary into the crop.

## 10.3 Providing cultivated but unsown areas within fields

Areas can be tilled but left unsown within the interior of the field to provide habitat for breeding birds that favour more open vegetation, and which avoid hedgerows and trees along field margins. Large plots (about 2 ha) of cultivated and unsprayed land can be created in suitable locations within arable on suitable chalky or sandy soil to provide nesting and feeding areas for Eurasian thick-knees and also northern lapwings. Small (e.g. 4 m × 4 m), un-drilled patches can be left in winter wheat to provide sparsely vegetated conditions for breeding Eurasian sky larks: so-called sky lark scrapes (Morris *et al.* 2004).

Plots for Eurasian thick-knees are usually cultivated in February/March to provide the open conditions they require for nesting. If a pair does nest on a plot, then the half of it without the nest can be cultivated in May to help maintain open conditions for chicks to feed in.

Sky lark scrapes are created by briefly turning off the seed drill during sowing and then managing these areas in the same way as the rest of the crop, including herbicide application, to minimize disruption to farming. Creation of two sky lark scrapes per hectare has been found to prolong the sky lark breeding season and increase the productivity of later nesting attempts, resulting in an estimated increase in overall productivity of 49% compared to in conventionally managed winter wheat fields (Morris *et al.* 2004). Because only a small proportion of the field is un-drilled, this method causes minimal loss of income. The patches probably work by increasing access to food, since birds did not preferentially nest within them and they do not support greater quantities of food compared to the rest of the crop.

## 10.4 Minimizing destruction of birds' nests during mechanical operations

Mechanical operations in fields can be damaging to ground-nesting birds, as in grasslands. Ploughing, rolling, and harrowing in late spring can cause high nest loss of earlier-breeding species (Berg *et al.* 1992). Some later-nesting birds, particularly crop-nesting harriers, *Circus* spp., in Europe, are vulnerable to loss of nestlings during harvesting (Corbacho *et al.* 1999; Millon *et al.* 2002). Organic farming usually relies to a greater extent on tillage, rather on pesticides, to reduce weed problems compared to conventional farming. Hence, there can be a greater potential for nest loss caused by mechanical operations in organic systems (Lokemoen and Beiser 1997).

Cultivation dates can be shifted to some extent so that they take place before or after the main breeding season. Dates of rolling and harrowing are not particularly crucial from an agricultural perspective. Individual nests can also be located and protected. This is obviously a time-consuming operation and only practical for small numbers of relatively conspicuous birds. It has successfully been used to protect eggs and young of Eurasian thick-knees from agricultural operations and to prevent destruction and successfully increase productivity of crop-nesting harriers. Eggs of birds such as northern lapwings can be lifted just before the machinery passes and returned to the nest, or placed in a newly created, artificial scrape, immediately afterwards. Alternatively, the area surrounding the nest can be spared from mechanical operations altogether. Nests remaining in an island of unploughed land in an otherwise ploughed field might be more conspicuous and thereby susceptible to predation. Any interventions to protect eggs must be continued throughout the nesting period. Saving a nest from ploughing soon after the eggs have been laid may be counterproductive if the clutch is then destroyed by farming operation immediately prior to hatching, and at that stage then too late for the bird to lay a successful, replacement clutch.

Nestlings of harriers can be transferred to artificial nest sites in nearby unharvested crops, or protected by fencing off the nest from harvesting and retaining the fence to protect them from ground predators. The second technique, though, can leave nestlings vulnerable to human persecution. These interventions have been shown to successfully increase harrier productivity (Corbacho *et al.* 1999; Millon *et al.* 2002).

## 10.5 Provided unharvested crops for birds to eat

Specific crops can be grown to provide seeds for birds during the non-breeding season. Strips of unharvested crops are often referred to as wild-bird cover or, if specifically intended for gamebirds, as game crops. The main considerations are:

- the types of crop;
- their location within the field.

Wild-bird cover provides a source of seeds both from the planted crop and from other arable weeds growing among it. Some wild-bird cover, for example kale and mixed crops containing turnips, can also attract high densities of insectivorous birds. Densities of seed-eating birds on wild-bird cover are typically even higher than on cereal stubbles (Henderson *et al.* 2004).

Maize and other plants that produce large seeds are favoured by larger birds, especially gamebirds, corvids, and pigeons. Quinoa is particularly favoured by a wide range of small, seed-eating songbirds. Kale, millet, and the wheat/rye hybrid triticale are used by both large and small seed-eaters. Kale appears to be the most widely used crop of all (Henderson *et al.* 2004; Stoate *et al.* 2004).

346 | Arable land

Different types of seed-bearing plants can be sown together, providing they have compatible sowing times and similar soil and management requirements. A favoured mix is maize and kale. Mixtures of cereals and kale can be used as a 2-year crop. The cereal provides grain for buntings and gamebirds that prefer larger seeds, whereas the kale provides an abundance of small seeds for other seed-eating songbirds in the second year. Otherwise, it is usually best to plant different seed-bearing crops in single-species strips and manage each optimally.

The quantities of seeds produced by wild-bird cover and game crops can be maximized by suitable fertilizer application and control of arable weeds that produce less seed. As with other crops, growing them as part of a suitable rotation will help to maintain soil fertility and help reduce seedbanks of unwanted plant species.

Positioning wild-bird cover adjacent to suitable hedgerows and wooded boundaries attracts highest overall densities of birds, because most species prefer feeding close to cover (Henderson *et al.* 2004).

Commonly used crops for geese include maize and soybeans (Figure 10.5). Maize, though, forms stands too dense for geese to access and has to be knocked

**Fig. 10.5** Sacrificial crops. Geese feed on a variety of crops in winter, both on the residue from harvesting, such as these snow geese feeding on maize, and on crops specifically grown and left unharvested for them. A crop rotation used to provide food for wintering Canada geese, *Branta canadensis*, and snow geese in the north-eastern USA involving soybeans, winter wheat, white clover, and maize is described in the text (Delaware Bay, Delaware, USA).

down to allow them access. Geese avoid areas close to cover and sources of potential disturbance. Provision of winter food for geese can be made more economic by contracting out farming operations, and paying for it by allowing the contractor to take a proportion of the crops for themselves.

A crop rotation used to provide food for wintering Canada geese, and snow geese in the north-eastern USA involves soybeans, winter wheat, white clover, *Trifolium repens*, and maize. Soybeans are planted during the first year and the contractor harvests 80% of them in the autumn, leaving the remainder as food for geese. Dry, hard soybeans are unattractive to geese. They are only eaten once the field has become wet or subject to shallow, artificial flooding. The areas harvested for soybeans are then direct-drilled with winter wheat to provide grazing for geese that winter. This direct-drilling (a form of non-inversion tillage) is cheaper than full cultivation and also helps prevent soil erosion. In late February white clover seed is broadcast on the areas sown with winter wheat. The winter wheat acts as a nurse crop for the white clover and the freezing and thawing of the soil surface helps the clover seed become buried within the soil and germinate. The white clover then establishes and is left as goose food for 3 years. It is mown two to three times during the second half of summer each year to maintain it at an optimal height (15–20 cm) for feeding geese. The final stage of the rotation involves planting maize, which is left unharvested for the geese. Strips around the edges of the maize are sequentially knocked down at intervals from mid-to-late January onwards, to provide a continual supply of newly accessible carbohydrate-rich cobs for the geese to feed on at a time of year when there is little other food available.

Crops such as buckwheat and sorghum can be grown and left unharvested to provide smaller seed for other wintering wildfowl. It can be worth planting such crops in widely spaced rows to allow other wild, seed-bearing plants to grow up between them. Conversely, these crops can be used to enhance the seed supply of land under moist-soil management (Section 8.4.1) that is failing to develop a sufficient cover of wild, annual plants. Crops can be established in these areas by lightly disking to break up the soil surface and then planting in wide rows.

## 10.6 Manipulating flooding regimes

Shallow flooding of croplands can provide valuable feeding habitat for wetland birds. It is used agriculturally:
- during rice cultivation;
- to control agricultural pest species, particularly nematodes on potatoes, flower bulbs, tomatoes, bell peppers, aubergines, corn, soybeans, and other crops and kill off flood-intolerant weed species.

**Fig. 10.6** Flooding rice stubble. Flooding rice stubble in winter (a) greatly increases its value for wintering waterbirds. Flooding also speeds up decomposition of rice straw left on fields following harvesting (b) and inhibits weed growth, so reducing preparation costs for the next crop. The foraging activities of the waterfowl themselves also increase straw decomposition (Bird *et al.* 2000). The greatest number of bird species in flooded rice fields are typically found in water 10–20 cm deep.

Different treatments of rice stubble (ploughing, burning, chopping, rolling, disking or cutting, and removing) have little or no effect on bird use following re-flooding, although rice-harvesting techniques that leave tall rice stems discourage some waterbird species, particularly lsmall waders/shorebirds (Day and Colwell 1998; Elphick and Oring 1998, 2003). Gradually lowering water levels in spring prior to planting the next crop exposes bare mud for feeding shorebirds/waders (photographs by Chris Elphick).

Rice fields provide valuable feeding habitat and, to a far lesser extent, breeding habitat for waterbirds (e.g. Fasola and Ruiz 1996; Elphick 2000; Fujioka *et al.* 2001; Maeda 2001; Czech and Parsons 2002; Elphick and Oring 2003). In the southern USA crawfish aquaculture is sometimes used in association with rice cultivation and also provides valuable feeding habitat for waterbirds (e.g. Huner *et al.* 2002). The value of rice fields or feeding waterbirds can be greatly increased by shallowly flooding them between harvesting and planting of the next crop, rather than leaving them dry during this period (Figure 10.6). This flooding reduces the suitability of fields for some small number of seed-eating landbirds, most raptors, and some other species, but benefits some insectivorous songbirds (Elphick and Oring 1998; Elphick 2004).

Flooding of normally dry fields controls agricultural pest species by creating anaerobic conditions. Because the rate of development of anaerobic conditions increases with soil temperature, longer periods of flooding are necessary when soil temperatures are lower. Typical flooding periods last for between 4 and 10 weeks to control nematodes, although far longer periods may be necessary for weed control. The typical depth of flooding is 10–40 cm.

Shallow flooding increases the accessibility of soil invertebrates to feeding waders/shorebirds and suspends seeds, making them available to dabbling ducks. Patches of shallow water also provide safe roost sites for waterbirds, thus enabling them to exploit food resources on surrounding farmland and elsewhere. Because the technique relies on the creation of anoxic conditions, it is likely to reduce populations of earthworms, at least those of non-aquatic species (Ausden *et al.* 2001). Hence, such flooding is only likely to provide a temporary abundance of accessible soil invertebrates.

## 10.7 Uncropped habitats

The presence of uncropped habitats in areas of arable land will greatly influence their overall value for birds (e.g. Best *et al.* 2001; Fuller *et al.* 2004; Jones *et al.* 2005; Moreira *et al.* 2005). Many songbirds and gamebirds that utilize crops also require adjacent uncropped habitats along field boundaries, such as hedgerows, scattered bushes, trees, and grassy margins for cover, for nesting, and to provide additional feeding habitat (Figure 10.7). Adjacent woodlands may also be used by birds utilizing farmland (e.g. Berg 2002b). For example, territories of yellowhammers, and ortolan buntings, *Emberiza hortulana*, in mixed farmland in Poland are both strongly associated with the presence of nearby woodlands. Both species use trees in these woodlands as song posts, while yellowhammers also nest in shrubs along the woodlands'

**Fig. 10.7** Untidy corners and field margins. These can greatly increase the density and diversity of birds found in an area otherwise dominated by crops. This small patch of rank grassland amongst arable was used for foraging by breeding red-backed shrikes *Lanius collurio* and great grey shrikes *L. excubitor*. In Europe shrikes are found mainly in unintensively-managed open habitats with scattered trees, where they feed mainly on large insects. The majority of European shrike species have undergone large population declines associated with intensification of agriculture (Mikashevice, Brest region, Belarus).

edge (Golawski and Dombrowski 2002). Alternatively, uncropped habitats can simply support species that are not found in arable habitats, for example perennial plants and woodland-edge small mammals along grassy field margins and hedgerows.

The highest bird diversity on farmland is associated with a high proportion of uncropped habitats, especially hedgerows, scrub, trees, and woodland (e.g. Berg 2002b; Moreira *et al.* 2005; Sanderson *et al.* 2007). However, increasing the proportion of trees and scrub will decrease the suitability of farmland for specialist, open-ground species, such as Eurasian golden-plovers, northern lapwings, Eurasian sky larks, calandra larks, *Melanocorypha calandra*, and corn buntings, *Miliaria calandra*, in Europe and grasshopper sparrows, dickcissels, *Spiza americana*, and meadowlarks, *Sturnella* spp., in North America (e.g. Grant *et al.* 2004; Moreira *et al.* 2005). In North America presence of woody vegetation also results in increased rates of brood parasitism by brown-headed cowbirds (e.g. Patten *et al.* 2006). Hedgerows, trees, and scrub also make areas less suitable for wintering geese (Gill 1996).

The main types of uncropped habitat that can be created or managed for wildlife along field boundaries are:

- hedgerows, shelterbelts, shrubby fencerows, and other scattered trees and scrub;
- grassy field edges (grass margins/field borders, grass filter strips, and beetle banks);
- water-filled ditches;
- nectar strips.

These can also be used in combination. Banks and walls can also provide some suitable habitat for wildlife, for example the former are important for subterranean-nesting bumblebees (Kells and Goulson 2003). Walls provide habitat for a range of species absent from cropped habitats, including a variety of plants and lizards.

### 10.7.1 Hedgerows, shelterbelts, shrubby fencerows, and other scattered trees and scrub

The range of bird species utilizing scrubby field margins/hedgerows will be influenced primarily by its height and width and the presence or absence of trees along it. Overall densities and numbers of species of breeding birds tend to be highest along tall and wide hedges with many trees, although this is mainly due to these supporting more species associated with woodland and woodland edge. Maintaining vegetation cover at the hedge base increases its value for wildlife (Hinsley and Bellamy 2000). Some bird species, though, prefer tall hedges with few trees, for example hedge accentors, *Prunella modularis*, and lesser whitethroat, *Sylvia curruca*, whereas others prefer shorter hedges with fewer trees, for example common whitethroat and yellowhammer (Green *et al.* 1994; Parish *et al.* 1994).

Hedgerows can also support small mammals typical of woodlands or woodland-edge habitat, for example yellow-necked mice, and hazel dormice in Europe. These tend to prefer, or are better able to disperse along, hedges without gaps. Hazel dormice also prefer taller, wider, less intensively managed hedges (Kotzageorgis and Mason 1997; Bright 1998; Bright and MacPherson 2002). Some older hedgerow trees, especially pollards, can contain reasonable quantities of dead and decaying wood for saproxylic species (Section 7.1.6).

The height of scrub will be influenced by the height and frequency of cutting. The density of trees can be increased by planting and avoiding cutting self-sown trees during hedge-trimming. Tagging saplings can help prevent them being accidentally cut. Like other forms of vegetation removal, cutting will have damaging effects in the short to medium term. Hence it is best to cut hedgerows as infrequently as is needed to maintain a dense-enough cover while keeping the hedge narrow enough so that it does not unduly interfer with agricultural operations. Cutting should obviously be avoided during the bird nesting season and ideally only carried out in late winter, so that it does not remove berries for birds in winter. As always, a rotation is best to minimize the risk of complete extinction of a species from a given area.

### 10.7.2 Grassy field edges (grass margins/field borders, grass filter strips, and beetle banks)

Grassy field edges can provide feeding and nesting habitat for a range of bird species that also utilize adjacent cropped habitats, together with a variety of grassland plants, invertebrates, and small mammals. The main bird species to benefit from these are breeding and wintering sparrows and other grassland species such as red-winged blackbirds, *Agelaius phoeniceus*, and dickcissels in North America and breeding buntings in Europe (e.g. Bryan and Best 1991; Bradbury *et al.* 2000; Marcus *et al.* 2000; Peach *et al.* 2001; Henningsen and Best 2005; Smith *et al.* 2005a). Fallow field borders have been shown to benefit northern bobwhites (e.g. Palmer *et al.* 2005). There is, though, some concern that nesting birds might suffer higher levels of predation along linear features, because predators often concentrate their activities along them. Grass margins are rarely botanically rich, having usually been surrounded by high levels of fertilizer use and subject to pesticide spray drift and often relatively unsympathetic past management. Raised, free-draining strips of tussocky grassland, known as beetle banks, can be provided for over-wintering predatory ground beetles, rove beetles, and spiders, that reduce numbers of crop pest species. They can be created using two-directional ploughing to create a ridge and then drilling this with tussock and mat-forming perennial grass species (Sotherton 1995; MacLeod *et al.* 2004). Tussocky grassland also provides valuable nesting habitat for some species of bumblebee (Svensson *et al.* 2000; Kells and Goulson 2003).

The key decisions regarding creating grass margins are their location and width, and whether to establish them by sowing or through natural regeneration. Locating grass margins adjacent to hedgerows and other scrub will increase their value for the majority of farmland birds that require scrub for nesting and cover, but make them less suitable for specialist, open-ground birds.

Wider grass margins will obviously provide a greater area of desirable grassland habitat per unit length of field boundary. There is evidence that at least some farmland bird species, for example cirl buntings and yellowhammers, prefer wider grass margins (e.g. 6 m as opposed to 2 m; Bradbury *et al.* 2000; Peach *et al.* 2001). Increasing the width of the margin will reduce the proportion of its area vulnerable to pesticide spray drift. The decision will be a trade-off between maximizing the width of the grass margin and minimizing loss of crop yield.

Sowing reduces the abundance of agriculturally undesirable weed species on the field margin and maximizes that of desired perennial plants. Sown perennial forbs are often included to provide nectar sources for insects, which are probably often in short supply on most farmland. When establishing beetle banks, sowing with tussock-forming grasses such as cock's-foot, *Dactylis glomerata*, and false oat-grass, *Arrhenatherum elatius*, results in higher densities of these groups than leaving these strips to regenerate naturally (Collins *et al.* 2003).

Any seed mix needs to be tailored to the specific conditions at the site. Because of the impracticality of grazing grass margins on otherwise cropped fields, there are likely to be few gaps available in the sward available for establishment of additional species once the sward has established. Therefore, it is best to include rapidly establishing, long-lived perennial plants, rather than species that require continual gap formation for them to persist.

Mowing can be used to maintain the sward at a desired height and prevent growth of woody vegetation, if required. As with fallow and set-aside, younger areas of grassland will contain more bare ground and seed-bearing ruderal plants, and support a different fauna. Burning and light disking can be used to set back succession and provide bare ground and suitable conditions for prolifically seed-producing ruderals plants and species that feed on them. Light disking is used to provide suitable habitat for northern bobwhites. As in other types of grassland, management should be restricted to times of the year that minimize its detrimental effect on nectar sources, larval foodplants, and nesting birds and only carried out on rotation (Chapter 5). Spraying with herbicides is also used. However, unwanted pesticide drift should be minimized. Herbicide drift tends to kill of perennial vegetation, some of which can be important larval food and provide nectar sources for adult insects. Leaving an unsprayed field margin adjacent to these boundary features will help reduce spray drift on to the grass margin.

### 10.7.3 Water-filled ditches/dykes

Water-filled ditches can also be important boundary features in some areas, but are rarely, if ever, specifically created to benefit wildlife on farmland. Existing ditches can, though, be managed to maximize their wildlife benefit.

Water-filled ditches along field boundaries provide important foraging habitat for several farmland songbirds in Europe: song thrushes, *Turdus philomelos*, Eurasian tree sparrows, *Passer montanus*, and reed buntings (Brickle and Peach

2004; Peach *et al.* 2004; Field and Anderson 2004). They will also support their own wetland flora and fauna. Wet ditches on arable field margins will be especially vulnerable to pesticide spray drift and nutrient enrichment from adjacent crops (Bradbury and Kirby 2006). Details of managing water-filled ditches to maximize their value for wildlife are discussed in Section 8.6. Management of water-filled ditches surrounded by arable fields differs from those on wet grassland in that the only option for managing vegetation on their margins is by periodic cutting, and that water levels are usually held at far lower levels, so that they still provide land drainage.

### 10.7.4 Nectar strips

These consist of sown strips of nectar-rich forbs for adult insects to feed on, mixed with non-aggressive grasses to help suppress unwanted weed species. Extending existing field margins from 0.5 to 2.0 m by sowing mixtures of non-invasive grasses and pollen- and nectar-producing forbs has been shown to substantially increase numbers of adult butterflies, bees, hoverflies, and other insects along them, and to establish breeding populations of some butterfly species (Lagerlof *et al.* 1992; Harwood *et al.* 1994; Feber and Smith 1995). The continuity of nectar sources can be extended by cutting a proportion of the strips to between 10 and 20 cm in mid-summer and autumn to extend the flowering season in these areas. No fertilizers or herbicides should be applied, other than spot-spraying or weed-wiping of specific injurious weeds. Nectar strips should be re-sown when necessary.

# 11
# Gardens, backyards, and urban areas

Urban areas, gardens, and backyards can support a surprisingly diverse range of wildlife, including a number of species rare in or absent from more semi-natural habitats (Figure 11.1). There are numerous guides to the specifics of managing these and other urban spaces for wildlife. This chapter will focus on the general principles of this management. It also briefly discusses management of mineral-extraction sites and developed land or buildings that are not currently in use, commonly known as brownfield or post-industrial sites. Mineral-extraction sites usually occur in the wider countryside, but are discussed here because of their similarity to brownfield sites.

In temperate areas the low-nutrient and well-drained ground of brownfield and some mineral-extraction sites can be of exceptional conservation value for invertebrates, which on the cool edges of their range require warm, open early-successional habitat (Harvey 2004; Figure 11.2). Some brownfield sites also retain relict areas of semi-natural habitat, including patches of wetland. Areas of industrial waste products can support very characteristic assemblages of plants. Notable among these are spectacular displays of orchids of the genus *Dactylorhiza* on the infertile, alkaline conditions of weathered pulverized fuel ash (PFA) and characteristic assemblages of rare lichens on metal-rich mine workings (Purvis 2001). Brownfield sites, though, are usually viewed as waste ground and subject to re-development, or pressure to tidy them up to create green spaces for recreation. Similarly, mineral-extraction sites are often 'restored' to other priority habitats or by replacing topsoil and land-forming to create gentle, grassy slopes. These activities destroy any existing conservation value.

Buildings can provide important nest sites for some bird species and roosts for bats. In the south-eastern USA a large proportion of the Atlantic coast population of least terns, *Sterna antillarum*, nest on gravel-covered roof-tops. In the state of Georgia 73% of least terns nest on rooftops, compared to only 1% on beaches (Krogh and Schweitzer 1999). Peregrine falcons, *Falco peregrinus*, nest on ledges in many cities in North America and Europe, while a large proportion of the

**Fig. 11.1** The value of UK gardens for plants and invertebrates.

Invertebrates: the most comprehensive survey of a garden for invertebrates has been a 15-year study of a 741-$m^2$ suburban garden in Leicestershire, England (Owen 1991). This garden was managed to provide good lhabitat for wildlife, but was not atypical in the features that it contained: lawns, herbaceous borders, vegetables, fruit bushes, rockeries, shrubberies, a compost heap, a few trees, and a small pond. Over a 15-year period 1602 species of insects and 121 species of other invertebrates were recorded.

Because some groups of insects were known to have been under-recorded, the total number of species visiting the garden over this period will have been considerably higher. Many of the species will undoubtedly have been only temporary colonists in the garden and a high proportion of the winged insects recorded must have only been passing through. However, the number of species recorded is still impressive! Of the parasitic wasps recorded, 20 were believed to be new records for Britain and a further four were previously undescribed to science. Of the insect groups thought to have been well recorded, an amazing 21% of the known British fauna were found in the garden.

Plants: a survey of 60 gardens in the city of Sheffield, England, found that they contained an amazing total of 1166 species of vascular plants. Thirty per cent of these were native species (Smith *et al.* 2006), this representing approximately a quarter of the UK's native vascular flora.

**Golden plusia moth,** *Polychrysia moneta* (pictured). This is one of the 28 species of moths in the UK that are more or less restricted to gardens, parks, orchards, and the outside walls of buildings (from Emmett and Heath 1991). Most are restricted to these habitats because their larval foodplants are alien/exotic species confined to these habitats. In the UK golden plusia moth caterpillars feed on cultivated delphiniums and larkspurs, *Delphinium* spp., in gardens.

**Fig. 11.2** 'Brownfield' sites and invertebrates. This 20-ha former ash field in East London contains a mixture of bare and sparsely vegetated ground, well-drained herb-rich grassland, seasonally wet areas, and scattered scrub. It supports an exceptional invertebrate fauna containing a high proportion of rare species. Twelve per cent of the approximately 800 invertebrate species recorded there are found in less than 4% of the 10-km squares in Great Britain.

populations of several swift species and even the threatened lesser kestrel, *Falco naumanni*, nest in roofs. Gravestones, walls, and old stone buildings can support rich assemblages of lichens that extend their range in lowland areas where other suitable rocks are rare or absent.

Overall, urban areas typically support a greater species richness of alien/exotic, and in some cases also native, plants compared to the surrounding landscape (Roy *et al.* 1999; Kühn *et al.* 2004). They can also contain far higher densities of ponds than in the wider countryside. In the city of Sheffield, England, 14% of dwellings are estimated to have ponds in their gardens, providing an estimated total of 25 200 garden ponds in the entire city (Gaston *et al.* 2005a)!

## 11.1 Managing urban areas, gardens, and backyards for wildlife

There is often a distinction made between nature conservation in urban areas and the wider countryside. Management of urban green spaces usually places a higher value on recreational and educational needs and community involvement. Habitats in urban areas have also rarely been subject to long periods of traditional management. Consequently, conservation of cultural habitats and consideration of the needs of specific rare species is less often a consideration. Most practical differences in management between habitats in urban areas, gardens, and backyards and those in the wider countryside are due to their small size, requiring them to be managed more intensively to maintain their interest, and the impracticality of using grazing or burning to arrest succession. Cutting and removal is generally used instead.

Despite these differences, there is still the potential to manage *large* areas of urban green space using similar techniques to those in the wider countryside (Figure 11.3). An example of a nature reserve in an urban setting, which is important for protecting endangered species, is the 250-ha Karori Wildlife Sanctuary in Wellington, New Zealand. This is surrounded by a fence to protect its inhabitants from alien/exotic predators.

The main techniques for managing habitats in urban spaces, gardens, and backyards or wildlife are:

- minimizing or avoiding harmful gardening practices, especially pesticide use;
- planting flowers, shrubs, and trees that provide good wildlife habitat;
- creating features that provide good habitat for wildlife such as ponds, marshy areas, wildflower meadows, and piles of logs and other plant material;
- providing artificial nest and hibernation sites.

In addition, artificial feeding can be used to attract birds and mammals to gardens and backyards and in many cases probably increase their population sizes. Predation by domestic cats is often an issue.

Another consideration when designing a garden or backyard to benefit wildlife is to ensure that fences do not unduly impede movement of animals between them, for example the annual migrations of toads to and from breeding ponds.

When creating and managing habitats in urban areas, gardens and backyards, it is also important to consider the effects of these activities on the wider environment. In particular, using peat and peat-based compost will encourage destruction of valuable peatlands and using weathered limestone for rockeries will encourage destruction of limestone pavement. Designing an area that requires frequent watering will increase pressure on often scarce water resources, which might impact on wetlands.

Managing for wildlife | 359

**Fig. 11.3** Stockholm's National City Park. This comprises a wedge of 27 km² of land and water stretching into the middle of Stockholm. Most is managed in a low-intensity manner, with plenty of tussocky grassland, dead wood, and other features normally associated with the wider countryside. These photographs were taken within 3 km of the city centre.

### 11.1.1 Minimizing or avoiding harmful gardening practices

A visit to most garden centres reveals the bewildering array of herbicides, insecticides, fungicides, acaricides, molluscicides, and rodenticides available for use in gardens. In Europe an estimated €560 million are spent each year on pesticides for homes and gardens (European Crop Protection Association figures for 2000). Any pesticide that is effective against its target groups, irrespective of how 'garden friendly' it is considered to be, will still be damaging.

Many types of garden plants will succumb to the effects of insects, snails, and slugs unless dosed with pesticides. If you want to avoid using pesticides, then it is best to choose hardy plants that are best suited to the climatic and soil conditions of your area, and just give up on trying to grow more sensitive species that require constant watering, fertilizing, and protection from invertebrates and disease. Doing so will also increase the amount of time you can spend enjoying your garden or backyard, rather than on maintaining it. A good starting point is to look at which types of attractive plants are growing well in surrounding gardens and backyards, particularly those that are regenerating naturally and becoming weeds themselves.

Potentially harmful effects of trimming and pruning can also be reduced. Bushes can be trimmed less frequently to minimize numbers of caterpillars removed from the plant with cut foliage. Some authors suggest leaving the clippings next to the plant for a day or two to increase the chances of caterpillars returning to it. Dead stems, shrivelled leaves, and flower heads should also be left in place, since these can be important food sources and over-wintering sites for insects. Conversely, judicious pruning can be used to increase and in some cases prolong the flowering period of some plants, and thereby potentially increase their value as nectar sources.

### 11.1.2 Planting flowers, shrubs, and trees that provide good wildlife habitat

The suitability of gardens and backyards for wildlife can be improved by selecting plants that provide:

- foliage for plant-eating insects;
- berries for birds and small mammals;
- shelter, cover, and variation in structure;
- nectar sources for insects (Figure 11.4).

These can be planted to create flower-rich borders in sunshine to attract warmth-loving insects and to create or enhance existing sheltered glades and

**Fig. 11.4** Flowery borders. This flowerbed is both beautiful and good for wildlife, and requires little maintenance. The plants have been chosen for their quality as nectar sources and insect foodplants, aesthetic appeal, structure, and suitability for conditions in the garden.

The plants in this border include, for example meadow crane's-bill, *Geranium pratense* (a good early-summer nectar source and which grows well at its sunny end), and dusky crane's-bill *Geranium phaeum* (a good spring nectar source that grows well in the shadier, drier conditions next to the hedge). Later in the year wild teasels, *Dipsacus fullonum*, grow up to provide nectar in mid–late summer, seeds for European goldfinches, *Carduelis carduelis*, and good winter structure in the garden. Other valuable nectaring plants in the border include golden margoram, *Origanum vulgare*, and foxgloves, *Digitalis purpurea*.

The only maintenance needed is occasionally cutting back of plants that are growing too well, and removal of seedling of plants that are seeding too successfully. Maintaining a high ground cover for most of year also reduces establishment of unwanted weeds.

woodland-edge habitat (see Section 7.4.1 regarding management of woodland edge, glades, and rides). Vegetation structure is an important consideration when designing woodland and woodland-edge habitat. Providing a canopy, understorey, and field layer will maximize the interest of the planting. Increasing the vegetation structure by mixed planting and creation of mosaics of shade and open areas will also probably increase the range of niches available for invertebrates.

It is widely stated that native plant species support a richer invertebrate fauna than introduced species, and consequently that there should be a presumption for planting native species in urban areas. While this may be true for insects associated with many tree species, it is not necessarily the case for herbaceous plants. Gardens and backyards can provide valuable habitat for a variety of insects dependent on alien/exotic plant species, or on plants that are local or scarce in the wild, but frequently planted in gardens and backyards. In the study by Owen (1991) there were 68 species of moths whose larvae fed on plants in the garden. Of these, 46 fed on native plants and 38 on alien/exotic ones. Overall, 27% of native plant species in the garden were used by moth larvae, compared to 35% of alien/exotic plant species. The best way to increase the range of breeding moths is to maximize the range of foodplants of both native and alien/exotic species.

The value of gardens and backyards for nectar-feeding insects can be maximized by providing a continuity of suitable nectar sources throughout the season. As a general rule, flowers that attract butterflies also tend to be attractive to moths, but the reverse is not necessarily true. White flowers that are fragrant at night are usually good for attracting moths. Although there is a tendency for wildlife gardeners to prefer natural forms of native plants, many garden cultivars of these species produce far more flowers and have substantially longer flowering periods. There is, though, the potential danger of introducing cultivars that have the potential to interbreed with native stock outside of the garden or backyard.

### 11.1.3 Creating specific features for wildlife

The value of urban areas, gardens, and backyards for wildlife can be increased by incorporating features described in the following sections.

*Piles of logs and other plant material, and compost heaps*

Piles of logs and other plant material provide food and cover for a wide range of wood and other detritivore-feeding invertebrates and their predators. They also provide cover, nest, and hibernation sites for small mammals, reptiles and amphibians. Logs will provide habitat for fungi. As with dead wood in general, it is probably best to position logs in both sun and shade to attract the maximum range of species (Section 7.1.6). There is little information on the effectiveness of log piles in providing habitat for saproxylic invertebrates in gardens and backyards. However, the results of one study found that small stacks of silver birch logs in gardens were poorly colonized by saproxylic species, although they did provide habitat for a wide range of other invertebrates (Gaston *et al.* 2005b).

*Ponds and marshy areas*

Standard methods for creating garden ponds involve using an impermeable liner to create the pond and adding subsoil or other nutrient-poor material to provide a substrate but without raising nutrient levels too highly. Ponds should only be planted with native species. Many alien/exotic aquatic plants introduced by the garden trade have spread into semi-natural wetlands where they out-compete native vegetation. Any plants bought at garden centres should be checked for small fragments of invasive alien/exotic plants attached to them. It is also worth introducing key invertebrate species that are unlikely to ever colonize the pond. Valuable groups are zooplankton such as water fleas, Cladocera, and copepods, Copepoda, to feed on algae in the water column, and aquatic snails to graze algae on plants.

The wildlife interest of a pond can be maximized by providing a variety of different water depths and vegetation types. Shallow, warm margins are especially valuable for invertebrates. It is also worth creating areas suitable for the establishment of emergent plants and deeper areas that remain as open water and provide habitat for submerged and floating plants. Most fish decimate invertebrates and greatly reduce the pond's wildlife interest. They may also cause it to become dominated by algae by eating the zooplankton that feed on these algae and by stirring up the bottom sediments and releasing nutrients. Fish also make ponds unsuitable for most amphibians by predating their larvae.

Most ponds in gardens and backyards are isolated from other water bodies, surrounded by undisturbed vegetation or paved or other hard surfaces and have relatively stable water levels. In contrast, most semi- or near-natural water bodies are part of larger wetland complexes, surrounded by other semi-natural habitats and have variable and often large seasonal variations in water levels. To increase the value of these ponds for wildlife it is therefore worth considering:

- providing a variety of both permanent ponds and temporary pools, rather than just one single, permanent pond;
- designing ponds to have lower water levels in summer that expose damp mud and emergent vegetation;
- regularly disturbing vegetation around the margins of the pond to maintain early-successional habitat.

Permanent and seasonal water bodies support quite different assemblages of species. Creating both will increase the range of species in the garden, backyard, or other green space. Some amphibians prefer temporary pools for breeding (Figure 11.5). Observing variations in the fauna resulting from periodic drying out and re-flooding of ponds can add to their enjoyment.

Seasonal variations in water levels are important features of most natural wetlands. Damp mud is an important habitat for many flies and beetles, while the drawdown zone of water bodies can support a quite different flora from that of

**Fig. 11.5** Permanent ponds and temporary pools. Many species prefer temporary water bodies rather than permanent ones.

This garden contains three ponds. Common frogs, *Rana temporaria*, only breed successfully in the shallow, largely unvegetated temporary one where their tadpoles are less heavily predated by smooth newts, *Lissotriton vulgaris*. Most smooth newts breed in the more vegetated, two permanent ponds.

more stable water margins. It is therefore worth designing a pond so that at least a proportion of it has gently enough sloping sides to create a drawdown zone. Creating variation in topography within the drawdown zone will increase the variety of different microhabitats provided as water levels fall: wet and dry mud, tussocky edges, and drying-out pools. Additional periodic disturbance, especially pulling up plants, can be used to further increase variations in conditions on the pond's margins and prevent them from becoming dominated by one or a small number of more competitive perennial plants. The ideal is to do this little and often during the growing season, thereby mimicking conditions created by grazing and poaching by herbivores. It is also worth including a steeper profile and more stable vegetation or hard surface around a proportion of the pond's perimeter, so that people can get close to the water's edge and look for animal life in the water.

An excellent method to provide water for a pond is to connect it to a water butt or drainage system that collects water from a nearby roof, so conserving water resources. Connecting the pond directly to the run-off from the roof, rather than via a water butt, will provide greater variation in water levels. It will mimic the situation in a more natural wetland where fluctuations in water levels vary depending on rates of inflow from its catchment—in this case the roof.

The principles of preventing succession in garden or backyard ponds are the same as those in more natural water bodies, by regularly clearing out a proportion of the vegetation. Again, little and often is best; a rule of thumb being to never clear out more than a third of the pond at any one time.

*Wildflower meadows and other grasslands*

Conventional, closely mown lawns provide good foraging habitat for many birds that feed on soil invertebrates. They can also support a surprisingly diverse flora. The lawns in 52 gardens in Sheffield, England, supported a total of 159 vascular plant species, with an average of 24 species per lawn (Thompson *et al.* 2004). In shorter swards, trampled, and other bare areas can provide nesting areas for solitary bees.

The visual, and probably also wildlife, interest of existing lawns can be increased by mowing at different heights and frequencies to vary their structural and floral composition. This can be done on existing areas of lawn to allow low-growing rosette-forming and other forbs in the lawn to flower and provide nectar sources for insects. As in other grasslands, though, managing by cutting catastrophically removes habitat used by invertebrates. There is, therefore, the potential to create an ecological trap by encouraging them to colonize the grassland and then suddenly destroying the entire patch of habitat by mowing. Again, cutting small patches at periodic intervals during the growing season to simulate patchy grazing is probably best, but obviously more time-consuming.

Wildflower meadows will provide an additional habitat and can be visually stunning. It is best to remove the topsoil to reduce nutrient levels in the upper soil before establishing the meadow. Creating small-scale variation in soil types and topography increases small-scale variation in vegetation composition and structure. It is cheapest to establish the majority of plants from seed, and then add plugs, mature plants, and bulbs of plants that do not establish well from seed. During the establishment phase vegetation can be cut around favoured individual plants to provide them with a competitive advantage. This can be important in helping forbs to establish, particularly if grass growth is vigorous.

The principles of managing botanically species-rich wildflower meadows in gardens and backyards are similar to those of managing other hay meadows (Section 5.5.1). However, the practicalities differ in several respects. In agriculturally managed meadows the date of cutting will often be a compromise between cutting late enough to achieve conservation objectives, and early enough to provide high-enough quality herbage for agricultural use. There will be no agricultural constraints when managing small meadows in urban areas, gardens, and backyards. Meadows can therefore be cut later in the season, ideally in autumn

# Gardens, backyards, and urban areas

**Fig 11.6** Garden wildflower meadows. This garden meadow contains an abundance of cowslips, *Primula veris*, and fritillaries in spring and was inspired by the meadow shown in Fig. 2.2.

In traditionally managed hay meadows gaps in the sward are important in providing places for plants to germinate in and are provided by aftermath grazing. This is impractical in gardens and backyards. A successful alternative method of gap creation is to open up the meadow to winter trampling by children (our garden in Histon, Cambridge, England).

or even winter. It is, though, important to not cut all of the meadow at the same time, in order to provide a continuity of habitat for invertebrates and any small mammals. Tussocks and seed heads provide important over-wintering sites for invertebrates.

As discussed in Section 5.5.1, aftermath grazing is important in maintaining the high botanical species richness of hay meadows by providing gaps for plants to germinate in and reducing re-growth of more competitive plant species following hay-cutting. Aftermath grazing will not be an option in most gardens, backyards, and urban spaces. Dominance by more competitive species can instead be prevented by selective cutting and removal, rather than by aftermath grazing. Germination gaps, otherwise created by trampling of large herbivores, can be created by other forms of soil disturbance (Figure 11.6).

## 11.1.4 Artificial nest and hibernation sites

There is a wide range of artificial nest and hibernation sites that can be used in gardens and backyards and on buildings. These include various designs of nestboxes for birds, bat boxes, hibernation boxes for West European hedgehogs, *Erinaceus europaeus*, and nest sites for bumblebees and solitary bees and wasps

(Figure 11.7). Artificial platforms can be erected for nesting white storks. Gravel-topped rooftops in the south-eastern USA can be modified to improve their suitability for nesting least terns. Mesh can be fitting over drains and rainspouts to prevent chicks from falling down them, and a low parapet attached to prevent chicks from falling off the roof and to provide shade from them.

The first consideration when providing artificial nest sites is whether the area is suitable for successful breeding. If unsuitable, then providing the artificial nest site might encourage the species to nest somewhere where it will have lower breeding success than elsewhere, thereby creating an ecological trap. The next decision is where to locate the nest site, to both maximize its suitability for the target species and, where relevant, minimize the risk of predation on it.

There is little information of the effectiveness of providing artificial nesting and hibernation sites in urban areas and gardens and backyards. A study by Gaston *et al.* (2005b) found high occupancy of artificial solitary bee and wasp nest sites in gardens, but no use of artificial bumblebee nest sites, possibly because they were not positioned in suitable locations.

**Fig. 11.7** Artificial solitary bee and wasp nest sites. This artificial nest bank is constructed from a mixture of sand and mud to provide a variety of different substrates for bees and wasps to excavate nest holes in. It also includes cut stems of butterfly-bush, *Buddleja davidii*, inserted into the mud for bees and wasps to nest in. This bank had 19 nests.

## 11.1.5 Minimizing predation on wildlife by domestic cats

Domestic cats kill a large number of birds, amphibians, reptiles, and mammals (Barratt 1997; Woods *et al.* 2003; Lepczyk *et al.* 2004). There is, though, little information on the extent to which this affects populations of these species. They may be mainly taking individuals that otherwise succumb to disease or starvation. Irrespective of this, cats can be a real nuisance for people wishing to attract birds and other wildlife to their garden or backyard. Individual cats vary greatly in the numbers of vertebrates they catch. Only a small proportion kill large numbers (Nelson *et al.* 2005).

Numbers of vertebrates killed by cats in gardens and backyards can be reduced by:

- owners imposing curfews on their cats;
- attaching warning devices to cats' collars;
- using ultrasonic deterrents to deter cats from entering particular areas.

Daylight curfews should reduce the numbers of birds killed by cats, particularly during periods when there are large numbers of vulnerable fledglings present. Night-time curfews result in smaller numbers of mammals taken. However, cats subject to nighttime curfews tend to catch greater numbers of reptiles and amphibians overall (Woods *et al.* 2003). Day- or night-time curfews are unlikely to be acceptable to many cat owners, though.

Attaching bells or electronic sonic bleepers to cats' collars alerts potential prey to their approach. The collars need to be attached using a quick-release mechanism to prevent cats from becoming caught on vegetation. Warning devices consisting of bells or electronic sonic bleepers have both been shown to reduce numbers of animals caught by cats by between a third and a half (Ruxton *et al.* 2002; Nelson *et al.* 2005). Electronic sonic bleepers are more expensive than bells, and a high proportion of both are lost when collars fall off. Bleepers also produce a more irritating noise to humans. Attaching bells on quick-release collars is the best option.

Other owners' cats can be discouraged from your garden or backyard using ultrasonic cat deterrents. These detect the presence of an animal using a motion sensor and then produce a high frequency ultrasonic alarm to scare it away. Experiments have shown these devices to be effective at deterring cats, with this effect appearing to *increase* with the length of time the device is in operation (Nelson *et al.* 2006).

## 11.2 Brownfield and mineral-extraction sites

Features of brownfield sites considered important for invertebrate assemblages in temperate areas are:

- a diversity of larval food plants including many ruderal species;
- a diversity and continuity of nectar sources;
- plants stressed by drought, pollutants such as high levels of heavy metals, and mineral deficiency;
- bare and sparsely vegetated ground, especially on friable substrates that invertebrates can burrow in;
- varied vegetation structure;
- a continuity of dead stems, leaves, flower heads, and seeds of open-ground vegetation which is not destroyed by vegetation management.

**Fig. 11.8** Former mineral-extraction sites. The early-successional habitats produced by some types of mineral extraction can be highly valuable for invertebrates. The best way to maintain these is by retaining and creating steep slopes and gullies that will continually erode to provide a continuity of bare and disturbed ground. This sand quarry supports a range of warmth-loving insects, otherwise rare in this area, and especially a wide range of solitary bees and wasps. The latter include the spider-hunting wasp, *Episyron gallicum*, at its first recorded UK site. The nearest other known colonies of this wasp are in central France (Sandy Heath Quarry, Bedfordshire, England).

If an area has been found to support an important invertebrate fauna, then the best way to conserve this is either through non-intervention, while the area continues to support the above-mentioned features, or if not by infrequent patchy disturbance. Motorbike scrambling by local youths is important in helping maintain areas of bare and sparsely vegetated ground at many sites. Although scrub can be an important component of these areas, it often requires patchy removal eventually to prevent it completely dominating.

Key features for important warmth-loving, edge-of-range invertebrates in mineral-extraction sites are similar to those in brownfield sites, but also include:

- vertical or near-vertical exposures for solitary bees and wasps to nest in;
- groundwater-fed seepages, similar to those on soft cliffs (Figure 9.1), and other seasonal pools and damp ground.

If a mineral-extraction site has been found to support an important early successional flora and fauna, then efforts should be made to retain this over at least a proportion of it (Figure 11.8). Little or no further management will usually be required. The value of these features for invertebrates will generally be greater if they face towards the sun.

Small areas of open, early-successional habitat can be created on the roofs of buildings. These green roofs have providing suitable breeding habitat for black redstarts, *Phoenicurus ochruros*, northern lapwings, and little ringed plovers, *Charadrius dubius* (Gedge and Kadas 2005). Creation of brownfield habitat on roofs has also been proposed to compensate for invertebrate habitat lost to development (Harvey 2004).

# References

Able, K.W., Nemerson, D.M., Bush, R., and Light, P. (2001) Spatial variation in Delaware Bay (USA) marsh creek fish assemblages. *Estuaries* **24**, 441–452.

Adamo, M.C., Puglisi, L., and Baldaccini, N.E. (2004) Factors affecting Bittern *Botaurus stellaris* distribution in a Mediterranean wetland. *Bird Conservation International* **14**, 153–164.

Albrecht, H. and Mattheis, A. (1998) The effects of organic and integrated farming on rare arable weeds on the Forschungsverbund Agrarökosysteme München (FAM) research station in southern Bavaria. *Biological Conservation* **86**, 347–356.

Aldridge, D.C. (2000) The impacts of dredging and weed cutting on a population of freshwater mussels (Bivalvia: Unionidae). *Biological Conservation* **95**, 247–257.

Allison, M. and Ausden, M. (2006) Effects of removing the litter and humic layers on heathland establishment following plantation removal. *Biological Conservation* **127**, 177–182.

Allombert, S., Gaston, A.J., and Martin, J.-L. (2005) A natural experiment on the impact of overabundant deer on songbird populations. *Biological Conservation* **126**, 1–13.

Andrén, H. (1995) Effects of landscape composition on predation rates at habitat edges. In Hansson, L., Fahrig, L., and Merriam, G. (eds), *Mosaic Landscapes and Ecological Processes*, pp. 225–255. Chapman and Hall, London.

Andresen, H., Bakker, J.P., Brongers, M., Heydemann, B., and Irmler, U. (1990) Long-term changes of salt-marsh communities by cattle grazing. *Vegetatio* **89**, 137–148.

Anderson, G.Q.A., Haskins, L.R., and Nelson, S.H. (2004) The effects of bioenergy crops on farmland birds in the UK: a review of current knowledge and future predictions. In Parris, K. and Poincet, T. (eds), *Biomass and Agriculture; Sustainability, Markets and Policies*, pp. 199–218. OECD, Paris.

Anderson, J.T. and Smith, L.M. (2000) Invertebrate response to moist-soil management of playa wetlands. *Ecological Applications* **10**, 550–558.

Annand, E.M. and Thompson, III, F.R. (1997) Forest bird response to regeneration practices in central hardwood forests. *Journal of Wildlife Management* **61**, 159–171.

Armstrong, A. and Rose, S. (1999) Ditch water levels managed for environmental aims: effects on field soil water regimes. *Hydrology and Earth System Sciences* **3**, 385–394.

Artman, V.L., Sutherland, E.K., and Downhower, J.F. (2001) Prescribed burning to restore mixed-oak communities in southern Ohio: effects on breeding bird populations. *Conservation Biology* **15**, 1423–1434.

Askins, R.A. (1994) Open corridors in a heavily forested landscape: impact on shrubland and forest-interior birds. *Wildlife Society Bulletin* **22**, 339–347.

Atkinson, I.A.E. (1985) The spread of commensal species of *Rattus* to oceanic islands and their effects on island avifaunas. In Moors, P.J. (ed.), *Conservation of Island Birds*. International Council for Bird Preservation Technical Publication no. 3, pp. 35–81. ICBP, Cambridge.

Atkinson, P.W., Fuller, R.J., and Vickery, J.A. (2002) Large-scale patterns of summer and winter bird distribution in relation to farmland type in England and Wales. *Ecography* **25**, 466–480.

Ausden, M. and Hirons, G.J.M. (2002) Grassland nature reserves for breeding waders in England and the implications for the ESA agri-environment scheme. *Biological Conservation* **106**, 279–291.

Ausden, M., Sutherland, W.J., and James, R. (2001) The effects of flooding lowland wet grassland on soil macroinvertebrate prey of breeding wading birds. *Journal of Applied Ecology* **38**, 320–338.

# References

Ausden, M., Rowlands, A., Sutherland, W.J., and James, R. (2003) Diet of breeding Lapwing *Vanellus vanellus* and Redshank *Tringa totanus* on coastal grazing marsh and implications for habitat management. *Bird Study* **50**, 285–293.

Ausden, M., Hall, M., Pearson, P., and Strudwick, T. (2005) The effects of cattle grazing on tall-herb vegetation and molluscs. *Biological Conservation* **122**, 317–326.

Baar, J. and Kuyper, T.W. (1998) Restoration of aboveground ectomycorrhizal flora in stands of *Pinus sylvestris* (Scots pine) in The Netherlands by removal of litter and humus. *Restoration Ecology* **6**, 227–237.

Baattrup-Pedersen, A., Larsen, S.E., and Riis, T. (2002) Long-term effects of stream management on plant communities in two Danish lowland streams. *Hydrobiologia* **481**, 33–45.

Baines, D. and Andrew, M. (2003) Marking of deer fences to reduce frequency of collisions by woodland grouse. *Biological Conservation* **110**, 169–176.

Baker, W.L. (2006a) Fire and restoration of sagebrush ecosystems. *Wildlife Society Bulletin* **34**, 177–185.

Baker, W.L. (2006b) Fire history in ponderosa pine landscapes of Grand Canyon National Park: is it reliable enough for management and restoration? *International Journal of Wildland Fire* **15**, 433–437.

Bakker, J.P. (1998) The impact of grazing on plant communities. In WallisDeVries, M.F., Bakker, J.P., and Van Wieren, S.E. (eds), *Grazing and Conservation Management*, pp. 137–184. Kluwer Academic Publishers, Dordrecht.

Bakker, J.P., De Bie, S., Dallinga, J.H., Pjaden, P., and De Vries, Y. (1983) Sheep-grazing as a management tool for heathland conservation and regeneration in the Netherlands. *Journal of Applied Ecology* **20**, 541–560.

Bamber, R.N., Batten, S.D., Sheader, M., and Bridgwater, N.D. (1992) On the ecology of brackish water lagoons in Great Britain. *Aquatic Conservation—Marine and Freshwater Ecosystems* **2**, 65–94.

Bamber, R.N., Batten, S.D., and Bridgwater, N.D. (1993) Design criteria for the creation of brackish lagoons. *Biodiversity and Conservation* **2**, 127–137.

Barker, A.M., Vinson, S.C., and Boatman, N.D. (1997) Timing and cultivation of rotational set-aside for grass weed control to benefit chick-food insects. In Kirkwood, R.C. (ed.), *The Brighton Crop Protection Conference—Weeds*, vol. 3, pp. 1191–1196. British Crop Protection Council, Brighton.

Barker, C.G., Power, S.A., Bell, J.N.B., and Orme, C.D.L. (2004) Effects of habitat management on heathland response to atmospheric nitrogen deposition. *Biological Conservation* **120**, 41–52.

Barratt, D.G. (1997) Predation by house cats, *Felis catus* (L), in Canberra, Australia. 1. Prey composition and preference. *Wildlife Research* **24**, 263–277.

Barrett, P.R.F., Littlejohn, J.W., and Curnow, J. (1999) Long-term algal control in a reservoir using barley straw. *Hydrobiologia* **415**, 309–313.

Bassett, P.A. (1980) Some effects of grazing on vegetation dynamics in the Camargue, France. *Vegetatio* **43**, 173–184.

Beasley, C.E., Green, R.E., Robson, R., Taylor, C.R., and Winspear, R. (1999) Factors affecting the numbers and breeding success of Stone Curlews *Burhinus oedicnemus* at Porton Down, Wiltshire. *Bird Study* **46**, 145–156.

Bedford, A.P. and Powell, I. (2005) Long-term changes in the invertebrates associated with the litter of *Phragmites australis* in a managed reedbed. *Hydrobiologia* **549**, 267–285.

Beintema, A.J. and Müskens, G.J.D.M. (1987) Nesting success of birds breeding in Dutch grasslands. *Journal of Applied Ecology* **24**, 743–758

Bell, J.R., Wheater, C.P., and Cullen, W.R. (2001) The implications of grassland and heathland management for the conservation of spider communities: a review. *Journal of Zoology* **255**, 377–387.

Beltman, B., Van den Broek, T., Barendregt, A., Bootsma, M.C., and Grootjans, A.P. (2001) Rehabilitation of acidified and eutrophied fens in the Netherlands: effects of hydrologic manipulation and liming. *Ecological Engineeering* **17**, 21–31.

Bengtsson, J., Ahnström, J., and Weibull, A.-C. (2005) The effects of organic agriculture on biodiversity and abundance: a meta-analysis. *Journal of Applied Ecology* **42**, 261–269.

Benoit, L.K. and Askins, R.A. (1999) Impact of the spread of *Phragmites* on the distribution of birds in Connecticut tidal marshes. *Wetlands* **19**, 194–208.

Benton, T.G., Vickery, J.A., and Wilson, J.D. (2003) Farmland biodiversity: is habitat heterogeneity the key? *Trends in Ecology and Evolution* **18**, 182–188.

Berdowski, J.J.M. and Zeilinga, R. (1987) Transition from heathland to grassland: damaging effects of the heather beetle. *Journal of Ecology* **75**, 159–251.

Berg, Å. (2002a) Breeding birds in short-rotation coppices on farmland in central Sweden—the importance of *Salix* height and adjacent habitats. *Agriculture, Ecosystems and Environment* **90**, 265–276.

Berg, Å. (2002b) Composition and diversity of bird communities in Swedish farmland-forest mosaic landscapes. *Bird Study* **49**, 153–165.

Berg, Å., Lindberg, T., and Kallebrink, K.G. (1992) Hatching success of lapwings on farmland: differences between habitats and colonise of different sizes. *Journal of Animal Ecology* **61**, 469–476.

Best, L.B., Campa, H., Kemp, K.E., Robel, R.J., Ryan, M.R., Savidge, J.A., Weeks, H.P., and Winterstein, S.R. (1997) Bird abundance and nesting in CRP fields and cropland in the Midwest: a regional approach. *Wildlife Society Bulletin* **25**, 864–877.

Best, L.B., Campa, H., Kemp, K.E., Robel, R.J., Ryan, M.R., Savidge, J.A., Weeks, H.P., and Winterstein, S.R. (1998) Avian abundance in CRP and crop fields during winter in the midwest. *American Midland Naturalist* **139**, 311–324.

Best, L.B., Bergin, T.M., and Freemark, K.E. (2001) Influence of landscape composition on bird use of rowcrop fields. *Journal of Wildlife Management* **65**, 442–449.

Bibby, C.J. (1979) Foods of the Dartford warbler *Sylvia undata* on southern English heathland (Aves: Sylviidae). *Journal of the Zoological Society of London* **188**, 557–576.

Bibby, C.J., Aston, N., and Bellamy, P.E. (1989) Effects of broadleaved trees on birds of upland conifer plantations in North Wales. *Biological Conservation* **49**, 17–29.

Binnie, R.C., Chestnutt, D.M.B., and Murdoch, J.C. (1980) The effects of time of initial defoliation on the productivity of perennial ryegrass swards. *Grass and Forage Science* **35**, 267–273.

Bird, J.A., Pettygrove, G.S., and Eadie, J.M. (2000) The impact of waterfowl foraging on the decomposition of rice straw: mutual benefits for rice growers and waterfowl. *Journal of Applied Ecology* **37**, 728–741.

Birdlife International (2004) *Birds in Europe: Population Estimates, Trends and Conservation Status*. Birdlife Conservation Series no. 12. Birdlife International, Cambridge.

Boatman, N.D., Dover, J.W., Wilson, P.J., Thomas, M.B., and Cowgill, S.E. (1989) Modification of farming practices at field margins to encourage wildlife and promote pest control. In Buckley, G.P. (ed.), *Biological Habitat Reconstruction*, pp. 299–311. Belhaven Press, London.

Bokdam, J. (2001) Effects of browsing and grazing on cyclic succession in nutrient-limited ecosystems. *Journal of Vegetation Science* **12**, 875–886.

Bokdam, J. and Gleichman, J.M. (2000) Effects of grazing by free-ranging cattle on vegetation dynamics in a continental north-west European heathland. *Journal of Applied Ecology* **37**, 415–431.

Bos, D., Bakker, J.P., de Vries, Y., and van Lieshout, S. (2002) Long-term vegetation changes in experimentally grazed and ungrazed back-barrier marshes in the Wadden Sea. *Applied Vegetation Science* **5**, 45–54.

Bouchard, V., Tessier, M., Digaire, F., Vivier, J.P., Valery, L., Gloaguen, J.C., and Lefeuvre, J.C. (2003) Sheep grazing as management tool in Western European saltmarshes. *Comptes Rendus Biologies* **326**, S148–S157.

Bouget, C. (2005) Short-term effects of windstorm disturbance on saproxylic beetles in broad-leaved temperate forests—Part 1: do environmental changes induce a gap effect? *Forest Ecology and Management* **216**, 1–14.

Bouget, C. and Duelli, P. (2004) The effects of windthrow on forest insect communities: a literature review. *Biological Conservation* **118**, 281–299.

Bowden, C.G.R. (1990) Selection of foraging habitats by Woodlarks (*Lullula arborea*) nesting in pine plantations. *Journal of Applied Ecology* **27**, 410–419.

Bracken, F. and Bolger, T. (2006) Effects of set-aside management on birds breeding in lowland Ireland. *Agriculture, Ecosystems and Environment* **117**, 178–184.

Bradbury, R.B. and Kirby, W.B. (2006) Farmland birds and resource protection in the UK: cross-cutting solutions for multi-functional farming? *Biological Conservation* **129**, 530–542.

Bradbury, R.B., Kyrkos, A., Morris, A.J., Clark, S.C., Perkins, A.J., and Wilson, J.D. (2000) Habitat associations and breeding success of yellowhammers on lowland farmland. *Journal of Applied Ecology* **37**, 789–805.

Bradbury, R.B., Browne, S.J., Stevens, D.K., and Aebischer, N.J. (2004) Five-year evaluation of the impact of the Arable Stewardship Pilot Scheme on birds. *Ibis* **146** (suppl. 2), 171–180.

Brambilla, M. and Rubolini, D. (2004) Water Rail *Rallus aquaticus* breeding density and habitat preferences in northern Italy. *Ardea* **92**, 11–17.

Brandeis, T.J., Newton, M., Filip, G.M., and Cole, E.C. (2002) Cavity-nester habitat development in artificially made Douglas-fir snags. *Journal of Wildlife Management* **66(3)**, 625–633.

Brawley, A.H., Warren, R.S., and Askins, R.A. (1998) Bird use of restoration and reference marshes within the Barn Island Wildlife Management Area, Stonington, Connecticut, USA. *Environmental Management* **22**, 625–633.

Bregnballe, T. and Madsen, J. (2004) Tools in waterfowl reserve management: effects of intermittent hunting adjacent to a shooting-free core area. *Wildlife Biology* **10**, 261–268.

Brickle, N.W. and Peach, W.J. (2004) The breeding ecology of Reed Buntings *Emberiza schoeniclus* in farmland and wetland habitats in lowland England. *Ibis* **146** (suppl. 2), 69–77.

Britton, A.J., Carey, P.D. Pakeman, R.J., and Marrs, R.H. (2000a) A comparison of regeneration dynamics following gap creation at two geographically contrasting heathland sites. *Journal of Applied Ecology* **37**, 832–844.

Britton, A.J., Marrs, R.H., Carey, P.D., and Pakeman, R.J. (2000b) Comparison of techniques to increase *Calluna vulgaris* cover on heathland invaded by grasses in Breckland, south east England. *Biological Conservation* **95**, 227–232.

Britton, R.H. and Johnson, A.R. (1987) An ecological account of a Mediterranean Salina: the Salin de Giraud, Camargue (S. France). *Biological Conservation* **42**, 185–230.

Brunsting, A.M.H. and Heil, G.W. (1985) The role of nutrients in the interactions between a herbivorous beetle and some competing plant species in heathlands. *Oikos* **44**, 23–44.

Bright, P.W. (1998) Behaviour of specialist species in habitat corridors: arboreal dormice avoid corridor gaps. *Animal Behaviour* **56**, 1485–1490.

Bright, P. and MacPherson, D. (2002) *Hedgerow management, dormice and biodiversity*. English Nature Research Report No. 454. English Nature, Peterborough.

Broadmeadow, S. and Nisbet, T.R. (2004) The effects of riparian forest management on the freshwater environment: a literature review of best management practice. *Hydrology and Earth System Sciences* **8**, 286–305.

Bryan, G.G. and Best, L.B. (1991) Bird abundance and species richness in grassed waterways in Iowa rowcrop fields. *American Midwest Naturalist* **126**, 90–102.

Bryant, J.C. and Chabreck, R.H. (1998) Effects of impoundment on vertical accretion of coastal marsh. *Estuaries* **21**, 416–422.

Brys, R., Jacquemyn, H., and De Blust, G. (2005) Fire increases aboveground biomass, seed production and recruitment success of *Molinia caerulea* in dry heathland. *Acta Oecologica-International Journal of Ecology* **28**, 299–305.

Buchsbaum, R.N., Catena, J., Hutchins, E., and James-Pirri, M.J. (2006) Changes in salt marsh vegetation, *Phragmites australis*, and nekton in response to increased tidal flushing in a New England salt marsh. *Wetlands* **26**, 544–557.

Buckingham, D.L. and Peach, W.J. (2006) Leaving final-cut grass silage in situ overwinter as a seed resource for declining farmland birds. *Biodiversity and Conservation* **15**, 3827–3845.

Buckingham, D.L., Evans, A.D., Morris, A.J., Orsman, C.J., and Yaxley, R. (1999) Use of set-aside land in winter by declining farmland bird species in the UK. *Bird Study* **46**, 157–169.

Buckingham, D.L., Peach, W.J., and Fox, D.S. (2006) Effects of agricultural management on the use of lowland grassland by foraging birds. *Agriculture, Ecosystems and Environment* **112**, 21–40.

Buffington, J.M., Kilgo, J.C., Sargent, R.A., Miller, K.V., and Chapman, B.R. (1997) Comparison of breeding bird communities in bottomland hardwood forests of different successional stages. *Wilson Bulletin* **109**, 314–319.

Bullock, D.J. (1985) Annual diets of hill sheep and feral goats in southern Scotland. *Journal of Applied Ecology* **22**, 423–433.

Bullock, J.M. and Webb, N.R. (1995) Responses to severe fires in heathland mosaics in southern England. *Biological Conservation* **73**, 207–214.

Bullock, J.M. and Pakeman, R.J. (1996) Grazing of lowland heath in England: management methods and their effects on heathland vegetation. *Biological Conservation* **79**, 1–13.

Bullock, D.J. and Oates, M.R. (1998) Rare and minority breeds in management for nature conservation: many questions and few answers? In Lewis, R.M., Alderson, G.L., and Mercer, J.T. (eds), *The Potential Use of Rare Livestock Breeds in UK Farming Systems*. British Society of Animal Science Meeting and Workshop Publication, Penicuik, Scotland.

Burgess, N.D. and Hirons, G.J.M. (1992) Creation and management of artificial nesting sites for wetland birds. *Journal of Environmental Management* **34**, 285–295.

Bury, R.B. (2004) Wildfire, fuel reduction, and herpetofaunas across diverse landscape mosaics in northwestern forests. *Conservation Biology* **18**, 968–975.

Butler, J., Currie, F., and Kirby, K. (2002) There's life in that dead wood so leave some in your woodland. *Quarterly Journal of Forestry* **96**, 131–137.

Byelich, J., DeCapita, M.E., Irvine, G.W., Radtke, R.E., Johnson, N.I., Jones, W.R., Mayfield, H., and Mahalak, W.J. (1985) *Kirtland's Warbler Recovery Plan*. US Fish and Wildlife Service, Rockville, MD.

Calladine, J., Baines, D., and Warren, P. (2002) Effects of reduced grazing on population density and breeding success of black grouse in northern England. *Journal of Applied Ecology* **39**, 772–780.

Campbell, L.H., Avery, M.I., Donald, P.F., Evans, A.D., Green, R.E., and Wilson, J.D. 1997. *A Review of the Indirect Effects of Pesticides on Birds*. Joint Nature Conservation Committee Report no. 227. Joint Nature Conservation Committee, Peterborough.

Capizzi, D. and Luiselli, L. (1996) Ecological relationships between small mammals and age of coppice in an oak-mixed forest in central Italy. *Revue d'Ecologie-la Terre et la vie* **51**, 277–291.

Carter, S.P. and Bright, P.W. (2003) Reedbeds as refuges for water voles (*Arvicola terrestris*) from predation by introduced mink (*Mustela vison*). *Biological Conservation* **111**, 371–376.

Cattin, M.-F., Blandenier, G., Banasek-Richter, C., and Bersier, L.-F. (2003) The impact of mowing as a management strategy for wet meadows on spider (Araneae) communities. *Biological Conservation* **113**, 179–188.

Centre for Ecology and Hydrology (2004) *Information Sheet 1: Control of Algae with Barley Straw*. CEH Wallingford, Oxon.

Chabreck, R.H., Joanen, T., and Paulus, S. L. (1989) Southern coastal marshes. In Smith, L.M., Pederson, R.L., and Kaminski, R.M. (eds), *Habitat Management for Migrating and Wintering Waterfowl in North America*, pp. 249–277. Texas Tech University Press, Lubbock.

Chamberlain, D.E., Fuller, R.J., Bunce, R.G.H., Duckworth, J.C., and Shrubb, M. (2000) Changes in the abundance of farmland birds in relation to the timing of agricultural intensification in England and Wales. *Journal of Applied Ecology* **37**, 771–788.

Christensen, M., Hahn, K., Mountford, E.P., Odor, P., Standovar, T., Rozenbergar, D., Diaci, J., Wijdeven, S., Meyer, P., Winter, S., and Vrska, T. (2005) Decaying wood in European beech (*Fagus sylvatica*) forest reserves. *Forest Ecology and Management* **210**, 267–282.

Christian, D.P., Collins, P.T., Hanowski, J.M., and Niemi, G.J. (1997) Bird and small mammal use of short-rotation hybrid plantations. *Journal of Wildlife Management* **61**, 171–182.

Churchfield, S., Hollier, J., and Brown, V.K. (1997) Community structure and habitat use of small mammals in grasslands of different successional age. *Journal of Zoology* **242**, 519–530.

Civil Aviation Authority (1998) *Aerodrome Bird Control*. CAP 680. Documedia Solutions, Cheltenham.

Clément, B. and Touffet, J. (1990) Plant strategies and secondary succession on Brittany heathlands after severe fire. *Journal of Vegetation Science* **1**, 195–202.

Collins, K.L., Botaman, N.D., Wilcox, A., and Holland, J.M. (2003) Effects of different grass treatments used to create overwintering habitat for predatory arthropods on arable farmland. *Agriculture, Ecosystems and Environment* **96**, 59–67.

Converse, S.J., White, G.C., and Block, W.M. (2006) Small mammal responses to thinning and wildfire in ponderosa pine-dominated forests of the southwestern United States. *Journal of Wildlife Management* **70**, 1711–1722.

Cooke, A.S. and Lakhani, K.H. (1996) Damage to coppice regrowth by muntjac deer *Muntiacus reevesi* and protection with electric fencing. *Biological Conservation* **75**, 231–238.

Cooke, A.S. and Farrell, L. (2001) Impact of muntjac deer (*Muntiacus reevesi*) at Monks Wood National Nature Reserve, Cambridgeshire, eastern England. *Forestry* **74**, 241–250.

Copeland, T.E., Sluis, W., and Howe, H.F. (2002) Fire season and dominance in an Illinois tallgrass prairie restoration. *Restoration Ecology* **10**, 315–323.

Corbacho, C., Sanchez, J.M., and Sanchez, A. (1999) Effectiveness of conservation measures on Montagu's Harriers in agricultural areas of Spain. *Journal of Raptor Research* **33**, 117–122.

Cowie, N.R., Sutherland, W.J., Ditlhogo, M.K.M., and James, R. (1992) The effects of conservation management of reed beds. II. The flora and litter disappearance. *Journal of Applied Ecology* **29**, 277–284.

Crawley, M.J., Johnston, A.E., Silvertown, J., Dodd, M., de Mazancourt, C., Heard, M.S., Henman, D.F., and Edwards, G.R. (2005) Determinants of species-richness in the park grass experiment. *American Naturalist* **165**, 179–192.

Creel, S., Winnie, J., Maxwell, B., Hamlin, K., and Creel, M. (2005) Elk alter habitat selection as an antipredator response to wolves. *Ecology* **86**, 3387–3397.

Critchley, C.N.R and Fowbert, J.A. (2000) Development of vegetation on set-aside land for up to nine years from a national perspective. *Agriculture, Ecosystems and Environment* **79**, 159–174.

Critchley, C.N.R., Chambers, B.J., Fowbert, J.A., Sanderson, R.A., Bhogal, A., and Rose, S.C. (2002) Association between lowland grassland plant communities and soil properties. *Biological Conservation* **105**, 199–215.

Cunningham, H.M., Chaney, K., Bradbury, R.B., and Wilcox, A. (2004) Non-inversion tillage and farmland birds: a review with special reference to the UK and Europe. *Ibis* **146** (suppl. 2), 192–202.

Cunningham, H.M., Chaney, K., Bradbury, R.B., and Wilcox, A. (2005) The effect of non-inversion on field usage by UK farmland birds in winter. *Bird Study* **52**, 173–179.

Czech, H.A. and Parsons, K.C. (2002) Agricultural wetlands and waterbirds: a review. *Waterbirds* **25** (special publication 2), 56–65.

Danell, K. and Sjöberg, K. (1982) Successional patterns of plants, invertebrates and ducks in a man-made lake. *Journal of Applied Ecology* **19**, 395–409.

Darvill, B., Knight, M.E., and Goulson, D. (2004) Use of genetic markers to quantify bumblebee foraging range and nest density. *Oikos* **107**, 471–478.

Davies, D.M., Graves, J.D., Elias, C.O., and Williams, P.J. (1997) The impact of *Rhinanthus* spp. on sward productivity and composition: implications for the restoration of species-rich grasslands. *Biological Conservation* **82**, 87–93.

Davies, Z.G., Wilson, R.J., Coles, S., and Thomas, C.D. (2006) Changing habitat associations of a thermally constrained species, the silver-spotted skipper butterfly, in response to climate warming. *Journal of Animal Ecology* **75**, 247–256.

Day, J.H. and Colwell, M.A. (1998) Waterbird communities in rice fields subjected to different post-harvest treatments. *Colonial Waterbirds* **21(2)**, 185–197.

Decleer, K. (1990) Experimental cutting of reedmarsh vegetation and its influence on the spider (Araneae) fauna in the Blankaart Nature Reserve, Belgium. *Biological Conservation* **52**, 161–185.

Decocq, G., Aubert, M., Dupont, F., Alard, D., Saguez, R., Wattez-Franger, A., De Foucault, B., Delelis-Dusollier, A., and Bardat, J. (2004) Plant diversity in a managed temperate deciduous forest: understorey response to two silvicultural systems. *Journal of Applied Ecology* **41**, 1065–1079.

Degn, H.J. (2001) Succession from farmland to heathland: a case for conservation of nature and historic farming methods. *Biological Conservation* **97**, 319–330.

Delgado, A. and Moreira, F. (2002) Do wheat, barley and oats provide similar habitat and food resources for birds in cereal steppes? *Agriculture, Ecosystems and Environment* **93**, 441–446.

Delibes, M. and Hiraldo, F. (1981) The rabbit as prey in the Iberian Mediterranean ecosystem. In Myers, K. and MacInnes, C.D. (eds), *Proceedings of the World Lagomorph Conference*, pp. 614–622. University of Guelph, Guelph.

Delisle, J.M. and Savidge, J.A. (1997) Avian use and vegetation characteristics of conservation reserve program fields. *Journal of Wildlife Management* **61**, 318–325.

Desmond, J.S., Zedler, J.B., and Williams, G.D. (2000) Fish use of tidal creek habitats in two southern Californian salt marshes. *Ecological Engineering* **14**, 233–252.

de Szalay, F.A. and Resh, V.H. (1997) Response of wetland invertebrates and plants important in waterfowl diets to burning and mowing of emergent vegetation. *Wetlands* **17**, 149–156.

Devries, B.W.L., Jansen, J., Vandobben, H.F., and Kuyper, T.W. (1995) Partial restoration of fungal and plant-species diversity by removal of litter and humus layers in stands of Scots pine in the Netherlands. *Biodiversity and Conservation* **4**, 156–164.

Diáz, M., Campos, P., and Pulido, F.J. (1997) The Spanish dehesas: a diversity in land-use and wildlife. In Pain, D.J. and Pienkowski, M.W. (eds), *Farming and birds in Europe. The Common Agricultural Policy and its Implications for Bird Conservation*, pp. 178–209. Academic Press, London.

Diemont W.H. and Linthorst Homan, H.D.M. (1989) Re-establishment of dominance by dwarf shrubs on grass heaths. *Vegetatio* **85**, 13–19.

Directorate General for Agriculture and Rural Development (2005) *Agri-environment Measures: Overview on General Principles, Types of Measures, and Application*. European Commission Directorate General for Agriculture and Rural Development.

Ditlhogo, M.K.M., James, R., Laurence, B.R., and Sutherland, W.J. (1992) The effects of conservation management of reed beds. I. The invertebrates. *Journal of Applied Ecology* **29**, 265–276.

Dodson, E.K. and Fiedler, C.E. (2006) Impacts of restoration treatments on alien plant invasion in *Pinus ponderosa* forests, Montana, USA. *Journal of Applied Ecology* **43**, 887–897.

Dolman, P.M. and Sutherland, W.J. (1992) The ecological changes of Breckland grass heaths and the consequences of management. *Journal of Applied Ecology* **29**, 402–413.

Dolman, P.M. and Sutherland, W.J. (1994) The use of soil disturbance in the management of Breckland grass heaths for nature conservation. *Journal of Environmental Management* **51**, 123–140.

Donald, P.F. and Evans, A.E. (1994) Habitat selection by corn buntings *Miliaria calandra* in Britain. *Bird Study* **41**, 199–210.

Donald, P.F., Haycock, D., and Fuller, R.J. (1997) Winter bird communities in forest plantations in western England and their response to vegetation, growth stage and grazing. *Bird Study* **44**, 206–219.

Donald, P.F., Fuller, R.J., Evans, A.D., and Gough, S.J. (1998) Effects of forest management and grazing on breeding bird communities in plantations of broadleaved and coniferous trees in western England. *Biological Conservation* **85**, 183–197.

Donald, P.F., Evans, A.D., Buckingham, D.L., Muirhead, L.B., and Wilson, J.D. (2001a) Factors affecting the territory distribution of Skylarks *Alauda arvensis* breeding on lowland farmland. *Bird Study* **48**, 271–278.

Donald, P.F., Green, R.E., Heath, M.F. (2001b) Agricultural intensification and the collapse of Europe's farmland bird populations. *Proceedings of the Royal Society of London Series B Biological Sciences* **268**, 25–29.

Donald, P.F., Sanderson, F.J., Burfield, I.J., and van Bommel, F.P.J. (2006) Further evidence of continent-wide impacts of agricultural intensification on European farmland birds 1990–2000. *Agriculture, Ecosystems and Environment* **116**, 189–196.

Donlan, J., Greene, H.W., Berger, J., Bock, C.E., Bock, J.H., Burney, D.A., Estes, J.A., Foreman, D., Martin, P.S., Roemer, G.W. *et al.* (2005) Re-wilding North America. *Nature* **436**, 913–914.

Dorland, E., van den Berg, L.J.L., Brouwer, E., Roelofs, J.G.M., and Bobbink, R. (2005) Catchment liming to restore degraded, acidified heathlands and moorland pools. *Restoration Ecology* **13**, 302–311.

Drake, C.M. (1998) The important habitats and characteristic rare invertebrates of lowland wet grassland in England. In Joyce, C.B. and Wade, P.M. (eds), *European Wet Grasslands: Biodiversity, Management and Restoration*, pp. 137–149. John Wiley and Sons, Chichester.

Duncan, P. and D'herbes, J.M. (1982) The use of domestic herbivores in the management of wetlands for waterbirds in the Camargue, France. In Scott, D.A. (ed.), *Management of Wetlands and Their Birds*, pp. 51–67. International Waterfowl Research Bureau, Slimbridge.

Dzwonko, Z. and Gawroński, S. (2002) Effect of litter removal on species richness and acidification of a mixed oak-pine woodland. *Biological Conservation* **106**, 389–398.

Elphick, C.S. (2000) Functional equivalency between rice fields and seminatural wetland habitats. *Conservation Biology* **14**, 181–191.

Elphick, C.S. (2004) Assessing conservation trade-offs: identifying the effects of flooding rice fields for waterbirds on non-target bird species. *Biological Conservation* **117**, 105–110.

Elphick, C.S. and Oring, L.W. (1998) Winter management of Californian rice fields for waterbirds. *Journal of Applied Ecology* **35**, 95–108.

Elphick, C.S. and Oring, L.W. (2003) Conservation implications of flooding rice fields on winter waterbird communities. *Agriculture, Ecosystems and Environment* **94**, 17–29.

Emmett, A.M. and Heath, J. (eds) (1991) *The Moths and Butterflies of Great Britain and Ireland*, vol. 7, part 2, *Lasiocampidae to Thyatiridae*. Harley Books, Colchester.

Esselink, P., Helder, G.J.F., Aerts, B.A., and Gerdes, K. (1997) The impact of grubbing by greylag geese (*Anser anser*) on the vegetation dynamics of a tidal marsh. *Aquatic Botany* **55**, 261–279.

Esselink, P., Zijlstra, W., Dijkema, K.S., and van Diggelen, R. (2000) The effects of decreased management on plant-species distribution patterns in a salt marsh nature reserve in the Wadden Sea. *Biological Conservation* **93**, 61–76.

Evans, A.D. (1996) The importance of mixed farming for seed-eating birds in the UK. In Pain, D.J. and Pienkowski, M.W. (eds), *Farming and Birds in Europe: The Common Agricultural Policy and its Implications for Bird Conservation*, pp. 331–357. Academic Press, London.

Evans, A.D. and Smith, K.W. (1994) Habitat selection of Cirl Buntings *Emberiza cirlus* wintering in Britain. *Bird Study* **41**, 81–87.

Evans, D.M., Redpath, S.M., Elston, D.A., Evans, S.A., Mitchell, R.J., and Dennis, P. (2006) To graze or not to graze? Sheep, voles, forestry and nature conservation in the British uplands. *Journal of Applied Ecology* **43**, 499–505.

Everall, N.C. and Lees, D.R. (1997) The identification and significance of chemicals released from decomposing barley straw during reservoir algal control. *Water Research* **31**, 614–620.

Eversham, B.C. and Telfer, M.G. (1994) Conservation value of roadside verges for stenotopic Carabidae: corridors or refugia? *Biodiversity and Conservation* **3**, 538–545.

Fasola, M. and Ruiz, X. (1996) The value of rice fields as substitutes for natural wetlands for waterbirds in the Mediterranean region. *Colonial Waterbirds* **19**, 122–128.

Feber, R.E. and Smith, H. (1995) Butterfly conservation on arable farmland. In Pullin, A.S. (ed.), *Ecology and Conservation of Butterflies*, pp 84–97. Chapman and Hall, London.

Feber, R.E., Brereton, T.M., Warren, M.S., and Oates, M. (2001) The impacts of deer on woodland butterflies: the good, the bad and the complex. *Forestry* **74**, 271–276.

Fernandes, P.M. and Botelho, H.S. (2003) A review of prescribed buring effectiveness in fire hazard reduction. *International Journal of Wildland Fire* **12**, 117–128.

Field, R.H. and Anderson, G.Q.A., (2004) Habitat use by breeding Tree Sparrows *Passer montanus*. *Ibis* **146** (suppl. 2), 60–68.

Fojt, W. and Harding, M. (1995) Thirty years of change in the vegetation communities of three valley mires in Suffolk, England. *Journal of Applied Ecology* **32**, 561–577.

Fletcher, A. (ed.) (2001) *Lichen Habitat Management*. Proceedings of a workshop held at Bangor, 3–6 September 1997. British Lichen Society, Natural History Museum, London.

Fowles, A.P., Alexander, K.N.A., and Key, R.S. (1999) The Saproxylic Quality Index: evaluating wooded habitats for the conservation of decaying-wood Coleoptera. *Coleopterist* **8(3)**, 121–141.

Freemark, K. and Boutin, C. (1995) Impacts of agricultural herbicide use on terrestrial wildlife in temperate landscapes—a review with special reference to North-America. *Agriculture, Ecosystems and Environment* **52**, 67–91.

Frid, C.L.J., Chandrasekara, W.U., and Davey, P. (1999) The restoration of mud flats invaded by common cord-grass (*Spartina anglica*, CE Hubbard) using mechanical disturbance and its effects on the macrobenthic fauna. *Aquatic Conservation: Marine and Freshwater Ecosystems* **9**, 47–61.

Friday, L. (ed.) (1997) *Wicken Fen, the Making of a Nature Reserve*. Harley Books, Colchester.

Fujioka, M., Armacost, J.W., Yoshida, H., and Maeda, T. (2001) Value of fallow farmlands as summer habitats for waterbirds in a Japanese rural area. *Ecological Research* **16**, 555–567.

Fuller, E. (2000) *Extinct Birds*. Oxford University Press, Oxford.

Fuller, R.J. (1992) Effects of coppice management on woodland breeding birds. In Buckley, G.P. (ed.), *Ecology and Management of Coppice Woodlands*, pp. 169–192. Chapman and Hall, London.

Fuller, R.J. (2001) Responses of woodland birds to increasing numbers of deer: a review of evidence and mechanisms. *Forestry* **74**, 289–298.

Fuller, R.J. and Moreton, B.D. (1987) Breeding bird populations of Kentish sweet chestnut (*Castanea sativa*) coppice in relation to age and structure of coppice. *Journal of Applied Ecology* **24**, 13–27.

Fuller, R.J. and Henderson, A.C.B. (1992) Distribution of breeding songbirds in Bradfield Woods, Suffolk, in relation to vegetation and coppice management. *Bird Study* **39**, 73–88.

Fuller, R.J. and Green, G.H. (1998) Effects of woodland structure on breeding bird populations in stands of coppiced lime (*Tilia cordata*) in western England over a 10-year period. *Forestry* **71(3)**, 199–215.

Fuller, R.J. and Gough, S.J. (1999) Changing patterns of sheep stocking in Britain and implications for bird populations. *Biological Conservation* **91**, 73–89.

Fuller, R.J., Hinsley, S.A., and Swetnam, R.D. (2004) The relevance of non-farmland habitats, uncropped areas and habitat diversity to the conservation of farmland birds. *Ibis* **146** (suppl. 2), 22–31.

Fuller, R.J., Atkinson, P.W., Garnett, M.C., Conway, G.J., Bibby, C.J., and Johnstone, I.G. (2006) Breeding bird communities in the upland margins (ffridd) of Wales in the mid-1980s. *Bird Study* **53**, 177–186.

Gabrey, S.W., Afton, A.D., and Wilson, B.C. (1999) Effects of winter burning and structural marsh management on vegetation and winter bird abundance in the Gulf Coast Chenier Plain, USA. *Wetlands* **19**, 594–606.

Gabrey, S.W., Afton, A.D., and Wilson, B.C. (2001) Effects of structural marsh management and winter burning on plant and bird communities during summer in the Gulf Coast Chenier Plain. *Wildlife Society Bulletin* **29**, 218–231.

Galbraith, H. (1988) Effects of agriculture on the breeding ecology of lapwings *Vanellus vanellus*. *Journal of Applied Ecology* **25**, 487–503.

Gaston, K.J., Warren, P.H., Thompson, K., and Smith, R.M. (2005a) Urban domestic gardens (IV): the extent of the resource and its associated features. *Biodiversity and Conservation* **14**, 3327–3349.

Gaston, K.J., Smith, R.M., Thompson, K., and Warren, P.H. (2005b) Urban domestic gardens (II): experimental tests of methods for increasing biodiversity. *Biodiversity and Conservation* **14**, 395–413.

Gedge, D. and Kadas, G. (2005) Green roofs and biodiversity. *Biologist* **52**, 161–169.

Gibb, H., Ball, J.P., Johansson, T., Atlegrim, O., Hjältén, J., and Danell, K. (2005) Effects of management on coarse woody debris volume and composition in boreal forest in northern Sweden. *Scandinavian Journal of Forest Research* **20**, 213–222.

Gilbert, G., Tyler, G., and Smith, K.W. (2003) Nestling diet and fish preference of Bitterns *Botaurus stellaris* in Britain. *Ardea* **91**, 35–44.

Gilbert, G., Tyler, G., and Smith, K.W. (2005a) Behaviour, home range size and habitat use by male Great Bittern *Botaurus stellaris* in Britain. *Ibis* **147**, 533–543.

Gilbert, G., Tyler, G.A., Dunn, C.J., and Smith, K.W. (2005b) Nesting habitat selection by bitterns *Botaurus stellaris* in Britain and the implications for wetland management. *Biological Conservation* **124**, 547–553.

Gilbert, L. and Anderson, P. (1998) *Habitat Creation and Repair*. Oxford University Press, Oxford.

Gill, J.A. (1996) Habitat choice in pink-footed geese: quantifying the constraints determining winter site use. *Journal of Applied Ecology* **33**, 884–892.

Gill, J.A., Watkinson, A.R., and Sutherland, W.J. (1996) The impact of sugar beet farming practice on wintering pink-footed goose *Anser brachyrhynchus* populations. *Biological Conservation* **76**, 95–100.

Gill, J.A., Norris, K., and Sutherland, W.J. (2001) Why behavioural responses may not reflect the population consequences of human disturbance. *Biological Conservation* **97**, 265–268.

Giller, K.E. and Wheeler, B.D. (1986) Past peat cutting and present vegetation patterns in an undrained fen in the Norfolk Broadland. *Journal of Ecology* **74**, 219–247.

Gillespie, I.G. and Allen, E.B. (2004) Fire and competition in a southern Californian grassland: impacts on the rare forb *Erodium macrophyllum*. *Journal of Applied Ecology* **41**, 643–652.

Gillings, S., Newson, S.E., Noble, D.G., and Vickery, J.A. (2005) Winter availability of cereal stubbles attracts declining farmland birds and positively influences breeding population trends. *Proceedings of the Royal Society of London Series B Biological Sciences* **272**, 733–739.

Gimmingham, C.H. (1972) *Ecology of Heathlands*. Chapman and Hall, London.

Giroux, J.-F. and Bédard, J. (1987) The effects of grazing by Greater Snow Geese on the vegetation of tidal marshes in the St Lawrence estuary. *Journal of Applied Ecology* **24**, 773–788.

Gloaguen J.C. (1993) Spatio-temporal patterns in post-burn succession on Brittany heathlands. *Journal of Vegetation Science* **4**, 561–566

Gołłwski, A. and Dombrowski, A. (2002) Habitat use of Yellowhammers *Emberiza citrinella*, Ortolan Buntings *E. hortulana*, and Corn Buntings *Miliaria calandra* in farmland in east-central Poland. *Ornis Fennica* **79**, 164–172.

Gordo, O. and Sanz, J.J. (2005) Phenology and climate change: a long-term study in a Mediterranean locality. *Oecologia* **146**, 484–495.

Gordon, I.J. (1989) Vegetation community selection by ungulates on the Isle of Rhum. III. Determinants of vegetation community selection. *Journal of Applied Ecology* **26**, 65–79.

Grant, S.A., Barthram, G.T., Lamb, W.I.C., and Milne, J.A. (1978) Effects of season and level of grazing on the utilisation of heather by sheep. I. Responses of the sward. *Journal of the British Grassland Society* **33**, 289–300.

Grant, S.A., Milne, J.A., Barthram, G.T., and Souter, W.G. (1982) Effects of season and level of grazing on the utilisation of heather by sheep. III. Longer-term responses and sward recovery. *Grass and Forage Science* **37**, 311–320.

Grant, S.A., Torvell, L., Smith, H.K., Suckling, D.E., Forbes, T.D.A., and Hodgson, J. (1987) Comparative studies of diet selection by sheep and cattle: blanket bog and heather moor. *Journal of Ecology* **75**, 947–960.

Grant, T.A., Madden, E., and Berkey, G.B. (2004) Tree and shrub invasion in northern mixed-grass prairie: implications for breeding grassland birds. *Wildlife Society Bulletin* **32**, 807–818.

Gratton, C. and Denno, R.F. (2005) Restoration of arthropod assemblages in a *Spartina* salt marsh following removal of the invasive plant *Phragmites australis*. *Restoration Ecology* **13**, 358–371.

Graveland, J. (1998) Reed die-back, water level management and the decline of the great reed warbler *Acrocephalus arundinaceus* in the Netherlands. *Ardea* **86**, 187–201.

Graveland, J. (1999) Effects of reed cutting on density and breeding success of Reed Warbler *Acrocephalus scirpaceus* and Sedge Warbler *A. schoenobaenus*. *Journal of Avian Biology* **30**, 469–482.

Gray, M.J., Kaminski, R.M., Weerakkody, G., Leopold, B.D., and Jensen, K.C. (1999) Aquatic invertebrate and plant responses following mechanical manipulations of moist soil habitat. *Wildlife Society Bulletin* **27**, 770–779.

Greatorex-Davies, J.N., Hall, M.L., and Marrs, R.H. (1992) The conservation of the pearl-bordered fritillary butterfly (*Boloria-Euphrosyne* L)—preliminary studies on the creation and management of glades in conifer plantations. *Forest Ecology and Management* **53**, 1–14.

Green, R.E. (1988) Effects of environmental factors on the timing and success of breeding common snipe *Gallinago gallinago* (Aves:Scolopacidae). *Journal of Applied Ecology* **25**, 79–93.

Green, R.E. (1995) Diagnosing causes of bird population decline. *Ibis* **137**, S47–S55.
Green, R.E. and Taylor, C.R. (1995) Changes in stone curlew *Burhinus oedicnemus* distribution and abundance and vegetation height on chalk grassland at Porton-Down, Wiltshire. *Bird Study* **42**, 177–181.
Green, R.E., Hirons, G.J.M., and Cresswell, B.H. (1990a) Foraging habits of female common snipe *Gallinago gallinago* during the incubation period. *Journal of Applied Ecology* **27**, 325–335.
Green, R.E., Hirons, G.J.M., and Kirby, J.S. (1990b) The effectiveness of nest defence by black-tailed godwits *Limosa limosa*. *Ardea* **78**, 405–413,
Green, R.E., Osborne, P.E., and Sears, E.J. (1994) The distribution of passerine birds in hedgerows during the breeding season in relation to characteristics of the hedgerow and adjacent farmland. *Journal of Applied Ecology* **31**, 677–692.
Green, R.E., Cornell, S.J., Scharlemann, J.P.W., and Balmford, A. (2005) Farming and the fate of wild nature. *Science* **307**, 550–555.
Green, T. (1996) Pollarding—origins and some practical advise. *British Wildlife* **8**, 100–105.
Greenwood, J.D. and Robinson, R.A. (2006) Principles of sampling. In Sutherland, W.J. (ed.), *Ecological Census Techniques: a Handbook*, 2nd edn, pp. 11–86. Cambridge University Press, Cambridge.
Griffis, K.L., Crawford, J.A., Wagner, M.R., and Moir, W.H. (2001) Understorey response to management treatments in northern Arizona ponderosa pine forests. *Forest Ecology and Management* **146**, 239–245.
Griffith, G.W., Bratton, J.H., and Easton, G. (2004) Charismatic magafungi—the conservation of waxcap grasslands. *British Wildlife* **16(1)**, 31–43.
Grootjans, A.P., Everts, H., Bruin, K., and Fresco, L. (2001) Restoration of wet dune slacks on the Dutch Wadden Sea Islands: recolonization after large-scale sod cutting. *Restoration Ecology* **9**, 137–146.
Gryseels, M. (1989a) Nature management experiments in a derelict reedmarsh. I: effects of winter cutting. *Biological Conservation* **47**, 171–193.
Gryseels, M. (1989b) Nature management experiments in a derelict reedmarsh. II: effects of summer mowing. *Biological Conservation* **48**, 85–99.
Gulati, R.D. and van Donk, E. (2002) Lakes in the Netherlands, their origin, eutrophication and restoration: state-of-the-art review. *Hydrobiologia* **478**, 73–106.
Guldemond, J.A., Parmentier, F., and Visbeen, F. (1993) Meadow birds, field management and nest protection in a Dutch peat soil area. *Wader Study Group Bulletin* **70**, 42–48.
Hågvar, S., Hågvar, G., and Mønnes, E. (1990) Nest site selection in Norwegian woodpeckers. *Holarctic Ecology* **13**, 156–165.
Hancock, M. (2000) Artificial floating islands for nesting Black-throated Divers *Gavia arctica* in Scotland: construction, use and effect on breeding success. *Bird Study* **47**, 165–175.
Hancock, M., Egan, S., Summers, R., Cowie, N., Amphlett, A., Rao, S., and Hamilton, A. (2005) The effect of experimental prescribed fire on the establishment of Scots pine *Pinus sylvestris* seedlings on heather *Calluna vulgaris* moorland. *Forest Ecology and Management* **212**, 199–213.
Hancock, M.H. and Wilson, J.D. (2003) Winter habitat associations of seed-eating passerines on Scottish farmland. *Bird Study* **50**, 116–130.
Hannerz, M. and Hanell, B. (1997) Effects on the flora in Norway spruce forest following clear-cutting and shelterwood cutting. *Forest Ecology and Management* **90**, 29–49.
Hanowski, J.M., Christian, D.P., and Nelson, M.C. (1999) Response of breeding birds to shearing and burning in wetland brush ecosystems. *Wetlands* **19**, 584–593.
Härdtle, W., Niemeyer, M., Niemeyer, T., Assmann, T., and Fottner, S. (2006) Can management compensate for atmospheric nutrient deposition in heathland ecosystems? *Journal of Applied Ecology* **43**, 759–769.

Harmon, M.E., Franklin, J.F., Swanson, F.J., Sollins, P., Gregory, S.V., Lattin, J.D., Anderson, N.H., Cline, S.P., Aumen, N.G., Sedell, J.R. et al. (1986) Ecology of coarse woody debris in temperate ecosystems. *Advances in Ecological Research* **15**, 133–302.

Hartley, S.E. and Mitchell, R.J. (2005) Manipulation of nutrients and grazing levels on heather moorland: changes in *Calluna* dominance and consequences for community composition. *Journal of Ecology* **93**, 990–1004.

Harvey, P. (2004) Brown roofs for invertebrates. *Essex Naturalist (New Series)* **21**, 79–88.

Harwood, R., Hickman, W.J., Macleod, A., and Sherrat, T.N. (1994) Managing field margins for hoverflies. *British Crop Protection Council Monograph* **58**, 147–152.

Haukos, D.A. and Smith, L.M. (1993) Moist-soil management of playa lakes for migrating and wintering ducks. *Wildlife Society Bulletin* **21**, 288–298.

Hautala, H., Jalonen, J., Laaka-Linberg, S., and Vanha-Majamaa, I. (2004) Impacts of retention felling on coarse woody debris (CWD) in mature boreal spruce forests in Finland. *Biodiversity and Conservation* **13**, 1541–1554.

Hawke, C.J. and José, P.V. (1996) *Reedbed Management for Commercial and Wildlife Interests*. RSPB, Sandy.

Heil, G.W. and Diemont, W.H. (1983) Raised nutrient levels change heathland into grassland. *Vegetatio* **53**, 113–120.

Heilmann-Clausen, J. and Christensen, M. (2003) Fungal diversity on decaying beech logs—implications for sustainable forestry. *Biodiversity and Conservation* **12**, 953–973.

Heilmann-Clausen, J., Aude, E., and Christensen, M. (2005) Cryptogam communities on decaying dead wood—does tree species diversity matter? *Biodiversity and Conservation* **14**, 2061–2078.

Henderson, A.G., Cooper, J., Fuller, R.J., and Vickery, J. (2000) The relative abundance of birds on set-aside and neighbouring fields in summer. *Journal of Applied Ecology* **37**, 335–347.

Henderson, I.G., Vickery, J.A., and Carter, N. (2004) The use of winter bird crops by farmland birds in lowland England. *Biological Conservation* **118**, 21–32.

Henningsen, J.C. and Best, L.B. (2005) Grassland bird use of riparian filter strips in Southeast Iowa. *Journal of Wildlife Management* **69**, 198–210.

Hernández, L. and Laundré, J.W. (2005) Foraging in the 'landscape of fear' and its implications for habitat use and diet quality of elk *Cervus elaphus* and bison *Bison bison*. *Wildlife Biology* **11**, 215–220.

Hickey, M.B.C. and Doran, B. (2004) A review of the efficiency of buffer strips for the maintenance and enhancement of riparian ecosystems. *Water Quality Research Journal of Canada* **39**, 311–317.

Hill, D., Fasham, M., Tucker, G., Shewry, M., and Shaw, P. (eds) (2005) *Handbook of Biodiversity Methods: Survey, Evaluation and Monitoring*. Cambridge University Press, Cambridge.

Hill, M.O., Evans, D.F., and Bell, S.A. (1992) Long-term effects of excluding sheep from hill pastures in North Wales. *Journal of Ecology* **80**, 1–13.

Hille, M. and den Ouden, J. (2004) Improved recruitment and early growth of Scots pine (*Pinus sylvestris* L.) seedlings after fire and soil scarification. *European Journal of Forest Research* **123**, 213–218.

Hilszczanski, J., Gibb, H., Hjalten, J., Atlegrim, O., Johansson, T., Pettersson, R.B., Ball, J.P., and Danell, K. (2005) Parasitoids (Hymenoptera, Ichneunionoidea) of Saproxylic beetles are affected by forest successional stage and dead wood characteristics in boreal spruce forest. *Biological Conservation* **126**, 456–464.

Hinsley, S.A. and Bellamy, P.E. (2000) The influence of hedge structure, management and landscape context on the value of hedgerows to birds: a review. *Journal of Environmental Management* **60**, 33–49.

## References

Hodder, K.H., Bullock, J.M., Buckland, P.C., and Kirby, K.J. (2005) *Large herbivores in the wildwood and modern naturalistic grazing systems.* English Nature Research Report Number 648. English Nature, Peterborough.

Hole, D.G., Perkins, A.J., Wilson, J.D., Alexander, I.H., Grice, P.V., and Evans, A.D. (2005) Does organic farming benefit biodiversity? *Biological Conservation* **122**, 113–130.

Holmes, P.M., Richardson, D.M., van Wilgen, B.W., and Gelderblom, C. (2000) Recovery of South African fynbos vegetation following alien woody plant clearing and fire: implications for restoration. *Austral Ecology* **25**, 631–639.

Hope, D., Picozzi, N., Catt, D.C., and Moss, R. (1996) Effects of reducing sheep grazing in the Scottish Highlands. *Journal of Range Management* **49**, 301–310.

House, S.M. and Spellerberg, I.F. (1983) Ecology and conservation of the sand lizard (*Lacerta agilis* L.) habitat in Southern England. *Journal of Applied Ecology* **20**, 417–437.

Howe, H.F. (1995) Succession and fire season in experimental prairie plantings. *Ecology* **76**, 1917–1925.

Hudson, R., Tucker, G.M., and Fuller, R.J. (1994) Lapwing *Vanellus vanellus* populations in relation to agricultural changes: a review. In: Tucker, G.M., Davies, S.M., and Fuller, R.J. *The Ecology and Conservation of Lapwings Vanellus vanellus*, UK Nature Conservation, no. 9, pp 1–33. Joint Nature Conservation Committee, Peterborough.

Hulme, P.D., Merrell, B.G., Torvell, L., Fisher, J.M., Small, J.L., and Pakeman, R.J. (2002) Rehabilitation of degraded *Calluna vulgaris* (L.) Hull-dominated wet heath by controlled sheep grazing. *Biological Conservation* **107**, 351–363.

Huner, J.V., Jeske, C.W., and Norling, W. (2002) Managing agricultural wetlands for waterbirds in the coastal regions of Louisiana, USA. *Waterbirds* **25** (special publication 2), 66–78.

Hurford, C. and Schneider, M. (2006) *Monitoring Nature Conservation in Cultural Habitats: a Practical Guide and Case Studies.* Springer, Dordrecht.

Hyvärinen, E., Kouki, J., and Martikainen, P. (2006) Fire and green-tree retention in conservation of red-listed and rare deadwood-dependent beetle in Finnish boreal forests. *Conservation Biology* **20**, 1711–1719.

Imbeau, L., Mönkkönen, M., and Desrochers, A. (2000) Long-term effects of forestry on birds of the eastern Canadian boreal forests: a comparison with Fennoscandia. *Conservation Biology* **15**, 1151–1162.

Isaksson, D., Wallander, J., and Larsson, M. (2007) Managing predation on ground-nesting birds: The effectiveness of nest exclosures. *Biological Conservation* **136**, 136–142.

Jacob, J. (2003) The response of small mammal populations to flooding. *Mammalian Biology* **68**, 102–111.

Jacobson, S.K., McDuff, M.D., and Monroe, M.C. (2006) *Conservation Education and Outreach Techniques.* Techniques in Ecology and Conservation Series. Oxford University Press, Oxford.

Jacquemart, A.L., Champluvier, D., and De Sloover, J. (2003) A test of mowing and soil-removal restoration techniques in wet heaths of the high Ardenne, Belgium. *Wetlands* **23**, 376–385.

Jansen, A.J.M., Fresco, L.F.M., Grootjans, A.P., and Jalink, M.H. (2004) Effects of restoration measures on plant communities of wet heathland ecosystems. *Applied Vegetation Science* **7**, 243–252.

Janssens, F., Peeters, A., Tallowin, J.R.B., Bakker, J.P., Bekker, R.M., Fillat, F., and Oomes, M.J.M. (1998) Relationship between soil chemical factors and grassland diversity. *Plant and Soil* **202**, 69–78.

Jenkins, R.K.B. and Ormerod, S.J. (2002) Habitat preferences of breeding Water Rail *Rallus aquaticus*. *Bird Study* **49**, 2–10.

Jewell, P.L., Gusewell, S., Berry, N.R., Kauferle, D., Kreuzer, M., and Edwards, P.J. (2005) Vegetation patterns maintained by cattle grazing on a degraded mountain pasture. *Botanica Helvetica* **115**, 109–124.

Johnson, M. and Oring, L.W. (2002) Are nest exclosures an effective tool in plover conservation? *Waterbirds* **25**, 184–190.

Jones, C.A., Basch, G., Baylis, A.D., Bazzoni, D., Biggs, J., Bradbury, R.B., Chaney, K., Deeks, L.K., Field, R., Gómez, J.A. *et al.* (2006) *Conservation Agriculture in Europe: An Approach to Sustainable Crop Production by Protecting Soil and Water?* SOWAP, Jealott's Hill, Bracknell.

Jones, G.A., Sieving, K.E., and Jacobson, S.K. (2005) Avian diversity and functional insectivory on north-central Florida farmlands. *Conservation Biology* **19**, 1234–1245.

Jones, J.C., Reynolds, J.D., and Raffaelli, D. (2006) Environmental variables. In Sutherland, W.J. (ed.), *Ecological Census Techniques: a Handbook*, 2nd edn, pp. 370–407. Cambridge University Press, Cambridge.

Jonsell, M., Weslien, J., and Ehnström, B. (1998) Substrate requirements of red-listed saproxylic invertebrates in Sweden. *Biodiversity and Conservation* **7**, 749–764.

Jonsell, M., Nittérus, K., and Stighäll, K. (2004) Saproxylic beetles in natural and man-made deciduous high stumps retained for conservation. *Biological Conservation* **118**, 163–173.

Jonsell, M., Schroeder, M., and Weslien, J. (2005) Saproxylic beetles in high stumps of spruce: fungal flora important for determining the species composition. *Scandinavian Journal of Forest Research* **208**, 54–62.

Joyce, C.B., Vina-Herbon, C., and Metcalfe, D.J. (2005) Biotic variation in coastal water bodies in Sussex, England: implications for saline lagoons. *Estuarine, Coastal and Shelf Science* **65**, 633–644.

Joys, A.C., Fuller, R.J., and Dolman, P.M. (2004) Influences of deer browsing, coppice history, and standard trees on the growth and development of vegetation structure in coppiced woods in lowland England. *Forest Ecology and Management* **202**, 23–37.

Kaminski, R.M. and Prince, H.H. (1981) Dabbling duck and aquatic macroinvertebrate responses to manipulated wetland habitat. *Journal of Wildlife Management* **45(1)**, 1–15.

Kampf, H. (2000) The role of large grazing animals in nature conservation—a Dutch perspective. *British Wildlife* **12**, 37–46.

Katsimanis, N., Dretakis, M., Akriotis, T., and Mylonas, M. (2006) Breeding bird assemblages of eastern Mediterranean shrublands: composition, organisation and patterns of diversity. *Journal of Ornithology* **147**, 419–427.

Keeley, J.E. (2002) Fire management of California shrubland landscapes. *Environmental Management* **29**, 395–408.

Keeley, J.E. (2005) Fire management impacts on invasive plants in the Western United States. *Conservation Biology* **20**, 375–384.

Keeley, J.E. and Fotheringham, C.J. (2000) Historic fire regime in Southern Californian shrublands. *Conservation Biology* **15**, 1536–1548.

Keeley, J.E. and McGinnis, T.W. (2007) Impact of prescribed fire and other factors on cheatgrass persistence in a Sierra Nevada ponderosa pine forest. *International Journal of Wildland Fire* **16**, 96–106.

Keeley, J.E., Fotheringham, C.J., and Morais, M. (1999) Re-examining fire suppression impacts on brushland fire regimes. *Science* **284**, 1829–1832.

Keeley, J.E., Pfaff, A.H., and Safford, H.D. (2005a) Fire suppression impacts on postfire recovery of Sierra Nevada chaparral shrublands. *International Journal of Wildland Fire* **14**, 255–265.

Keeley, J.E., Baer-Keeley, M., and Fotheringham, C.J. (2005b) Alien plant dynamics following fire in Mediterranean-climate California shrublands. *Ecological Applications* **15**, 2109–2125.

Kells, A.R. and Goulson, D. (2003) Preferred nesting sites of bumblebee queens (Hymenoptera: Apidae) in agroecosystems in the UK. *Biological Conservation* **109**, 165–174.

King, J.L., Simovich, M.A., and Brusca, R.C. (1996) Species richness, endemism and ecology of crustacean assemblages in northern California vernal pools. *Hydrobiologia* **328**, 85–116.

Kirby, K.J. (1992a) Accumulation of decaying wood—a missing ingredient in coppicing? In Buckley, G.P. (ed.), *Ecology and Management of Coppice Woodlands*, pp. 99–112. Chapman and Hall, London.

Kirby, K.J., Wester, S.D., and Anctzak, A. (1991) Effects of forest management on stand structure and the quantity of fallen decaying wood: some British and Polish examples. *Forest Ecology and Management* **43**, 167–174.

Kirby, K.J., Thomas, R.C., Key, R.S., McClean, I.F.G., and Hodgetts, N. (1995) Pasture-woodland and its conservation in Britain. *Biological Journal of the Linnean Society* **56**, 135–153.

Kirby, P. (1992b) *Habitat Management for Invertebrates: a Practical Handbook*. RSPB, Sandy.

Kleijn, D. and van der Voort, L.A.C. (1997) Conservation headlands for rare arable weeds: the effects of fertiliser application and light penetration on plant growth. *Biological Conservation* **81**, 57–67.

Kleijn, D. and Sutherland, W.J. (2003) How effective are European agri-environment schemes in conserving and promoting biodiversity? *Journal of Applied Ecology* **40**, 947–969.

Knick, S.T. and Rotenberry, J.T. (1997) Landscape characteristics of disturbed shrubsteppe habitats in southwestern Idaho (U.S.A.). *Landscape Ecology* **12**, 287–297.

Knick, S.T., Dobkin, D.S., Rotenberry, J.T., Schroeder, M.A., Vander Haegen, W.M., and van Riper, C. (2003) Teetering on the edge or too late? Conservation and research issues for avifauna of sagebrush habitats. *The Condor* **105**, 611–634.

Knight, M.E., Martin, A.P., Bishop, S., Osborne, J.L., Hale, R.J., Sanderson, A., and Goulson, D. (2005) An interspecific comparison of foraging range and nest density of four bumblebee (*Bombus*) species. *Molecular Ecology* **14**, 1811–1820.

Knights, B. and White, E.M. (1998) Enhancing migration and recruitment of eels: the use of passes and associated trapping systems. *Fisheries Management and Ecology* **4**, 311–324.

Kost, M.A. and De Steven, D. (2000) Plant community response to prescribed burning in Wisconsin sedge meadows. *Natural Areas Journal* **20**, 36–45.

Kostecke, R.M., Smith, L.M., and Hands, H.M. (2004) Vegetation response to cattail management at Cheyenne Bottoms, Kansas. *Journal of Aquatic Plant Management* **42**, 39–45.

Kostecke, R.M., Smith, L.M., and Hands, H.M. (2005) Macroinvertebrate response to cattail management at Cheyenne Bottoms, Kansas, USA. *Wetlands* **25**, 758–763.

Kotzageorgis, G.C. and Mason, C.F. (1997) Small mammal populations in relation to hedgerow structure in an arable landscape. *Journal of Zoology* **242**, 425–434.

Krogh, M.G. and Schweitzer, S.H. (1999) Least terns nesting on natural and artificial habitats in Georgia, USA. *Waterbirds* **22**, 290–296.

Kruk, M., Noordervliet, A.A.W., and ter Keurs, W.J. (1996) Hatching dates of waders and mowing dates in intensively exploited grassland areas in different years. *Biological Conservation* **77**, 213–218.

Kruk, M., Noordervliet, M.A.W., and ter Keurs, W.J. (1997) Survival of black-tailed godwit chicks *Limosa limosa* in intensively exploited grassland areas in the Netherlands. *Biological Conservation* **80**, 127–133.

Kuffer, N. and Senn-Irlet, B. (2005) Influence of forest management on the species-richness and composition of wood-inhabiting basidiomycetes in Swiss forests. *Biodiversity and Conservation* **14**, 2419–2435.

Kühn, I., Brandl, R., and Klotz, S. (2004) The flora of German cities is naturally species-rich. *Evolutionary Ecological Research* **6**, 749–764.

Laaksonen, T., Ahola, M., Eeva, T., Vaisanen, R.A., and Lehikoinen, E. (2006) Climate change, migratory connectivity and changes in laying date and clutch size of the pied flycatcher. *Oikos* **114**, 277–290.

Lagerlof, J., Starck, J., and Svenson, B. (1992) Margins of agricultural fields as habitats for pollinating insects. *Agriculture, Ecosystems and Environment* **40**, 117–124.

Laiolo, P. (2005) Spatial and seasonal patterns of bird communities in Italian agroecosystems. *Conservation Biology* **19**, 1547–1556.

Laiolo, P., Rolando, A., and Valsania, V. (2004) Avian community structure in sweet chestnut coppiced woods facing natural restoration. *Revue d'Ecologie-la Terre et la Vie* **59**, 453–463.

Lambley, P. (2001) Management of lowland grassland for lichens. In Fletcher, A. (ed.), *Lichen Habitat Management*. Proceedings of a workshop held at Bangor, 3–6 September 1997, pp. 12-1–12-6. British Lichen Society, Natural History Museum, London.

Lanham, J.D., Keyser, P.D., Brose, P.H., and Van Lear, D.H. (2002) Oak regeneration using the shelterwood-burn technique: management options and implications for songbird conservation in the southeastern United States. *Forest Ecology and Management* **155**, 143–152.

Lepczyk, C.A., Mertig, A.G., and Liu, J.G. (2004) Landowners and cat predation across rural-to-urban landscapes. *Biological Conservation* **115**, 191–201.

Lever, C. (1994) *Naturalized Animals*. T. & A.D. Poyser Natural History, London.

Lilja, S. L., De Chantal, M., Kuuluvainen, T., Vanha-Majamaa, I., and Puttonen, P. (2005) Restoring natural characteristics in managed Norway spruce [*Picea abies* (L.) Karst.] stands with partial cutting, dead wood creation and fire: immediate effects. *Scandinavian Journal of Forest Research* **20**, 68–78.

Lindhe, A. and Lindelow, A. (2005) Cut high stumps of spruce, birch, aspen and oak as breeding substrates for saproxylic beetles. *Forest Ecology and Management* **203**, 1–20.

Lindhe, A., Asenblad, N., and Toresson, H.G. (2004) Cut logs and high stumps of spruce, birch, aspen and oak—nine years of saproxylic fungi succession. *Biological Conservation* **119**, 443–454.

Lindhe, A., Lindelow, A., and Asenblad, N. (2005) Saproxylic beetles in standing dead wood density in relation to substrate sun-exposure and diameter. *Biodiversity and Conservation* **14**, 3033–3053.

Linz, G.M., Blixt, D.C., Bergman, D.L., and Bleier, W.J. (1996) Response of ducks to glyphosate-induced habitat alterations of wetlands. *Wetlands* **16**, 38–44.

Llusia, D. and Oñate, J.J. (2005) Are the conservation requirements of pseudo-steppe birds adequately covered by Spanish agri-environmental schemes? An *ex-ante* assessment. *Ardeola* **52**, 31–42.

Lokemoen, J.T. and Beiser, J.A. (1997) Bird use and nesting in conventional, minimum-tillage and organic cropland. *Journal of Wildlife Management* **61**, 644–655.

Lonzarich, D.G. and Smith, J.J. (1997) Water chemistry and community structure of saline and hypersaline salt evaporation ponds in San Francisco Bay, California. *California Fish and Game* **83**, 89–104.

Loucougaray, G., Bonis, A., and Bouzillé, J.-B. (2004) Effects of grazing by horses and/or cattle on the diversity of coastal grasslands in western France. *Biological Conservation* **116**, 59–71.

MacArthur, R.H. and MacArthur, J.W. (1961) On bird species diversity. *Ecology* **42**, 594–598.

Mack, R.N. and Thompson, J.N. (1982) Evolution in steppe with few, large hooved mammals. *American Naturalist* **119**, 757–773.

MacLeod, A., Wratten, S.D., Sotherton, N.W., and Thomas, M.B. (2004) 'Beetle banks' as refuges for beneficial arthropods in farmland: long-term changes in predator communities and habitat. *Agricultural and Forest Entomology* **6**, 147–154.

Madsen, M., Neilsen, B.O., Holter, P., Pederson, O.C., Jespersen, J.B., Vagn Jespen, K-M., Nansen, P., and Gronvold, L. (1990) Treating cattle with Ivermectin and effects on the fauna and decomposition of dung pats. *Journal of Applied Ecology* **27**, 1–15.

Maeda, T. (2001) Patterns of bird abundance and habitat use in rice fields of the Kanto Plain, central Japan. *Ecological Research* **16**, 569–585.

Manning, P., Putwain, P.D., and Webb, N.R. (2004) Identifying and modelling the determinants of woody plant invasion of lowland heath. *Journal of Ecology* **92**, 868–881.

Manuwal, D.A. and Huff, M.H. (1987) Spring and winter bird populations in a Douglas fir sere. *Journal of Wildlife Management* **51**, 586–595.

Marage, D. and Lemperiere, G. (2005) The management of snags: a comparison in managed and unmanaged ancient forest of the Southern French Alps. *Annals of Forest Science* **62**, 135–142.

Marcus, J.F., Palmer, W.E., and Bromley, P.T. (2000) The effects of farm field borders on overwintering sparrow densities. *Wilson Bulletin* **112**, 517–523.

Martin, J.-L., Thibault, J.-C., and Bretagnolle, V. (2000) Black rats, island characteristics, and colonial nesting birds in the Mediterranean: consequences of an ancient introduction. *Conservation Biology* **14**, 1452–1466.

Martin, P.A. and Forsyth, D.J. (2003) Occurrence and productivity of songbirds in prairie farmland under conventional versus minimum tillage regimes. *Agriculture, Ecosystems and Environment* **96**, 107–117.

Mason, C.F., Heath, D.J., and Gibbs, D.J. (1991) Invertebrate assemblages of Essex salt marshes and their conservation importance. *Aquatic Conservation: Marine and Freshwater Ecosystems* **1**, 123–137.

McCoy, T.D., Ryan, M.R., Kurzejeski, E.W., and Burger, L.W. (1999) Conservation Reserve Program: source or sink habitat for grassland birds in Missouri? *Journal of Wildlife Management* **63**, 530–538.

McCoy, T.D., Kurzejeski, E.W., Burger, L.W., and Ryan, M.R. (2001a) Effects of conservation practice, mowing, and temporal changes on vegetation structure on CRP fields in northern Missouri. *Wildlife Society Bulletin* **29**, 979–987.

McCoy, T.D., Ryan, M.R., Burger, L.W., and Kurzejeski, E.W. (2001b) Grassland bird conservation: CP1 vs. CP2 plantings in Conservation Reserve Program fields in Missouri. *American Midwest Naturalist* **145**, 1–17.

Menard, C., Duncan, P., Fleurance, G., Georges, J.-Y., and Lila, M. (2002) Comparative foraging and nutrition of horses and cattle in European wetlands. *Journal of Applied Ecology* **39**, 120–133.

Merriam, K.E., Keeley, J.E., and Beyers, J.L. (2006) Fuel breaks affect nonnative species abundance in Californian plant communities. *Ecological Applications* **16**, 515–527.

Merrill, S.B., Cuthbert, F.J., and Oehlert, G. (1998) Residual patches and their contribution to forest-bird diversity on northern Minnesota aspen clearcuts. *Conservation Biology* **12**, 190–199.

Middleton, B. (1999) *Wetland Restoration, Flood Pulsing, and Disturbance Dynamics*. John Wiley and Sons, New York.

Middleton, B.A., Holsten, B., and van Diggelen, R. (2006) Biodiversity management of fens and fen meadows by grazing, cutting and burning. *Applied Vegetation Science* **9**, 307–316.

Millenbah, K.F., Winterstein, S.R., Campa, H., Furrow, L.T., and Minnis, R.B. (1996) Effects of conservation reserve program field age on avian relative abundance, diversity and productivity. *Wilson Bulletin* **108**, 760–770.

Millon, A., Bourrioux, J.-L., Riols, C., and Bretagnolle, V. (2002) Comparative breeding biology of Hen Harrier and Montagu's Harrier: an 8-year study in north-eastern France. *Ibis* **144**, 94–105.

Milsom, T.P., Ennis, D.C., Haskell, D.J., Langton, S.D., and McKay, H.V. (1998) Design of grassland feeding areas for waders during winter: the relative importance of sward, landscape factors and human disturbance. *Biological Conservation* **84**, 119–129.

Milsom, T.P., Langton, S.D., Parkin, W.K., Peel, S., Bishop, J.D., Hart, J.D., and Moore, N.P. (2000) Habitat models of bird species' distribution: an aid to the management of coastal grazing marshes. *Journal of Applied Ecology* **37**, 706–727.

Milsom, T.P., Hart, J.D., Parkin, W.K., and Peel, S. (2002) Management of coastal grazing marshes for breeding waders: the importance of surface topography and wetness. *Biological Conservation* **103**, 199–207.

Milsom, T.P., Sherwood, A.J., Rose, S.C., Town, S.J., and Runham, S.R. (2004) Dynamics and management of plant communities in ditches bordering arable fenland in eastern England. *Agriculture, Ecosystems and Environment* **103**, 85–99.

Minello, T.J., Zimmerman, R.J., and Medina, R. (1994) The importance of edge for natant macrofauna in a created salt-marsh. *Wetlands* **14**, 184–198.

Mitchell, L.R., Gabrey, S., Marra, P.P., and Erwin, R.M. (2007) *Impacts of Marsh Management on Coastal Marsh Bird Habitats in the Southeastern United States*. Studies in Avian Biology. Cooper Ornithological Society, Camirollo, California (in press).

Mitchell, R.J., Marrs, R.H., Le Duc, M.G., and Auld, M.H.D. (1999) A study of the restoration of heathland on successional sites: changes in vegetation and soil chemical properties. *Journal of Applied Ecology* **36**, 770–783.

Mönkkönen, M. and Helle, P. (1989) Migratory habits of birds breeding in different stages of forest succession: a comparison between the Palearctic and Nearctic. *Annales Zoologici Fennici* **26**, 323–330.

Moorcroft, D., Whittingham, M.J., Bradbury, R.B., and Wilson, J.D. (2002) The selection of stubble fields by wintering granivorous birds reflects vegetation cover and food abundance. *Journal of Applied Ecology* **39**, 535–547.

Moreira, F., Beja, P., Morgado, R., Reino, L., Gordinho, L., Delgado, A., and Borralho, R. (2005) Effects of field management and landscape context on grassland wintering birds in Southern Portugal. *Agriculture, Ecosystems and Environment* **109**, 59–74.

Moretti, M. and Barbalat, S. (2004) The effects of wildfires on wood-eating beetles in deciduous forests on the southern slope of the Swiss Alps. *Forest Ecology and Management* **187**, 85–103.

Moritz, M.A., Keeley, J.E., Johnson, E.A., and Schaffner, A.A. (2004) Testing a basic assumption of shrubland fire management: how important is fuel age? *Frontiers in Ecology and the Environment* **2**, 67–72.

Morris, A.J., Holland, J.M., Smith, B., and Jones, N.E. (2004) Sustainable Arable Farming for an Improved Environment (SAFFIE): managing winter wheat sward structure for Skylarks *Alauda arvensis*. *Ibis* **146** (suppl. 2), 155–162.

Morris, M.G. (2000) The effects of structure and its dynamics on the ecology and conservation of arthropods in British grasslands. *Biological Conservation* **95**, 129–140.

Moseby, K.E. and Read, J.L. (2006) The efficacy of feral cat, fox and rabbit exclusion fence designs for threatened species. *Biological Conservation* **127**, 429–437.

Moss, B., Madgwick, J., and Phillips, G. (1996) *A Guide to the Restoration of Nutrient-Enriched Shallow Lakes*. W.W. Hawes.

Musters, C.J.M., Kruk, M., De Graaf, H.J., and Ter Keurs, W.J. (2001) Breeding birds as a farm product. *Conservation Biology* **15**, 363–369.

Nelson, S.H., Evans, A.D., and Bradbury, R.B. (2005) The efficacy of collar-mounted devices in reducing the rate of predation of wildlife by domestic cats. *Applied Animal Behaviour Science* **94**, 273–285.

Nelson, S.H., Evans, A.D., and Bradbury, R.B. (2006) The efficacy of an ultrasonic cat deterrent. *Applied Animal Behaviour Science* **96**, 83–91.

Newton, I. (1998) *Population Limitation in Birds*. Academic Press, London.

Nilsen, L.S., Johansen, L., and Velle, L.G. (2005) Early stages of *Calluna vulgaris* regeneration after burning of coastal heath in central Norway. *Applied Vegetation Science* **8**, 57–64.

Nilsson, S.G., Niklasson, M., Hedin, J., Aronsson, G., Gutowski, J.M., Linder, P., Ljungberg, H., Mikusiński, G., and Ranius, T. (2003) Densities of large living and dead trees in old-growth temperate and boreal forests. *Forest Ecology and Management* **178**, 353–370.

Nitterus, K., Gunnarsson, B., and Axelsson, E. (2004) Insects reared from logging residue on clear-cuts. *Entomologica Fennica* **15**, 53–61.

Nogales, M., Martín, A., Tershy, B.R., Donlan, C.J., Veitch, D., Puerta, N., Wood, B., and Alonso, J. (2004) A review of feral cat eradication on islands. *Conservation Biology* **18**, 310–319.

Norden, B., Ryberg, M., Gotmark, F., and Olausson, B. (2004) Relative importance of coarse and fine woody debris for the diversity of wood-inhabiting fungi in temperate broadleaf forests. *Biological Conservation* **117**, 1–10.

Norris, K., Cook, T., O'Dowd, B., and Durdin, C. (1997) The density of redshank *Tringa totanus* breeding on the salt-marshes of the Wash in relation to habitat and its grazing management. *Journal of Applied Ecology* **34**, 999–1013.

Noss, R. and Soulé, M. (1998) Rewilding and biodiversity: complementary goals for continental conservation. *Wild Earth* **3 (8)**, 18–28.

Olff, H., Vera, F.W.M., Bokdam, J., Bakker, E.S., Gleichman, J.M., de Maeyer, K., and Smit, R. (1999) Shifting mosaics in grazed woodlands driven by the alternation of plant facilitation and competition. *Plant Biology* **1**, 127–137.

Oliver, C.D. and Larsen, B.C. (1990) *Forest Stand Dynamics*. McGraw-Hill, New York.

Osborne, J.L., Clark, S.J., Morris, R.J., Williams, I.H., Riley, J.R., Smith, A.D., Reynolds, D.R., and Edwards, A.S. (1999) A landscape-scale study of bumble bee foraging range and constancy, using harmonic radar. *Journal of Applied Ecology* **36**, 519–533.

Ovenden, G. N., Swash, A.R.H., and Smallshire, D. (1998) Agri-environment schemes and their contribution to the conservation of biodiversity in England. *Journal of Applied Ecology* **35**, 955–960.

Owen, J. (1991) *The Ecology of a Garden: The First Fifteen Years*. Cambridge University Press, Cambridge.

Owen, M. (1975) Cutting and fertilising grassland for winter goose management. *Journal of Wildlife Management* **39**, 163–167.

Pakeman, R.J., Hulme, P.D., Torvell, L., and Fisher, J.M. (2003) Rehabilitation of degraded dry heather [*Calluna vulgaris* (L.) Hull] moorland by controlled sheep grazing. *Biological Conservation* **114**, 389–400.

Palmer, W.E., Wellendorf, S.D., Gillis, J.R., and Bromley P.T. (2005) Effects of field borders and nest-predator reduction on abundance of northern bobwhites. *Wildlife Society Bulletin* **33**, 1398–1405.

Parish, T., Lakhani, K.H., and Sparks, T.H. (1994) Modelling the relationship between bird population variables and hedgerow and other field margin attributes. I. Species richness of winter, summer and breeding birds. *Journal of Applied Ecology* **31**, 764–775.

Patten, M.A., Shochat, E., Reinking, D.L., Wolfe, D.H., and Sherrod, S.K. (2006) Habitat edge, land management, and rates of brood parasitism in tallgrass prairie. *Ecological Applications* **16**, 687–695.

Patterson, M.P. and Best, L.B. (1996) Bird abundance and nesting success in Iowa CRP fields: the importance of vegetation structure and composition. *American Midland Naturalist* **135**, 153–167.

Peach, W.J., Lovett, L.J., Wotton, S.R., and Jeffs, C. (2001) Countryside stewardship delivers cirl buntings (*Emberiza cirlus*) in Devon, UK. *Biological Conservation* **101**, 361–373.

Peach, W.J., Denny, M., Cotton, P.A., Hill, I.F., Gruar, D., Barritt, D., Impey, A., and Mallord, J. (2004) Habitat selection by song thrushes in stable and declining farmland populations. *Journal of Applied Ecology* **41**, 275–293.

Pearce-Higgins, J.W. and Yalden, D.W. (2004) Habitat selection, diet, arthropod availability and growth of a moorland wader: the ecology of European Golden Plover *Pluvialis apricaria* chicks. *Ibis* **146**, 335–346.

Pearce-Higgins, J.W. and Grant, M.C. (2006) Relationships between bird abundance and the composition and structure of moorland vegetation. *Bird Study* **53**, 112–125.

Peck, K.M. (1989) Tree species preferences shown by foraging birds in a forest plantation in northern England. *Biological Conservation* **48**, 41–57.

Percival, S.M. (1993) The effects of reseeding, fertilizer application and disturbance on the use of grasslands by barnacle geese, and the implications for refuge management. *Journal of Applied Ecology* **30**, 437–443.

Perkins, A.J., Whittingham, M.J., Bradbury, R.B., Wilson, J.D., Morris, A.J., and Barnett, P.R. (2000) Habitat characteristics affecting use of lowland agricultural grassland by birds in winter. *Biological Conservation* **95**, 279–294.

Perrow, M., and Davy, A.J. (eds) (2002a) *Handbook of Ecological Restoration. Volume 1: Principles of Restoration.* Cambridge University Press, Cambridge.

Perrow, M., and Davy, A.J. (eds) (2002b) *Handbook of Ecological Restoration. Volume 2: Restoration in Practice.* Cambridge University Press, Cambridge.

Peterken, G.F. (1981) *Woodland Conservation and Management*. Chapman and Hall, London.

Peterken, G.F. (1996) *Natural Woodland*. Cambridge University Press, Cambridge.

Petersen, P.M. (2002) Importance of site conditions and time since abandonment for coppice vegetation on Langeland, Denmark. *Nordic Journal of Botany* **22**, 463–481.

Piessens, K., Aerts, N., and Hermy, M. (2006) Long-term (1978–2003) effects of an extensive grazing regime on plant species composition of a heathland reserve. *Belgian Journal of Botany* **139**, 49–64.

Piha, M., Pakkala, T., and Tiainen, J. (2003) Habitat preferences of the skylark *Alauda arvensis* in southern Finland. *Ornis Fennica* **80**, 97–110.

Plumb, G.E. and Dodd, J.L. (1993) Foraging ecology of bison and cattle on a northern mixed prairie: implications for natural area management. *Ecological Applications* **3**, 631–43.

Potts, G.T. (1986) *The Partridge: Pesticides, Predation and Conservation*. London: Collins.

Poulin, B. and Lefebvre, G. (2002) Effect of winter cutting on the passerine breeding assemblage in French Mediterranean reedbeds. *Biodiversity & Conservation* **11**, 1567–1581.

Poulin, B., Lefebvre, G., and Mauchamp, A. (2002) Habitat requirements of passerines and reedbed management in Southern France. *Biological Conservation* **107**, 315–325.

Poulsen, B.O. (2002) Avian richness and abundance in temperate Danish forest: tree variables important to birds and their conservation. *Biodiversity and Conservation* **11**, 1551–1566.

Power, S.A., Ashmore, M.R., Cousins, D.A., Sheppard, L.J., (1998) Effects of nitrogen addition on the stress sensitivity of *Calluna vulgaris*. *New Phytologist* **138**, 663–673.

Pratt, R.M., Putman, R.J., Ekins, J.R., and Edwards, P.J. (1986) Use of habitat by free-ranging cattle and ponies in the New Forest, Southern England. *Journal of Applied Ecology* **23**, 539–557.

Provencher, L., Gobris, N.M., Brennan, L.A. Gordon, D.R., and Hardesty, J.L. (2002) Breeding bird response to midstorey hardwood reduction in Florida sandhill longleaf pine forests. *Journal of Wildlife Management* **66**, 641–661.

Purvis, W. (2001) Metal-rich habitats. In Fletcher, A. (ed.), *Lichen Habitat Management*. Proceedings of a workshop held at Bangor, 3–6 September 1997, pp. 17-1–17-10. British Lichen Society, Natural History Museum, London.

Putnam, R.J., Pratt, R.M., Ekins, J.R., and Edwards, P.J. (1987) Food and feeding behaviour of cattle and ponies in the New Forest, Hampshire. *Journal of Applied Ecology* **24**, 369–380.

Pyke, C.P. and Marty, J. (2004) Cattle grazing mediates climate change impacts on ephemeral wetlands. *Conservation Biology* **19**, 1619–1625.

Pywell, R. (2006) *A Literature Review and Gap Analysis of Grassland Restoration Research in the UK and Europe*. Report to the Department for Environment, Food and Rural Affairs (BD1458), NERC Centre for Ecology and Hydrology, Monks Wood.

Pywell, R.F., Bullock, J.M., Walker, K.J., Coulson, S.J., Gregory, S.J., and Stevenson, M.J. (2004) Facilitating grassland diversification using the hemiparasitic plant *Rhinanthus minor*. *Journal of Applied Ecology* **41**, 880–887.

Rands, M.R.W. (1985) Pesticide use on cereals and the survival of grey partridge chicks: a field experiment. *Journal of Applied Ecology* **22**, 49–54.

Ranius, T., Kindvall, O., Kruys, N., and Jonsson, B.G. (2003) Modelling dead wood in Norway spruce stands subject to different management regimes. *Forest Ecology and Management* **182**, 13–29.

Ranius, T., Ekvall, H., Jonsson, M., and Bostedt, G. (2005) Cost-efficiency of measures to increase the amount of coarse woody debris in managed Norway spruce forests. *Forest Ecology and Management* **206**, 119–133.

Ratcliffe, N., Schmitt, S., and Whiffin, M. (2005) Sink or swim? Viability of a black-tailed godwit population in relation to flooding. *Journal of Applied Ecology* **42**, 834–843.

Read, J.M., Birch, C.P.D., and Milne, J.A. (2002) HeathMod: a model of the impact of seasonal grazing by sheep on upland heaths dominated by *Calluna vulgaris* (heather). *Biological Conservation* **105**, 279–292.

Reddersen, J. (2001) SRC-willow (*Salix viminalis*) as a resource for flower-visiting insects. *Biomass and Energy* **20**, 171–179.

Redpath, S.M. and Thirgood, S.J. (1997) *Birds of Prey and Red Grouse*. The Stationery Office, London.

Redpath, S.M. and Thirgood, S.J. (1999) Numerical and functional responses of generalist predators: harriers and peregrines on grouse moors. *Journal of Animal Ecology* **68**, 879–892.

Redpath, S.M., Madders, M., Donnelly, E., Anderson, B., Thirgood, S., Martin, A., and McLeod, D. (1998) Nest site selection by Hen Harriers in Scotland. *Bird Study* **45**, 51–61.

Reinecke, K.J., Kaminski, R.M., Moorhead, D.J., Hodges, J.D., and Nassar, J.R. (1989) Mississippi alluvial valley. In Smith, L.M., Pederson, R.L., and Kaminski, R.M. (eds), *Habitat Management for Migrating and Wintering Waterfowl in North America*, pp. 203–247. Texas Tech University Press, Lubbock.

Ripple, W.J. and Beschta, R.L. (2003) Wolf re-introduction, predation risk, and cottonwood recovery in Yellowstone National Park. *Forest Ecology and Management* **184**, 299–313.

Robbins, C.S., Dawson, D.K., and Dowell, B.A. (1989) Habitat area requirements of breeding forest birds of the middle Atlantic states. *Wildlife Monographs* **103**, 1–34.

Roberts, P. (1982) Foods of the chough on Bardsey Island, Wales. *Bird Study* **29**, 155–161.

Robertson, P.A. (1993) *The Management of Artificial Coastal Lagoons in Relation to Invertebrates and Avocets Recurvirostra avosetta (L.)*. PhD thesis, University of East Anglia, Norwich.

Roem, W.J., Klees, H., and Berendse, F. (2002) Effects of nutrient addition and acidification on plant species diversity and seed germination in heathland. *Journal of Applied Ecology* **39**, 937–948.

Roman, C.T., Niering, W.A., and Warren, R.S. (1984) Salt marsh vegetation change in response to tidal restriction. *Environmental Management* **8**, 141–150.

Rook, A.J., Dumont, B., Isselstein, J., Osoro, K., WallisDeVries, M.F., Parente, G., and Mills, J. (2004) Matching type of livestock to desired biodiversity outcomes—a review. *Biological Conservation* **119**, 137–150.

Rosenstock, S.S. (1996) Shrub-grassland small mammal and vegetation responses to rest from grazing. *Journal of Range Management* **49**, 199–203.

Ross, S., Adamson, H., and Moon, A. (2003) Evaluating management techniques for controlling *Molinia caerulea* and enhancing *Calluna vulgaris* on upland wet heathland in Northern England, UK. *Agriculture, Ecosystems and Environment* **97**, 39–49.

Roy, D.B., Hill, M.O., and Rothery, P. 1999. Effects of urban land cover on the local species pool in Britain. *Ecography* **22**, 507–515.

RSPB, EN & ITE (1997) *The Wet Grassland Guide: Managing floodplain and coastal wet grasslands for wildlife*. RSPB, Sandy.

Ruxton, G.D., Thomas, S., and Wright, J.W. (2002) Bells reduce predation of wildlife by domestic cats (*Felis catus*). *Journal of Zoology* **256**, 81–83.

Ryan, M.R., Burger, L.W., and Kurzejeski, E.W. (1998) The impact of CRP on avian wildlife: a review. *Journal of Production Agriculture* **11**, 61–66.

Sage, R., Cunningham, M., and Boatman, N. (2006) Birds in willow short-rotation coppice compared to other arable crops in central England and a review of bird census data from energy crops in the UK. *Ibis* **148** (suppl. 1), 184–197.

Sage, R.B. and Robertson, P.A. (1996) Factors affecting songbird communities using new short rotation coppice habitats in spring. *Bird Study* **43**, 201–213.

Sanderson, F.J., Kloch, A., Sachanowicz, K., and Donald, P.F. (2007) Predicting the effects of agricultural change on farmland bird populations in Poland. (in press).

Sansen, U. and Koedam, N. (1996) Use of sod cutting for restoration of wet heathlands: revegetation and establishment of typical species in relation to soil conditions. *Journal of Vegetation Science* **7**, 483–486.

Schmidt, M.H., Lefebvre, G., Poulin, B., and Tscharntke, T. (2005a) Reed cutting affects arthropod communities, potentially reducing food for passerine birds. *Biological Conservation* **121**, 157–166.

Schmidt, N.M, Olsen, H., Bildsoe, M., Sluydts, V., and Leirs, H. (2005b) Effects of grazing intensity on small mammal population ecology in wet meadows. *Basic and Applied Ecology* **6**, 57–66.

Schurbon, J.M and Fauth, J.E. (2003) Effects of prescribed buring on amphibian diversity in a southeastern U.S. national forest. *Conservation Biology* **17**, 1338–1349.

Sedlakova, I. and Chytry, M. (1999) Regeneration patterns of Central European dry heathland: effects of burning, sod-cutting/turf-stripping and cutting. *Plant Ecology* **143**, 77–87.

Self, M. (2005) A review of management for fish and bitterns, *Botaurus stellaris*, in wetland reserves. *Fisheries Management and Ecology* **12**, 387–394.

Sinicrope, T.L., Hine, P.G., Warren, R.S., and Niering, W.A. (1990) Restoration of an impounded salt marsh in New England. *Estuaries* **13**, 25–30.

Sippola, A.L., Siitonen, J., and Kallio, R. (1998) Amount and quality of coarse woody debris in natural and managed coniferous forests near the timberline in Finnish Lapland. *Scandinavian Journal of Forest Research* **13**, 204–214.

Sitters, H.P., Fuller, R.J., Hoblyn, R.A., Wright, M.T., Cowie, N., and Bowden, C.G.R. (1996) The Woodlark *Lullu la arborea* in Britain: population trends, distribution and habitat occupancy. *Bird Study* **43**, 172–187.

Skidmore, P. (1991) *Insects of the British Cow-Dung Community*. Field Studies Council Occasional Publication no. 21. Field Studies Council, Shrewsbury.

Slavin, P. and Shisler, J.K. (1983) Avian utilization of a tidally restored salt hay farm. *Biological Conservation* **26**, 271–285.

Smart, J., Gill, J.A., Sutherland, W.J., and Watkinson, A.R. (2006) Grassland-breeding waders: identifying key habitat requirements for management. *Journal of Applied Ecology* **43**, 454–463.

Smith, A., Redpath, S., and Campbell, S. (2000a) *The Influence of Moorland Management on Grouse and their Populations*. Her Majesty's Stationery Office, Norwich.

Smith, L.M. and Kadlec, J.A. (1983) Seed banks and their role during drawdown of a North American marsh. *Journal of Applied Ecology* **20**, 673–684.

Smith, L.M., Haukos, D.A., and Prather, R.M. (2004a) Avian response to vegetative pattern in playa wetlands during winter. *Wildlife Society Bulletin* **32**, 474–480.

Smith, M.D., Barbour, P.J., Burger, L.W., and Dinsmore, S.J. (2005a) Density and diversity of overwintering birds in managed field borders in Mississippi. *Wilson Bulletin* **117**, 258–269.

Smith, R.K., Jennings, N.V., Robinson, A., and Harris, S. (2004b) Conservation of European hares *Lepus europaeus* in Britain: is increasing habitat heterogeneity in farmland the answer? *Journal of Applied Ecology* **41**, 1092–1102.

Smith, R.K., Jennings, N.V., and Harris, S. (2005b) A quantitative analysis of the abundance and demography of European hares *Lepus europaeus* in relation to habitat type, intensity of agriculture and climate. *Mammal Review* **35**, 1–24.

Smith, R.M., Thompson, K., Hodgson, J.G., Warren, P.H., and Gaston, K.J. (2006) Urban domestic gardens (IX): composition and richness of the vascular plant flora, and implications for native biodiversity. *Biological Conservation* **129**, 312–322.

Smith, R.S. and Rushton, S.P. (1994) The effects of grazing management on the vegetation of mesotrophic (meadow) grassland in Northern England. *Journal of Applied Ecology* **31**, 13–24.

Smith, R.S., Shiel, R.S., Millward, D., and Corkhill, P. (2000b) The interactive effects of management on the productivity and plant community structure of an upland meadow: an 8-year field trial. *Journal of Applied Ecology* **37**, 1029–1043.

Smith, R.S., Shiel, R.S., Bardgett, R.D., Millward, D., Corkhill, P., Rolph, G., Hobbs, P.J., and Peacock, S. (2003) Soil microbial community, fertility, vegetation and diversity as targets in the restoration management of a meadow grassland. *Journal of Applied Ecology* **40**, 51–64.

Sotherton, N.W. (1995) Beetle banks: helping nature to control pests. *Pest Outlook* **6**, 13–17.

Spellerberg, I.F. (2002) *Ecological Effects of Roads*. Science Publishers, Enfield, New Hampshire.

Stammel, B., Kiehl, K., and Pfadenhauer, J. (2003) Alternative management on fens: response of vegetation to grazing and mowing. *Applied Vegetation Science* **6**, 245–254.

Stoate, C., Henderson, I.G., and Parish, D.M.B. (2004) Development of an agri-environment scheme option: seed-bearing crops for farmland birds. *Ibis* **146** (suppl. 2), 203–209.

Stoneham, G., Chaudhri, V., Ha, A., and Strappazzon, L. (2003) Auctions for conservation contracts: an empirical examination of Victoria's BushTender trial. *Australian Journal of Agricultural and Resource Economics* **47**, 477–500.

Suárez, F., Naveso, M.A., and De Juana, E. (1997) Farming in the drylands of Spain: birds of the pseudosteppe. In Pain, D.J. and Pienkowski, M.W. (eds), *Farming and Birds in Europe. The Common Agricultural Policy and its Implications for Bird Conservation*, pp. 297–330. Academic Press, London.

Summers, R.W. (2004) Use of pine snags by birds in different stand types of Scots pine. *Bird Study* **51**, 212–221.

Summers, R.W., Proctor, R., Thorton, M., and Avey, G. (2004) Habitat selection and diet of the capercaillie *Tetrao urogallus* in Abernethy Forest, Strathspey, Scotland. *Bird Study* **51**, 58–68.

Sutherland, W.J. (2000) *The Conservation Handbook: Research, Management and Policy*. Blackwell Science, Malden, MA.

Sutherland, W.J. (ed.) (2006a) *Ecological Census Techniques: a Handbook*, 2nd edn. Cambridge University Press, Cambridge.

Sutherland, W.J. (2006b) Planning a research programme. In Sutherland, W.J. (ed.), *Ecological Census Techniques: a Handbook*, 2nd edn, pp. 1–10. Cambridge University Press, Cambridge.

Svensson, B., Lagerlof, J., and Svensson, B.G. (2000) Habitat preferences of nest-seeking bumble bees (Hymenoptera: Apidae) in an agricultural landscape. *Agriculture, Ecosystems and Environment* **77**, 247–255.

Swamy, V., Fell, P.E., Body, M., Keaney, M.B., Nyaku, M.K., McIlvain, E.C., and Keen, A.L. (2002) Macroinvertebrate and fish populations in a restored impounded salt marsh 21 years after the reestablishment of tidal flooding. *Environmental Management* **29**, 516–530.

Swengel, S.R. and Swengel, A.B. (2001) Relative effects of litter and management on grassland bird abundance in Missouri, USA. *Bird Conservation International* **11**, 113–128.

Symes, N.C. and Day, J. (2003) *A Practical Guide to the Restoration and Management of Lowland Heathland*. RSPB, Sandy.

Syphard, A.D., Franklin, J., and Keeley, J.E. (2006) Simulating the effects of frequent fire on southern California coastal shrublands. *Ecological Applications* **16**, 1744–1756.

Taft, O.W., Colwell, M.A., Isola, C.R., and Safran, R.J. (2002) Waterbird responses to experimental drawdown: implications for the multispecies management of wetland mosaics. *Journal of Applied Ecology* **39**, 987–1001.

Tainton, N. (ed.) (1999) *Veld Management in South Africa*. University of Natal Press, Pietermaritzburg.

Tattershall, F.H., Avundo, A.E., Manley, W.J., Hart, B.J., and MacDonald, D.W. (2000) Managing set-aside for field voles (*Microtus agrestis*). *Biological Conservation* **96**, 123–128.

Taylor, I.P. (2004) Foraging ecology of the black-fronted plover on saline lagoons in Australia: the importance of receding water levels. *Waterbirds* **27**, 270–276.

Ter Heerdt, G.N.J. and Drost, H. (1994) Potential for the development of marsh vegetation from the seed bank after a drawdown. *Biological Conservation* **67**, 1–11.

Tharme, A.P., Green, R.E., Baines, D., Bainbridge, I.P., and O'Brien, M. (2001) The effect of management for red grouse shooting on the population density of breeding birds on heather-dominated moorland. *Journal of Applied Ecology* **38**, 439–457.

Thirgood, S., Redpath, S., Newton, I., and Hudson, P. (1999) Raptors and red grouse: conservation conflicts and management solutions. *Conservation Biology* **14**, 95–104.

Thomas, C.D. and Jones, T.M. (1993) Partial recovery of a skipper butterfly (*Hesperia comma*) from population refuges: lessons for conservation in a fragmented landscape. *Journal of Animal Ecology* **62**, 472–481.

Thomas, C.D., Bodsworth, E.J., Wilson, R.J., Simmons, A.D., Davies, Z.G., Musche, M., and Conradt, L. (2001) Ecological and evolutionary processes at expanding range margins. *Nature* **511**, 577–581.

Thomas, J.A. (1983) The ecology and conservation of *Lysandra bellargus* (Lepidoptera: Lycaenidae) in Britain. *Journal of Applied Ecology* **20**, 59–83.

Thomas, J.A. (1994) Why small cold-blooded insects pose different conservation problems to birds in modern landscapes. *Ibis* **137**, S112–S119.

Thompson, D.B.A., MacDonald, A.J., and Hudson, P.J. (1995) Upland heaths and moors. In Sutherland, W.J. and Hill, D.A. (eds), *Managing Habitats for Conservation*, pp. 292–326. Cambridge University Press, Cambridge..

Thompson, K., Hodgson, J.G., Smith, R.M., Warren, P.H., and Gaston, K.J. (2004) Urban domestic gardens (III): composition and diversity of lawn floras. *Journal of Vegetation Science* **15**, 373–378.

Trager, M.D., Wilson, G.W.T., and Hartnett, D.C. (2004) Concurrent effects of fire regime, grazing and bison wallowing on tallgrass prairie vegetation. *American Midland Naturalist* **152**, 237–247.

Tucker, G. (1992) Effects of agricultural practices on field use by invertebrate-feeding birds in winter. *Journal of Applied Ecology* **29**, 779–790.

Tucker, G.M. and Heath, M.F. (1994) *Birds in Europe: Their Conservation Status*. Birdlife Conservation Series no. 3. Birdlife International, Cambridge.

Tucker, J.W., Robinson, W.D., and Grand, J.B. (2004) Influence of fire on Bachman's sparrow, an endemic North American songbird. *Journal of Wildlife Management* **68**, 1114–1123.

Twisk, W., Noordervliet, M.A.W., and ter Keurs, W.J. (2003) The nature value of the ditch vegetation in peat areas in relation to farm management. *Aquatic Ecology* **37**, 191–209.

Tyler, G.A., Green, R.E., and Casey, C. (1998) Survival and behaviour of Corncrake *Crex crex* chicks during the mowing of agricultural grassland. *Bird Study* **45**, 35–50.

Tyson, K.C., Garwood, E.A., Armstrong, A.C., and Scholefield, D. (1992) Effects of field drainage on the growth of herbage and the liveweight gain of grazing beef cattle. *Grass and Forage Science* **47**, 290–301.

Usher, M.B. (1992) Management and diversity of arthropods in *Calluna* heathland. *Biodiversity and Conservation* **1**, 63–79.

van den Berg, L.J.L., Bullock, J.M., Clarke, R.T., Langston, R.H.W., and Rose, R.J. (2001) Territory selection by the Dartford warbler (*Sylvia undata*) in Dorset, England: the role of vegetation type, habitat fragmentation and population size. *Biological Conservation* **101**, 217–228.

Van der Hut, R.M.G. (1986) Habitat choice and temporal differentiation in reed passerines of a Dutch marsh. *Ardea* **74**, 159–176.

Van der Putten, W.H. (1997) Die-back of *Phragmites australis* in European wetlands: an overview of the European Research Programme on Reed Die-back and Progression (1993–94). *Aquatic Botany* **59**, 263–275.

Vandvik, V., Heegaard, E., Maren, I.E., and Aarrestad, P.A. (2005) Managing heterogeneity: the importance of grazing and environmental variation on post-fire succession in heathlands. *Journal of Applied Ecology* **42**, 139–149.

Van Dyke, F. and Darragh, J.A. (2006) Short- and long-term changes in elk use and forage production in sagebrush communities following prescribed burning. *Biodiversity and Conservation* **15**, 4375–4398.

Van't Hul, J.T., Lutz, R.S., and Mathews, N.E. (1997) Impact of prescribed burning on vegetation and bird abundance at Matagorda Island, Texas. *Journal of Range Management* **50**, 346–350.

van Wilgen, B.W., Richardson, D.M., and Seydack, A.H.W. (1994) Managing fynbos for biodiversity—constraints and options in a fire-prone environment. *South African Journal of Science* **90**, 322–329.

Vera, F.W.M. (2000) *Grazing Ecology and Forest History*. CABI International, Wallingford.

Verdú, J.R., Crespo, M.B., and Galante, E. (2000) Conservation strategy of a nature reserve in Mediterranean ecosystems: the effects of protection from grazing on biodiversity. *Biodiversity and Conservation* **9**, 1707–1721.

Vereecken, H., Baetens, J., Viaene, P., Mostaert, F., and Meire, P. (2006) Ecological management of aquatic plants: effects in lowland streams. *Hydrobiologia* **570**, 205–210.

Vickery, J.A. and Gill, J.A. (1999) Managing grassland for wild geese in Britain: a review. *Biological Conservation* **89**, 93–106.

Vickery, J.A., Sutherland, W.J., and Lane, S.J. (1994) The management of grass pastures for brent geese. *Journal of Applied Ecology* **31**, 282–290.

Vickery, J.A., Tallowin, J.R., Feber, R.E., Asterak, E.J., Atkinson, P.W., Fuller, R.J., and Brown, V.K. (2001) The management of lowland neutral grasslands in Britain: effects of agricultural practices on birds and their food resources. *Journal of Applied Ecology* **38**, 647–664.

Vickery, P.D., Hunter, M.L., and Melvin, S.M. (1994) Effects of habitat area on the distribution of grassland birds in Maine. *Conservation Biology* **8**, 1087–1097.

Vulink, J.T. and Van Eerden, M.R. (1998) Hydrological conditions and herbivory as key operators for ecosystem development in Dutch artificial wetlands. In Wallis DeVries, M.F., Bakker, J.P., and Van Wieren, S.E. (eds), *Grazing and Conservation Management*, pp. 217–252. Kluwer Academic Publishers, Dordrecht..

Vulink, J.T., Drost, H.J., and Jans, L. (2000) The influence of different grazing regimes on *Phragmites-* and shrub vegetation in the well-drained zone of a eutrophic wetland. *Applied Vegetation Science* **3**, 73–80.

Vulliamy, B., Potts, S.G., and Wilmer, P.G. (2006) The effects of cattle grazing on plant-pollinator communities in a fragmented Mediterranean landscape. *Oikos* **114**, 529–543.

Wall, R. and Strong, L. (1987) Environmental consequences of treating cattle with the antiparasitic drug Ivermectin. *Nature* **327**, 418–421.

Wallace, K.J., Callaway, J.C., and Zedler, J.B. (2005) Evolution of tidal creek networks in a high sedimentation environment: a 5-year experiment at Tijuana Estuary, California. *Estuaries* **28**, 795–811.

WallisDeVries, M.F. (1995) Large herbivores and the design of large-scale nature reserves in Western Europe. *Conservation Biology* **9**, 25–33.

WallisDeVries, M.F. (2002) A quantitative conservation approach for the endangered butterfly *Maculinea alcon*. *Conservation Biology* **18**, 489–499.

Wallis DeVries, M. and Raemakers, I. (2001) Does extensive grazing benefit butterflies in coastal dunes? *Restoration Ecology* **9**, 179–188.

Waring, P. (2001) Grazing and cutting as conservation management tools—the need for a cautious approach, with some examples of rare moths which have been adversely affected. *The Entomologist's Record and Journal of Variation* **113**, 193–200.

Warnock, N., Page, G.W., Ruhlen, T.D., Nur, N., Takekawa, J.Y., and Hanson, J.T. (2002) Management and conservation of San Francisco Bay salt ponds: effects of pond salinity, area, tide and season on Pacific flyway waterbirds. *Waterbirds* 25 (special publication 2), 79–92.

Warren, M.S. (1985) The influence of shade on butterfly numbers in woodland rides, with special reference to the wood white, *Leptidea sinapis*. *Biological Conservation* **33**, 147–64.

Warren, M.S. (1987) The ecology and conservation of the heath fritillary butterfly, *Mellicta athalia*. III. Population dynamics and the effect of habitat management. *Journal of Applied Ecology* **24**, 499–513.

Warren, M.S. and Thomas, J.A. (1992) Butterfly response to coppicing. In Buckley, G.P. (ed.), *Ecology and Management of Coppice Woodlands*, pp. 249–270. Chapman and Hall, London.

Warren, M.S., Thomas, C.D., and Thomas, J.A. (1984) The status of the heath fritillary butterfly, *Mellicta athalia* Rott., in Britain. *Biological Conservation* **29**, 287–305.

Warren, R.S., Fell, P.E., Rozsa, R., Brawley, A.H., Orsted, A.C., Olson, E.T., Swamy, V., and Niering, W.A. (2002) Salt marsh restoration in Connecticut: 20 years of science and management. *Restoration Ecology* **10**, 497–513.

Watson, A.M. and Ormerod, S.J. (2004) The distribution of three uncommon freshwater gastropods in the drainage ditches of British grazing marshes. *Biological Conservation* **118**, 455–466.

Watson, A.M. and Ormerod, S.J. (2005) The distribution and conservation of threatened Sphaeriidae on British grazing marshland. *Biodiversity and Conservation* **14**, 2207–2220.

Webb, N.R. (1998) The traditional management of European heathlands. *Journal of Applied Ecology* **35**, 987–990.

Wheeler, B.D. and Giller, K.E. (1982) Species richness of herbaceous fen vegetation in broadland, Norfolk in relation to the quantity of above-ground plant material. *Journal of Ecology* **70**, 179–200.

Whisenant, S.G. (1999) *Repairing Damaged Wildlands: a Process-Oriented, Landscape-Scale Approach*. Cambridge University Press, Cambridge.

Whittingham, M.J. and Evans, K.L. (2004) The effects of habitat structure of predation risk of birds in agricultural landscapes. *Ibis* **146** (suppl. 2), 210–220.

Whittingham, M.J., Percival, S.M., and Brown, A.F. (2000) Time budgets and foraging of breeding golden plover *Pluvialis apricaria. Journal of Applied Ecology* **37**, 632–646.

Williams, D.D. (1997) Temporary ponds and their invertebrate assemblages. *Aquatic Conservation: Marine and Freshwater Ecosystems* **7**, 105–117.

Williams, G.D and Zedler, J.B. (1999) Fish assemblage composition in constructed and natural tidal marshes of San Diego Bay: relative influence of channel morphology and restoration history. *Estuaries* **22**, 702–716.

Wilson, A.M., Vickery, J.A., and Browne, S.J. (2001) Numbers and distribution of Northern Lapwings *Vanellus vanellus* breeding in England and Wales in 1988. *Bird Study* **48**, 2–17.

Wilson, C.W., Masters, R.E., and Bukenhofer, G.A. (1995) Breeding bird response to pine-grassland community restoration for red-cockaded woodpeckers. *Journal of Wildlife Management* **59**, 56–67.

Wilson, J. (1978) The breeding bird community of willow scrub at Leighton Moss, Lancashire. *Bird Study* **25**, 239–244.

Wilson, J.D., Taylor, R., and Muirhead, L.B. (1996) Field use by farmland birds in winter: an analysis of field type preferences using resampling methods. *Bird Study* **43**, 320–332.

Wilson, J.D., Evans, J., Browne, S.J., and King, J.R. (1997) Territory distribution and breeding success of skylarks *Alauda arvensis* on organic and intensive farmland in southern England. *Journal of Applied Ecology* **34**, 1462–1478.

Wilson, J.D., Morris, A.J., Arroya, B.E., Clark, S.C., and Bradbury, R.B. (1999) A review of the abundance and diversity of plant foods of seed-eating birds in northern Europe in relation to agricultural change. *Agriculture, Ecosystems and Environment* **75**, 13–50.

Wilson, P.J. and Aebischer, N.J. (1995) The distribution of dicotyledonous arable weeds in relation to distance from the field edge. *Journal of Applied Ecology* **32**, 295–310.

Woods, M., McDonald, R.A., and Harris, S. (2003) Predation of wildlife by domestic cats *Felis catus* in Great Britain. *Mammal Review* **33**, 174–188.

van Andel J. and Aronson, J. (2006) *Restoration Ecology: the New Frontier*. Blackwell Publishing, Malden, MA.

Yahner, R.H. (1995) Forest-dividing corridors and neotropical migrant birds. *Conservation Biology* **9**, 476–477.

# Index

Abernethy (Scotland) *212*
*Acacia* spp. 135
acacias 135
accentor, hedge 351
acidification 145–6, 148
*Acinonyx jubatus* 20
*Acrocephalus paludicola* 24
*Acrocephalus palustris* 205
*Acrocephalus schoenobaenus* 205
*Aedes sollicitans* 304, 317
*Agelaius phoeniceus* 352
agri-environment schemes 44, 53–5, 207, 233, 333
*Agrostis stolonifera* 289, 296
*Aimophila aestivalis* 215
*Aix sponsa* 283
*Alauda arvensis* 205, 336–7, 344, 350
alder *176*, 282
Aldringham Walks (England) *163*
algae 234, *234*, 320, 363
　blue-green 234
*Allium ursinum* 199
*Alnus glutinosa 176*, 282
*Ambystoma californiense* 243
*Ammodramas maritimus* 309
*Ammodramas nelsoni* 309
*Ammodramus henslowii* 121
*Ammodramus savannarum* 121, 350
*Ammophila arenaria* 329
amphibians 60, 67, 144, 174, 214, 242, 260, 267, 283, 288, 328, 330, 362–3, 368
amphipods 322
*Anas penelope* 308
*Anas platyrhynchos* 283
anenome, starlet sea 319
*Anisus vorticulus* 245
*Anser anser* 255–6, 305
*Anser brachyrhynchus* 338
*Anthus campestris* 143
*Anthus pratensis* 170–1
*Anthus trivialis* 211, 224
anti-parasitic drugs 75–7, 209
antlion 163
Apaj (Hungary) *337*
Apeldoorn (Netherlands) 66
*Apis mellifera* 123
*Apodemus flavicollis* 200, 351
*Apodemus sylvaticus* 200

*Aquila adalberti* 4
*Aquila heliaca* 14
arable 331–54
arable weeds 332–3, 338, 340–*3*, 345–6
Arguébanes (Spain) *111*
*Argynnis pandora* 111
Arnhem (Netherlands) 66
*Arrhenatherum elatius* 353
*Artemia* spp. 322
Arun Valley (England) *245*
*Arvicola terrestris* 247–8, 262
*Asio flammeus* 165, 170
ass, wild 74
Asteraceae 337
*Athene cunicularia* 103
*Atriplex portulacoides* 305
attributes (as measures of species of habitat condition) 36–40
auctioning conservation contracts 55
audits 52
aurochs 72
avens, mountain *89*
avermectins 75–7
avocets 319

backyards 355–68
badger, European 67, 299
*Baeolophus bicolor* 214
barley 337
　straw 234
*Bartramia longicauda* 64
bats 97, 180, 208–9, 355, 366
beaches 326
beak-sedge, brown 157
beaver, American 3
beavers 18
bed-lowering 257–62
bee wolf, European *123*
bee, honey 123
bees 62, 205, 354
　solitary 61, 92, 123, 144, 162, 302, 365, 366, 367, 369–70
beetle banks 352–3
beetle, heather 145
beetles 61–2, 91, 244, 249, 302, 319, 326, 341, 352, 363
　dead-wood 62, 180, 225
　dung 75–6, 97, 328

beetles (contd)
 ground 61, 248, 319, 352
 jewel 194
 longhorn 194
 rove 319, 352
Beloe Fish Ponds (Belarus) 236
Ben Lawers (Scotland) 108
bent, creeping 289, 296
*Betula pendula* 145, *189*, 362
*Betula pubescens* 145, *176*
*Betula* spp. 145–6, 150, 154–5, 160
Białowieża National Park (Poland) *184*
bilberry 222
biobridges – *see* green bridges
bioenergy 186, 205, 331
biofuel – *see* bioenergy
birch 145–6, 150, 154–5, 160
 downy 145, *176*
 silver 145, *189*, 362
*Bison bison* 102–3
bison, American 102–3
bittern, great 236, 254, 261
blackbird, red-winged 352
Blackwater National Wildlife Refuge (USA) *239, 310*
bladderwort, common 37, 40
blue-border management 276, *277–8*
blue butterfly
 Alcon 158
 silver-studded 151
boar, wild 3, 66, 103, 209
bobwhite, northern 214, 333, 352–3
bogs 142, 164, 166, 168–9, 229, 279–81
 raised – *see* mires, raised
*Boloria euphrosyne* 171
Borrowdale Woods (England) *210*
Bortenicha (Belarus) *24*
*Bos taurus primigensis* 72
*Botaurus stellaris* 236, 254, 261
box-junctions *195*
bracken 142, 145–6, 149, 160, 164, 166, 169, 171
Bradfield Woods (England) *199*
Bradshaw bucket, use of for cleaning out ditches 244
bramble 189
branchiopods 243
*Branta canadensis* 346–7
Brassicaceae 337
brassicas 337
Breckland (England) *61, 224, 343*
breeds
 differences between 74–5
 rare 75
broom 141

brownfield sites 355, *357*, 369–71
bryophytes 57, 67, 150, 168, 210
buckwheat 347
*Buddleja davidii* 367
buffalo, water 269, 273
buffer strips 233
bugs – *see* true bugs
buildings 355, *357*, 366–7, 370
bulldozing 137, 160, 162, 281
bulrush 316
 alkali 77, 273–5, 296, 325
 lesser 316
 spp. 229, 253, 255, 268, 273
bumblebees 63, 120, 351–2, 366–7
bunting
 cirl 338, 352
 corn 350
 ortolan 349
 reed 205, 336, 353
buntings 337, 346, 352
Buprestidae 194
*Burhinus oedicnemus* 102, 122, 344–5
burning 63, 85, 94–7, 117–21, 132–43, 146, 148–52, 158, 164–9, 192–3, 211–15, 219, 225, 235, 240, 255, 262–3, 266–9, 279, 281, 283–4, 309–10, 325, 353
 hazard-reduction 133–4, 139, 212–13
bustard
 great 332
 little 332
*Buteo regalis* 103
butterflies 17, 51, 62, 194, 207–8, 343, 354, 362
 white 343
butterfly-bush 367

*Calamagrostis epigejos* 157
*Calendula arvensis* 187
*Calluna vulgaris* 67, 142–3, *146–7*, 149–52, 154–7, 164–70, 279
Campbell Island (New Zealand) 80
*Canis lupus* 19
capercaillie, western 222
*Capreolus capreolus* 72
*Caprimulgus europaeus* 143, 158, 160, 224
cardinal butterfly *111*
*Carduelis cannabina* 336
*Carduelis carduelis* 361
*Carduelis flavirostris* 165
*Carex* spp. 101, 148, 275, 277, 289
*Carlina acanthifolia* 95
carrion 4, 92, 97, 155
Caryophyllaceae 337
*Castor canadensis* 3
Castros Las Cerras (Spain) *141*

cat, domestic  79–82, 368
cattail
  broadleaf  316
  narrowleaf  316
  spp.  229, 253, 255, 268, 273
cattle
  grazing by  71, 97–100, 103–4, 106–7, 114, 140, 153, 156, 169, 247–9, 269, 273, 275–6, 294–6, 306, 328
  Galloway  75, 153
  hardy breeds of  74–5
  Heck  *72*, 75
  Highland  *75*, 273
  trampling of nests by  298
Cerambycidae  194
*Cervus elaphus 19*, 155, 164, 166–7, 169–70, 176
chaparral  131, *134*, 135, 138, *139*
*Charadrius dubius*  370
*Charadrius montanus*  103
*Charadrius vociferus*  299
Charophyta  233, 246, 320
cheetah  20
*Chen caerulescens*  305, *346–7*
Chenopodiaceae  337
Cherry Hill (England)  *343*
chew sticks  81
chick loss, minimising it during agricultural operations  113–14, 116–18, 299, 344–5
children, trampling by  366
Chironomidae  241, 322–3, 325
chough, red-billed  97, 151
*Ciconia ciconia*  335, 367
*Circus cyaneus*  165, 169–71
*Circus* spp.  344–5
*Cirsium* spp.  101, 112, 117, 123, 177
*Cistothorus palustris*  309
*Cistothorus platensis*  309
*Cladium mariscus*  15, *266*, 268
Cladocera  363
clear-felling  192, 216–*17*, 219, 221, 223
*Clethrionomys glareolus*  200
cliffs  88, 92, 102, 108, 301–*2*
climate change  82–5, 243
clover, white  347
clubmoss, marsh  157
club-rush, sea  77, 273–5, 296, 325
coarse woody debris (CWD)  180, 214, 224–5
coastal squeeze  *312*
cock's-foot  353
cockle shells, on bird nesting islands  234
cocklebur  240
*Colchicum autumnale*  285
*Colinus virginianus*  214, 333, 352–3
compaction, soil  121, 123, 162, 342

composites  337
connectivity, habitat  64, 66–7
conservation
  corridors – *see* habitat corridors
  headlands  207, 342–*3*
  programs  53–4, 233
  tillage – *see* non-inversion tillage
Conservation
  Reserve Enhancement Program (CREP)  53–*4*
  Reserve Program (CRP)  53, 333–4
continuous-cover forestry (CCF)  216
Copepoda  363
copepods  363
coppice, short rotation (SRC)  186, 192, 205–7
coppicing  185, 192, 197–205, 281, 283
cordgrass  303–4, 316
  big  325
  common  302
  English  302
  saltmeadow  317
  smooth  *310*, 317
Corixidae  322
corncrake  118, 333
*Coronella austriaca*  143, *144*
*Corophium volutator*  322
corvids  68, 116, 345
*Corylus avellana*  89
cottongrass, hare's-tail  164, 169–71
couch, sea  305, *307–8*
cowbird, brown-headed  68, 91, 191, 350
cowslip  *366*
cow-wheat, common  6
coypu  303, 310
crakes  240
crane, common  4
crane's-bill
  dusky  361
  meadow  361
*Crataegus monogyna*  209
crawfish  349
Creagh Meagaidh (Scotland)  176
*Crex crex*  118, 333
*Crocidura* spp.  200
crows  216, 287, 297, 299
crustaceans  242, 314
cultural habitats  4–5, 13–16, 24–5, 73, 192
curlew, Eurasian  167, 295
cutting  63, 77–9, 94–7, 113–17, 142, 148–52, 160, 166–7, 194, 196, 235, 248, 254, 262–70, 276, 286, 288, 296, 299, 304, 325, 340, 354, 358, 365–6
cyanobacteria, control of  234
cycling, soil disturbance created by  121, 162

*Cygnus cygnus* 18
*Cynomys* spp. 18, 91, 102, *103*
cypress, bald 284

*Dactylis glomerata* 353
dams 231, 279–*80*
damselflies 145, 258
De
   Hoop Nature Reserve (Republic of South Africa) *133*
   Pine (Netherlands) *294*
   Weeribben National Park (Netherlands) *259*
dead wood 174, 180–1, 183–4, 192–4, 196–8, 202–4, 208–9, 211, 214–15, 219, 221, 223–7, 362
deer 3, 60, 66, 103, 155, 157, 186, 190–1, 196, 198, 201, 204, 209–11
   red *19*, 155, 164, 166–7, 169–70, 176
   roe 72
   white-tailed 210
dehesas – *see* wooded dehesas
Delaware Bay (USA) 316, *346*
*Delphinium* spp. 356
delphiniums 356
*Dendroica kirtlandii* 215
*Deschampsia flexuosa* 145, 147, 152, 157
dickcissel 350, 352
*Digitalis purpurea* 361
dikes – *see* ditches
*Dipsacus fullonum* 361
direct drilling 340, 347
disking 128, 239–40, 255, 268, 325, 347, 353
disturbance
   human 68–9
   mechanical soil 121–4, 160–4, 325, 328, 330
ditches 230, 242–50, 260–2, 271, 279, 284–5, 287–8, 291–3, 304, 316, 353–4
docks 114
donkeys, grazing and browsing by 74, 101, 328
dormouse, hazel 202, 351
dragonflies 145, 258
drain blocking – *see* grip-blocking
drawdowns 238–41, 249, 253–7, 276–8, 323–5, 363–4, *277*–*8*
dredging, of rivers 249, 252
*Dryas octopetala* 89
duck, wood *283*
ducks
   dabbling 241, 289, 349
   diving 319
dunes 102, 131, 142, 143, 327–30
   creation of inland *161*–2
dung 62, 75–7, 90, 92, 94, 96–8, 100–2, 114, 142, 152–5, 177, 267

Dungeness (England) *235*
dykes – *see* ditches

eagle
   imperial 14
   Spanish imperial 4
earthworms 125, 167, 289–90, 294, 296, 340, 349
eat-outs, by geese 305
ecoducts – *see* green bridges
ecological traps 341, 367
ecoplugs 79
ECOtillage – *see* non-inversion tillage
edge effects 67–8
Edwin B. Forsythe National Wildlife Refuge (USA) *324*
egrets 238, 241, 260, 281, 320, 325
*Eichhornia crassipes* 78
eider, common 67
*Elaphus maximus* 20
*Eleocharis parvula* 323, *324*
elephant
   African 20
   Asian 20
elk *19*, 176
Elmley Marshes (England) *295*
*Elytrigia atherica* 305, 307–8
Embalse del Tozo (Spain) *4*
*Emberiza cirlus* 338, 352
*Emberiza citrinella* 115, 349, 351–2
*Emberiza hortulana* 349
*Emberiza schoeniclus* 205, 336, 353
Ephydridae 322
*Episyron gallicum* 369
equines, grazing by, 99–101, 106 – *see also* ponies, effects of grazing by; and donkeys, effects of grazing by
*Equus africanus* 74
*Equus ferus ferus* 72
*Eresus cinnaberinus* 161
*Erica cinerea* 156
*Erica tetralix* 156, 164
*Erinaceus europaeus* 366
*Eriophorum vaginatum* 164, 169–71
*Eucalyptus* 211
*Euphorbia characias* 95
*Euroleon nostras* 163
eutrophication 125, 232–3, 238, 242, 246
Extremadura (Spain) *4, 251, 332*

Fabaceae 337
*Falco cherrug* 14
*Falco columbarius* 165, 169
*Falco naumanni* 357
*Falco peregrinus* 355

falcon
  peregrine 355
  saker 14
farming 53, 331–54
  mixed 334–6
  organic 334–6, 344
farmyard manure 125
*Fasciola hepatica* 76
*Felis catus* 79–82, 368
fences
  anti-predator 358
  reducing bird collisions with 211–*12*
  underwater *237*
fens 2, 14–15, 58, 229, 243–4, 252–78, 281–2, 287–8
fen-sedge, great 15, *266,* 268
ferret, black-footed 103
fertilizers 53, 59, 68, 87, 90, 112, 114, 124–5, 186, 205, 232–3, 250, 252, 284, 286–7, 321, 331, 336, 339, 341–2, 346, 352, 354
fescue, red 305, 308
*Festuca ovina* 145
*Festuca rubra* 305, 308
*Ficedula hypoleuca* 211
finches 206, 337
fir, douglas 184, 197
fire – *see* burning
  back- 118, 135, 150, 267
  crown- 133, 196, 211, 215
  head- 118, 150
  surface- 133, 211–14
firebreaks 120, 137, 150, 163, *213,* 267
fish 60, 66–7, 145, 231–2, 234, 236, 238, 241–2, 250, 260–1, 289, 303–4, 314–16, 320, 322, 325, 363
  passes 67
  salmonid 251
flamingos 319
fleabane, saltmarsh *324*
flies 61–2, 76, 97, 180, 319, 326, 363
  brine 322
  crane 302
flood alleviation 286, 314
floodplains – *see* river floodplains
fluke, liver 76
flycatcher, European pied 211
folding (livestock) 96–7
foot-drains 291, *294*
forage harvester 151
Ford Moss (England) *280*
forest
  old-growth 57, 179, 183–*4,* 191–2, 196
  primary 182–3
  virgin 182–3

forests 5, 67, 73, 173–227
fox
  red 67, 216, 293, 297, 299
  swift 103
foxglove *361*
Frampton Marsh (England) *302*
*Fritillaria meleagris* 15, *366*
fritillary (plant) *15, 366*
fritillary butterflies 200, 202
fritillary
  heath *6*
  pearl-bordered 171
frog, common *364*
frogbit 37, 40
frogs 267
fuel breaks – *see* firebreaks
fungi 58–9, 90, 110, 116, 180–1, 183, 193, 196–7, 203, 208–9, 225, 362
fungicides 342, 360
fynbos 131–5, *133,* 137–8, 140

*Gallinago gallinago* 290–1, 294–5
gamebirds 333, 341–2, 345–6, 349
gardens 355–68
Gargano Peninsula (Italy) *136*
garrigue 131, 135
*Garrulus glandarius* 179
*Gaultheria shallon* 146
*Gavia arctica* 237
geese 116, 125, 240, 255, 260, 289, 305, 308–9, 346–7, 351
gentian, marsh 158
*Gentiana pneumonanthe* 158
Georgia (USA) 355
*Geranium phaeum* 361
*Geranium pratense* 361
germination gaps 58, 77, 90, 95, 114, 116, 126, 128, 170, 188, 191, 297, 366
girdling, to provide dead wood 197
glasswort 305
*Glyceria maxima* 268, 273
goat
  dwarf 75
  feral 75
  pygmy 75
goats, grazing and browsing by 74–5, 97–8, 101–2, 108, 134, 140, 155, 170, 269, 328
godwit, black-tailed 290, 291, 294, 298–9
golden-plover, Eurasian 165, 167, 171, 350
goldfinch, European 361
Goor-Asbroek (Belgium) *189*
goose
  Canada 346–7
  greylag 255–6, 305

goose (contd)
  pink-footed 338
  snow 305, *346–7*
goosefoots 337
Goosemoor (England) *315*
gorse 143, 145–6, 153, 159
Grange Heath (England) *159*
grass
  filter strips 352
  margins 338, 352–3
grasshopper, blue-winged 161
grasslands
  dry 87–130
  wet 167, 229–30, 246, *248–50*, 263, *265*, 271–2, 276–7, 283–99
grazing
  aftermath 95, 114, 126, 186, 366
  extensive 69–71, 73, 94, 101, 112, 271, 328
  naturalistic 18–19, 69–74, 101, 112, 271
grazing pressure, method of estimating 106–7
grebes 319
green bridges *66*
greentree reservoir 283
green-tree retention (GTR) 219
grip-blocking 166, 279
grips 279, 291, *294*
grosbeak, blue 214
ground squirrels 91
group-selection system (in forestry) 216, 219
grouse
  black *165,* 170
  red 164, 166–8, 170–1
  woodland 211–12
*Grus grus* 4
gull
  herring 69
  lesser black-backed 69
gulls 116, 299, 319, 326

habitat corridors 64
*Haematopus ostralegus* 298
hair-grass, wavy 145, 147, 152, 157
*Hakea* spp. 135, 138
Hanstholm (Denmark) *146*
hare, European 98, 334
hares 103
harrier, northern 165, 169–71
harriers 344–5
hawk, ferruginous 103
hawthorn 209
hay meadows *96*, 113–14, 117, 125, 288
hay
  green 126–9
  marsh 263, 265
  salt 304, 316
haylage 113–14
hazel 89
heath
  cross-leaved 156, 164
  Atlantic upland – *see* moorland
  wet 142–3, 145–6, 148, 150–3, 155–8, 162, 164, 166, 168–70
heather 67, 142–3, *146–7*, 149–52, 154–7, 164–70, 279
  burning 148–52, 167–9
  bell 156
  growth phases 143, *147*
heathland, European Atlantic lowland *71*, 131–2, 141–64, 301, 327–8
hedgehog, West European 366
hedgerows 207, 342, 344, 346, 349–52
*Hediste diversicolor* 322
hefted flocks and herds 107, 169
hemlock 184
herbicide
  systemic 79, 129, 296
  translocated – *see* herbicide, systemic
herbicides 77–9, 129, 159–60, 190–1, 197, 203, 205, 281, 296, 303, 336, 340–4, 353–4, 360
herons 238, 241, 260, 281, 320, 325
*Herpestes auropunctatus* 79
hibernation sites, artificial 362, 366–7
  for reptiles 60
high-cut stumps 219, *221–2,* 223–5
*Hippuris vulgaris* 325
Histon (England) *366*
*Holcus lanatus* 157
hop 189
horse-riding, soil disturbance created by 121, 162–3
horses, grazing by 74, 97, 104, 186 – *see also* equines, grazing by
hoverflies 354
hummingbird, rufous 210
*Humulus lupulus* 189
hyacinth, water *78*
hydrology, basic principles of 230–1
*Hygrocybe* 58, 90, 116

ibises 281
icterids 309
in-bye 164
indicators, performance 36
intolerant tree species, definition of 176
islands 67, 79–82
  creation and management for nesting birds 234–7, 238, 240, 241, 317, 319, 321, 323–6

jay, Eurasian  179
Jug Bay Wetlands Sanctuary (USA)  *3, 210*
*Juncus* spp.  38, 40, 45–6, 101, 112, 289, 296

kale  345–6
Kalmthoutse Heide (Belgium)  *147*
Karori Wildlife Sanctuary (New Zealand)  358
kestrel, lesser  357
keystone species  18–20, 67
killdeer  299
kingbird, eastern  205
Kinguissie (Scotland)  168
knotgrasses  337
Kröller-Müller Museum (Netherlands)  *23*
kwongan  131

*Lacerta agilis*  143, 148, 156, 161
Lady Park Wood (England/Wales)  *204*
*Lagopus lagopus scoticus*  164, 166–8, 170–1
Lake
    Fausse Pointe State Park (USA)  *284*
    Hornborga (Sweden)  *256*
    Kvismaren (Sweden)  *9*
    Tåkern (Sweden)  *277*
lakes  233–42, 276–8
land-sparing  331
Langdonken (Belgium)  *7*
Laniidae  188
*Lanius collurio*  350
*Lanius excubitor*  350
Lanslebourg-Mont-Cenis (France)  *96*
lapwing
    black-winged  *119*
    northern  37, 40, 122, 167, 205, 290–1, 293, 295, 298–9, 336, 344–5, 350, 370
lark, calandra  350
    Eurasian sky  205, 336–7, 344, 350
    wood  143, 151, 160, 224
larks  332, 337
larkspurs  356
*Larus argentatus*  69
*Larus fuscus*  69
lawns, garden  365
leatherjackets  290, 294, 296
Leicestershire (England)  356
Leipsic (USA)  *54*
*Lepus europaeus*  98, 334
lichens  59, 90, 100, 102, 105, 122, 143, 151, 180–1, 183, 208, 327, 328, 355, 357
lime and liming  124, 148, 286, 290
limestone pavement  *89*, 358
limits of acceptable change (LACs)  36
*Limosa limosa*  290, 291, 294, 298–9
linnet, Eurasian  336

linseed  337
lion  20*, 66*
*Liriodendron tulipifera*  219
*Lissotriton vulgaris*  364
litter-stripping  193
Little Creek Wildlife Area (USA)  316*–317*
liverworts  57, 180–1, 183
livestock units, calculation of  106
lizard, sand  143, 148, 156, 161
lizards  103, 351
*Lochmaea suturalis*  145
*Lolium perenne*  114, 124
London, East (England)  *357*
loon, Arctic  237
lousewort, marsh  *266*
*Loxodonta africana*  20
*Lullula arborea*  143, 151, 160, 224
*Lycopodiella inundata*  157

machair  332–3, 335
*Maculinea alcon*  158
Maine (USA)  64
maize  345, *346,* 347
mallard  283
mallee  131
managed re-alignment  85, 311*–12,* 314
management
    planning  27–44
    costs of  65
managing change  1, 13–16
maquis  131, 135
mare's-tail  325
Margherita di Savoia (Italy)  *318*
margoram, golden  361
marigold, field  *187*
Marion Island (Republic of South Africa)  80
Market Weston Fen (England)  *266*
marram  329
Masai Mara (Kenya)  *3, 19*
Matalascañas (Spain)  329
mat-grass  *99*
Mattinata (Italy)  *136, 187*
mattoral  131
meadow birds  46, 54, 85, 243, 246, 248–9, 267, 286, 290–9
meadowlarks  350
meadows, garden wildflower  365–6
*Melampyrum pratense*  6
*Melanocorypha calandra*  350
*Meles meles*  67, 299
*Mellicta athalia*  6
*Melospiza melodia*  210
merlin  165, 169
Mersehead (Scotland)  *326*

Michigan (USA) 215
Mickfield Meadow (England) *15*
micro-habitats 60–1, 194, 244, 253, 263
*Micromys minutus* 339
*Microtus agrestis* 170–1, 339
midge larvae, non-biting 241, 322–3, 325
Mikashevice (Belarus) *350*
*Miliaria calandra* 350
millet 345
min till – *see* non-inversion tillage
mine workings, metal-rich 355
mineral-extraction sites 355, 369–70
minimum tillage – *see* non-inversion tillage
mink, American 67, 262
Minsmere (England) *104, 274*
mire, raised 229
mires 142–3, 145–6, 148, 150–1, 153, 155–7, 162, 229
Mississippi (USA) 283
Missouri (USA) 121
moist-soil management 17, 238–41, 255, 289, 320, 324–5
*Molinia caerulea* 145, 147, 150–3, 156, 164, 166, 169–70, 258
molluscs 322
    bivalve 323, 325
*Molothrus ater* 68, 91, 191, *350*
Mondhuie (Scotland) *280*
mongoose, small Indian 79
monitoring 16, 27–8, 36, 38–42, 44–54, 81, 106, 156
Monroy (Spain) 332
montados 4–5, 192, 207–9
moor-grass, purple 145, 147, 150–3, 156, 164, 166, 169–70, 258
moorland 131–2, 164–71
mosquito, salt marsh 304, 317
mosses 57, 102, 143, 145, 150–1, 180–1, 183
moth, golden plusia *356*
moths 62, 269, 356, 362
motorbikes, soil disturbance created by 121, 162, 370
Mount Tamalpais State Park (USA) *134*
mountain-pine, dwarf 145–6
mouse
    Eurasian harvest 339
    wood 200
    yellow-necked 200, 351
mowing – *see* cutting
muirburning – *see* heather burning
mulcher, forestry 77, 158–9, 190
*Muscardinus avellanarius* 202, 351
muskrat 255
mussels, freshwater 252

*Mustela nigripes* 103
*Mustela vison* 67, 262
*Mycocastor coypus* 303, 310

Nairobi National Park (Kenya) 66
*Nardus stricta* 99
National Park de Hoge Veluwe (Netherlands) 23, *161*
natural processes 2, 5, 13, 18–22, 25, 38, 181
naturalness 17–22, 25, 69–74
nectar strips 354
*Nematostella vectensis* 319
*Neomysis* spp. 322
nest
    exclosures – *see* nest protectors
    loss, minimising it during agricultural operations 113–14, 116–18, 299, 344–5
    protectors 297, 299
    trampling 109–10, 297–8, 305
nettle, common *93*
New Forest (England) *71*, 153, *157*
newt, smooth 364
nightjar, Eurasian 143, 158, 160, 224
nitrate 125, 232–3, 287
nitrogen 39–40, 59, 88, 90, 124–5, 145–6, 149, 160, 166, 169, 190, 193, 232–3, 341
    atmospheric deposition 145, 149, 166, 169, 193
non-inversion tillage 340–2
no-till – *see* non-inversion tillage
*Numenius arquata* 167, 295
nutria 303, 310

oak
    cork 4
    live 4
    pedunculate *208*
Oakens Wood (England) *196*
oaks/oak woodland *210*, 211, 219, 283
oat-grass, false 353
oats 337
objectives, setting site conservation 36–42
*Odocoileus virginianus* 210
*Oedipoda caerulescens* 161
oilseed rape 336–7
*Olea europaea* 187
olive groves 186–*7*, 192, 208
Omberg (Sweden) *222*
*Ondatra zibethicus* 255
Oostvaardersplassen (Netherlands) 16, 71–*2*, 255
open-marsh water management (OMWM) 304
*Ophrys bertolonii 136*
*Ophrys garganica 136*

orchid
  Bertoloni's *136*
  early spider *Ophrys sphegodes* *136*
  Gargano *136*
orchids, *Dactylorhiza* spp. 355
*Origanum vulgare* 361
Oriñón (Spain) *327*
*Ornithogalum umbellatum* 187
*Oryctolagus cuniculus* 18, 102–4, 155–6, 328
osier 186
*Otis tarda* 332
Ouse Washes (England) *18, 286*
ouzel, ring 165
oversowing, to diversify grassland 128–9
owl
  burrowing 103
  short-eared 165, 170
oxlip *199*
oystercatcher, Eurasian 298

*Panthera leo* 20, *66*
Parque National de Doñana (Spain) *272*
Parque Natural des Dunas (Spain) *329*
partridge, grey 333, 338, 340, 342
*Passer montanus* 353
*Passerculus sandwichensis* 64, 206
*Passerella iliaca* 210
*Passerina caerulea* 214
pasture-woodland – *see* wood pasture
peas 337
*Pedicularis palustris* 266
Pembury Walks (England) *154*
*Perdix perdix* 333, 338, 340, 342
persicarias 337
pesticides 5, 64, 68, 232, 252, 331, 333, 336, 341–4, 352–4, 358, 360
*Philanthus triangulum* 123
*Phoenicurus ochruros* 370
*Phoenicurus phoenicurus* 211
phosphorus 39–40, 58, 88, 124–6, 160, 232–3
*Phragmites australis* 9, 67, 77, 229, 233, 249, 252–78, 302–3, 305, 309, 316, 325
phrygana 131, 135
*Phylloscopus sibilatrix* 211
*Picea abies* 182, 184, 215, 223
Pickering Creek Audubon Center (USA) *54*
pickleweed *305*
*Picoides borealis* 2, 14
Picos de Europa (Spain) *108*
Pieridae 343
pigeons 345
pigs 4, 71, 75, 103, 186, 209
pine
  jack 215
  lodgepole 145
  maritime 145
  ponderosa 214
  Scots 145, 159, 197, 212, *222*
pines 2, 135, 138, 146, 155, 161, 211, 214
pink family 337
*Pinus banksiana* 215
*Pinus contorta* 145
*Pinus mugo* 145, *146*
*Pinus pinaster* 145
*Pinus ponderosa* 214
*Pinus* spp. 2, 135, 138, 146, 155, 161, 211, 214
*Pinus sylvestris* 145, 159, 197, 212, *222*
pipit
  meadow 170–1
  tawny 143
  tree 211, 224
planning, reverse 42
Planorbidae 241
Plateaux (Netherlands) *258, 285*
playa lakes 239, 241
*Plebejus argus* 151
ploughing 73, 121–3, 137, 162, 340–1, 344–5, 352
  chisel *122*
plover
  little ringed 370
  mountain 103
plovers 116, 234, 241, 323
*Pluchea odorata* *324*
*Pluvialis apricaria* 165, 167, 171, 350
pocket gophers 102
Point Lobos State Reserve (USA) *139*
poisoning 80–2
pollarding 186, 209
*Polychrysia moneta* 356
Polygonaceae 337
*Polygonum* spp. 239
ponds
  fish 234, *236*
  garden 357, 362–5
  permanent 150, 242–9
  temporary – *see* pools, temporary
  turf 257–9
pondweed, sharp-leaved 245
ponies, grazing and browsing by 71, 97–100, 153–4, 156–7 169, 186, 248–9, 269, 271, 273–5, 306, 328
pony
  Carmargue 75
  Exmoor 75
  hardy breeds of 74–5
  Icelandic 75
  Konik Polski or Konik *72*, 75, 274

pony (contd)
  New Forest 71, 75
  Przewalski 75
  Shetland 75
  Welsh Mountain 75
pools
  temporary 60, 62, 92, 103, 182, 242, *250*, 288, 328, 363–4
  vernal 242–*3*
poplar 186, 206
*Populus* spp. 186, 206
Poroszló (Hungary) *14*
post-industrial sites – *see* brownfield sites
*Potamogeton acutifolius* 245
potassium 124
prairie dogs 18, 91, 102, *103*
prairies 87, 102, 119, 121, 341
*Primula elatior* 199
*Primula veris* 366
proteas 133, 138
*Protea* spp. 133, 138
*Prunella modularis* 351
pseudosteppes *332,* 335, 340
*Pseudotsuga menziesii* 184, 197
*Pteridium aquilinum* 142, 145–6, 149, 160, 164, 166, 169, 171
*Pterocles alchata* 332
*Pterocles orientalis* 332
*Puccinellia maritima* 305–6, *307,* 308
Pulborough Brooks (England) *123*
pulverised fuel ash (PFA) 355
*Pyrrhocorax pyrrhocorax* 97, 151

*Quercus robur* 208
*Quercus rotundifolia* 4
*Quercus* spp. *210*, 211, 219, 283
*Quercus suber* 4
quinoa 345

rabbit, European 18, 102–4, 155–6, 328
rabbits 91, 122, 327
  encouraging 102, 104
Rabinówka (Poland) *3*
rafts, for nesting waterbirds 237
ragworm 322
ragwort 112
Rainham Marshes (England) *93*
ramsons *199*
*Rana temporaria 364*
rangelands 87
rat, black 80, 200
rats, eradication of 79–82
rattle, yellow 126–*7*
rattlesnakes 103
*Rattus rattus* 80, 200

*Rattus* spp., eradication of 79–82
redshank, common 37, 40, 290–1, 295, 298–9, 308
redstart
  black 370
  common 211
reduced tillage – *see* non-inversion tillage
reed cutting 3, 9, 262–9
reed, common, and reedbeds 9, 67, 77, 229, 233, 249, 252–78, 302–3, 305, 309, 316, 325
reed-cutters, amphibious 267, *270*
regulated tidal exchange (RTE) 85, 311, 313–18
reptiles 60, 79, 123, 143, 148, 151, 162, 214, 309, 328, 330, 362, 368
Retamosa (Spain) *4*
rewilding 19–20
*Rhinanthus* spp. 126–*7*
rhododendron 145–6
*Rhododendron ponticum* 145–6
*Rhynchospora fusca* 157
rice fields 347–9
ring-barking – *see* girdling
Rio Almonte (Spain) *251*
river floodplains 249, 251, *264–5*
  re-naturalisation of 20–*1*
River
  Prypyat (Belarus) *264–5*
  Skjern (Denmark) *21*
rivers 249–52
roads 66, 144, 303, 309, 316
roofs
  green 370
  use of by nesting birds 355, 257, 370
  using run-off from to supply garden ponds 365
rotational management 40, 120–1, 139, 149, 151–2, 163, 168, 196, 198–202, 205–7, 214, 216–25, 238–41, 244–6, 263, 268–9, 281, 309, 332, 335, 338–40, 346–7, 352–3
rotovation 121–3, 150, 162–3, 325, 328, 330
rowan *176*
rowcrops – *see* arable
*Rubus* spp. 189
*Rumex* spp. 114
*Ruppia maritima* 320, 323–4
rushes 38, 40, 45–6, 101, 112, 289, 296
Russian Federation 335
rye 345
Rye Harbour (England) *237*
rye-grass, perennial 114, 124

saffron, meadow *285*
sage scrub, coastal 131
sagebrush steppe 131, 134–5, 137, 140
salamander, Californian tiger 243

*Salicornia* spp. 305
salinas 304, *318*, 322
saline lagoons 317–26
*Salix* spp. 145, 186, 205–6, 240, 255, 282, 287
*Salix viminalis* 186
salt evaporation ponds – *see* salinas
saltmarsh 288, 301–17
   creation 310–15
   restoration 315–18
saltmarsh-grass, common 305–6, *307*, 308
San Luis National Wildlife Refuge (USA) *243*
sand dunes – *see* dunes
sandgrouse
   black-bellied 332
   pin-tailed 332
sandpiper, upland 64
Sandy Heath Quarry (England) *369*
saproxylic – *see* dead-wood
savannah 87
sawfly larvae and pupae 340
*Saxicola torquata* 165
*Scirpus maritimus* 77, 273–5, 296, 325
scrub 74, 88–9, 91, 93, 96, 101, 142–3, 145–6, 148–50, 152–5, 158–9, 165, 168, 173–9, 187–91, 194, 196, 209, 279, 287, 328, 334, 342, 350–2, 370
   riverine 252
   wet 252–3, 262–4, 269, 275, 281–3
seabirds 80, 82
sea-level rise 82–5, *312*
sea-purslane 305
sedge-cutting 3, 15, 263
sedges 101, 148, 275, 277, 289
sediment
   recharge 311
   trickle-charging 311
seedling plugs, use of to diversify grassland 129
*Selasphorus rufus* 210
selection system (in forestry) 216, 219, *220*
*Senecio* spp. 112
sequoia, giant 213
*Sequoiadendron giganteum* 213
set-aside 207, 338–40
shallon 146
shearing 281
sheep
   grazing and browsing by 74–5, 97–101, 103–8, 114, 134, 140, 154–6, 164, 166–7, 169–70, 186, 249, 269, 275, 288, 296–7, 306, 328
   Hebridean 75, *154*
   Manx Loaghtan 75
   trampling of nests by 298
   primitive 74, 101, 155, 188

Shetland 75
Soay 75
sheep's-fescue 145
Sheffield (England) 356–7, 365
shelterwood system 216, *218*–19
shingle
   coastal 326–7
   on bird nesting islands and rafts 234–7
shorebirds 68, 143, 165, 240–1, 303, 317–20, 323–5, 333, 348–9
shredding 186
shrews, white-toothed 200
shrike
   great grey 350
   red-backed 350
shrikes 188
shrimp
   brine 322
   mud *322*
   opossum 322
shrimps 241, 315, 322
silage 113–15
site action planning – *see* management planning
sky lark scrapes 344
slot-seeding 129
sluices 231
slurry 124
small mammals 90–2, 116, 120–1, 180, 200, 205, 210, 214, 248, 288, 303, 309, 333, 339, 350–2, 360, 362, 366
small-reed, wood 157
smartweeds 239
snags 180, 183, 197, 219, *222*
snail, little ramshorn whirlpool *245*
snails 360, 363
   ramshorn 241
snake, smooth 143, *144*
snipe, common 290–1, 294–5
sod-cutting 7, 58, 147–9, 160–2, 193, 257, *258*, 328, 330
solar ponds – *see* salinas
Solway Firth (Scotland) *326*
*Somateria mollissima* 67
*Sorbus aucuparia* 176
sorghum 239, 347
sousliks 91
soybean 346–7
sparrow
   Bachman's 215
   clay-colored 206
   Eurasian tree 353
   field 206
   fox 210
   grasshopper 121, 350
   Henslow's 121

sparrow (contd)
  Nelson's sharp-tailed  309
  savannah  64, 206
  seaside  309
  song  210
sparrows  206, 337, 352
*Spartina alterniflora* 310, 317
*Spartina anglica*  302
*Spartina cynosuroides*  325
*Spartina patens*  317
*Spartina* spp.  303–4, 316
*Spermophilus* spp.  91
*Sphagnum* moss  148, 164, 169, 279
spider, ladybird  161
spiders  63, 144, 263, 341, 352
spikerush, dwarf  323–4
*Spiza americana*  350, 352
*Spizella pallida*  206
*Spizella pusilla*  206
spraying
  boom  79
  spot  79, 160, 354
spruce, Norway  182, 184, 215, 223
spurge, large Mediterranean  95
star of Bethlehem  187
starling, common  116
steppe  87, 332
*Sterna antillarum*  355, 367
stilts  319
Stockholm National City Park (Sweden)  359
stonechat, common  165
stoneworts  233, 246, 320
stork, white  335, 367
straw sausages  234
strip-felling  216
structural marsh management  304
stubble
  rice  348
  winter  336, 337, 340, 345
*Sturnella* spp.  350
*Sturnus vulgaris*  116
sub-soiling  121, 123
sugar beet  338
sulphur, atmospheric deposition  145–6
surveillance  39, 42, 44–52
*Sus scrofa*  3, 66, 103, 209
swamps  2, 229–30, 240, 243–4, 248–9, 252–78, 281–2, 287–9, 325
  *see also* wet woodland
swan, whooper  18
sweet-grass, reed  268, 273
swifts  357
*Sylvia communis*  336, 351
*Sylvia curruca*  351
*Sylvia undata*  143, 151, 159

Sylviidae  179, 205
*Sympetrum piedmontani*  258

targets, setting conservation  17, 36–41
tarpan  72
tasselweed, beaked  320, 323–4
Tautra Island (Norway)  67
*Taxodium distichum*  284
teasel, wild  361
tern, least  355, 367
terns  69, 234, 237, 319, 326
*Tetrao tetrix*  165, 170
*Tetrao urogallus*  222
*Tetrax tetrax*  332
The
  Burren (Ireland)  89
  Wash (England)  302
Thelnetham Fen (England)  266
Thetford Forest (England)  224
thick-knee, Eurasian  102, 122, 344–5
thistle, acanthus-leaved carline  95
thistles  101, 112, 117, 123, 177
Thompson Common (England)  190
thrush, song  353
tidegates  311
Tipulidae  290, 294, 296, 302
titmouse, tufted  214
toads  358
tolerant tree species, definition of  176–7
topping  113, 116–17
transhumance  89, 108
*Trifolium repens*  347
*Tringa totanus*  37, 40, 290–1, 295, 298–9, 308
triticale  345
*Troglodytes troglodytes*  210
true bugs  62, 144
Trujillo (Spain)  332
Tryggelev Nor (Denmark)  321
*Tsuga*  184
Tudeley Woods (England)  189
tulip poplar  219
*Turdus philomelos*  353
*Turdus torquatus*  165
turf stripping – *see* sod cutting
turnips  345
twite  165
two-storied high forest system  217
*Typha angustifolia*  316
*Typha latifolia*  316
*Typha* spp.  229, 253, 255, 268, 273
*Tyrannus tyrannus*  205

*Ulex europaeus*  143, 145–6, 153, 159
Umbelliferaceae  93, 145, 209

umbellifers 93, 145, 209
underpasses 66
undersowing 340
Unionidae 252
Upton and Woodbastwick Marshes (England) *282*
*Urtica dioica* 93

*Vaccinium myrtillus* 222
Vanellus melanopterus 119
*Vanellus vanellus* 37, 40, 122, 167, 205, 290–1, 293, 295, 298–9, 336, 344–5, 350, 370
variable-retention forestry 216, 219, *221–2*
vegetated filter strips 233
Veijlerne (Denmark) *295*
Vera hypothesis 73, 177–9
*Vermivora celata* 210
vetches 337
veteran trees 5, 194, *207–8*
vole
  bank *200*
  field 170–1, 339
  water 247–8, 262
voles 91
*Vulpes velox* 103
*Vulpes vulpes* 67, 216, 293, 297, 299
vultures 4

waders 68, 240–1, 303, 317–20, 323–5, 333, 348–9
  breeding 37, 40, 46, 54, 85, 143, 165, 243, 246, 248–9, 267, 286, 290–9
Walbeswick (England) *84, 319*
warbler
  aquatic *24*
  Dartford 143, 151, 159
  Kirtland's 215
  marsh 205
  orange-crowned 210
  sedge 205
  wood 211
warblers 269
  Old World 179, 205
washlands 286
wasps
  parasitic 356
  solitary 61, 92, 123, 144, 162, 302, 366–7, 369–70
  spider-hunting 369
water
  boatmen 322
  fleas 363
  meadows 285
  quality 41, 50, 232–4, 250
waxcap fungi 58, 90, 116

wedge-felling 216
weed-cutters, amphibious 244, 262, *270*
weed-wiping 79, 354
wheat 338, 344–5, 347
whitethroat
  common 336, 351
  lesser 351
Wicken Fen (England) 14–15
widgeon grass 320, 323–4
wigeon, Eurasian 308
wild-bird cover 345–6
wildflower meadows, garden 365–6
wildfowl 17–18, 43, 68, 234, 237–41, 243, 248, 252, 255, 260, 267, 286–7, 289, 291, 296, 303–4, 308–9, 316, 320, 323, 325–6, 347
wildlife
  bridges – *see* green bridges
  corridors – *see* habitat corridors
  overpasses – *see* green bridges
willows 145, 186, 205–6, 240, 255, 282, 287
Windsor Great Park (England) *208*
wolf, gray 19
wooded dehesas *4–5*, 192, 207–9
woodland
  ancient 183
  glades 73, 142, 181, 187, 193–7, 210
  rides 158, 193–8, 200, 202, 205, 207, 210, 215
  theories of regeneration 174–9
  wet *264, 282–4*
woodlands 5, 173–227
wood-meadow 186
wood-pasture 71, 73, 179, 186, 192, *204*, *208–9*
woodpecker, red-cockaded *2*, 214
woodpeckers 3, 180–1, 222
worms, polychaete 322–3, 325
Worth Matravers (England) *302*
wren
  marsh 309
  sedge 309
  Stephen Island 79
  winter 210

*Xanthium* spp. 240
*Xenicus lyalli* 79

yellowhammer *115*, 349, 351–2
Yellowstone National Park (USA) 19
Ynys-Hir (Wales) *122*
Yorkshire-fog 157
Yosemite National Park (USA) *213*
Yremossen (Sweden) *182, 222*

Żywkowo (Poland) *335*

**DISCARDED**
CONCORDIA UNIVERSITY LIBRARIES
CONCORDIA UNIV. LIBRARY
MONTREAL